Venture Capital

Springer
Berlin
Heidelberg
New York
Hong Kong
London
Milan
Paris
Tokyo

Stefano Caselli
Stefano Gatti
Editors

Venture Capital

A Euro-System Approach

**With 52 Figures
and 46 Tables**

 Springer

Prof. Stefano Caselli
Prof. Stefano Gatti
IEMIF Bocconi
Via Sarfatti 25
20136 Milan
Italy
stefano.caselli@uni-bocconi.it
stefano.gatti@uni-bocconi.it

ISBN 3-540-40234-9 Springer-Verlag Berlin Heidelberg New York

Cataloging-in-Publication Data applied for
A catalog record for this book is available from the Library of Congress.
Bibliographic information published by Die Deutsche Bibliothek
Die Deutsche Bibliothek lists this publication in the Deutsche Nationalbibliografie; detailed bibliographic
data is available in the Internet at <http://dnb.ddb.de>.

Springer-Verlag Berlin Heidelberg New York
a member of Springer Science+Business Media

http://www.springer.de

© Springer-Verlag Berlin · Heidelberg 2004
Printed in Germany

The use of general descriptive names, registered names, trademarks, etc. in this publication does not
imply, even in the absence of a specific statement, that such names are exempt from the relevant protective laws and regulations and therefore free for general use.

Hardcover-Design: Erich Kirchner, Heidelberg

SPIN 11379768   43/3111/DK-5 4 3 2 1 – Printed on acid-free paper

Preface

Josh Lerner

Jacob H. Schiff Professor of Investment Banking Harvard Business School and National Bureau of Economic Research

During much of the 1970s and 1980s, venture capital and private equity remained largely a United States phenomena. Over the past decade, however, private equity has spread around the globe, taking particularly firm root in Western Europe-indeed, growing 4,700% since 1984 through 2001. Today, Europe is the dominant private equity market outside the United States. Despite this tremendous growth and the current recessionary climate, there is ample room for attractive expansion in European private equity; both for venture capital and private equity.

There have been several reasons for this growth. The first has been the internationalization of capital sources. The key sources of capital for European private equity funds have traditionally been segmented by national boundaries: historically, the pattern in Europe has been for private equity groups to raise funds from banks, insurance companies, and government bodies in their own country, with little involvement from foreign investors. These barriers are now breaking down for two principal reasons. First, institutional investors, particularly in the United States, have become increasingly interested in European funds. Second, many international private equity firms have become more active in Europe.

A second driver of growth has been the entry of new talent into the industry. Traditionally, many European private equity investors had come from financial or consulting backgrounds, rather than from operating roles. Perhaps as a result of these backgrounds, the relationships between private equity investors and portfolio companies have tended to be much more distant in Europe, with a greater emphasis on an assessment of financial performance than on "hands-on" scrutiny. Today, we are seeing groups increasingly emulating U.S. funds, and seeking to play an ongoing role in managing these firms. The track records of a number of groups that have adopted these approaches attest to the benefits that influx of talent can bring.

A third change relates to the geographic distribution of investments. European private equity investors have historically evinced a strong tendency to invest in the same country in which the fund is located. This reluctance to invest across national boundaries reflects both the legacy of legal and regulatory barriers to such transnational investments (now greatly reduced) and the very distinct business cultures that characterize any European nation. Progress is being made in creating truly pan-European private equity market, though much work remains to be done here.

A final critical reason for the growth of private equity investment in Europe has been a change in its focus. At the beginning of the decade, private equity investing outside the United States was targeted at buyout or expansion financing of already-profitable businesses. These investments provided an opportunity for companies to expand production or reorganize operations. During the past decade, however, private equity investing has expanded to include venture capital. The same underlying technological innovations that drove the U.S. venture capital revolution unleashed a similar burst of entrepreneurial activity in Europe. Recent experience for both venture capital and private equity in many countries points to potentially attractive opportunities for both investors and businesses requiring capital in the Europe.

Thus, despite continued cultural and regulatory challenges—as well as the "boom and bust" pattern that has characterized the industry globally since its inception—European private equity represents a diverse and vibrant market. But while European private equity has made considerable strides, academic research has lagged. Published works on venture capital and private equity remain largely focused on the U.S. market.

Thus, this set of papers plays an important role. By collecting a set of research papers into the nature of private equity in Europe, including ones that employ a variety of methodologies, we gain a better understanding of what makes this market distinctive. In particular, the distinctive financial structures, regulatory environments, and fundraising processes that characterize the European market are explicated here with admirable clarity.

To be sure, even after reading the volume, the reader will surely note that important questions still remain unanswered. Many relate to the future of the European market. For instance, to what extent will these distinctive elements survive, or will the industry converge to the U.S. template? But these issues will doubtless stimulate research by the authors in the years to come!

March 15, 2003

Foreword

Stefano Caselli and Stefano Gatti

"Giordano Dell'Amore" Institute of Financial Markets and Financial Intermediation – L. Bocconi University

It was about march 2000 when we first started thinking about a research covering the present state of the art of the Venture Capital (VC) industry in Europe and Italy. The industry had passed through a period of strong increase in both the number of deals and the amount of funds invested in high tech or newly born firms. This scenario was somewhat different from the previous ten years situation. Apart from UK, Continental Europe had never had a strong VC industry. Most part of the financial intermediation was rooted in commercial banks; in an international comparison with the anglosaxon countries, equity capital markets – particularly those dedicated to small, fast-growing, high-tech firms – were small or in their beginnings and this fact had limited for a long time a full development of a mature VC industry.

A fortunate coincidence of institutional reforms, of a new technology revolution (the internet wave), of the sharp decrease in the Government bond yields had led to strongly increased savings available for the financing of new companies and new ideas. A large number of SME's were facing the opportunity of a fast access to equity capital markets, much faster than before. In this situation, it was quite natural to launch in Italy a research project focused on the new challenges faced by the relatively new industry of VC.

What happened after march 2000 is well know and doesn't deserve further comments. From our point of view, however, the correction in stock prices and the smashing of the Stock Exchange bubble has had a beneficial effect and has driven Venture Capitalists back to the fundamentals of the firm analysis and to more careful screening policies of funded companies.

The book is articulated into three parts.

The first section of the book (How Does A Venture Capital Work And Why Is Venture Capital Needed In Modern Economies?), is dedicated to the analysis of the investment process of a Venture Capital Investor/Closed End Fund; a thorough description of the financing stages and investment process and a careful study of the main determinants of the demand and offer of venture capital is conducted in chapters 1 and 2. Chapter 3 focuses on the different funding policies available in the Venture Capital Industry. The topic of company valuation and of the correct appraisal of the price of an equity stake in young or high-tech firms is analyzed in chapter 4. The high tech bubble that boomed in march 2000 has shown how important it is to apply an appropriate valuation method, especially to high tech companies, usually lacking a good and reliable historical track record. As a matter of fact, market multiples and traditional real option models are progressively disappearing from the scene, and a critical analysis of the alternatives available to the analysts seems necessary. Chapter 5 focuses on the principles underpinning asset allocation, portfolio management and liquidity management in Venture Capital industry: the chapter points out the special features of portfolio management in Venture Capital in comparison with the asset allocation principles used by open-end funds investing in quoted (usually larger) firms. Finally, chapters 6. and 7. show exactly how a Venture Capital investor approaches the market in practice; the two chapters have been contributed by the CEO's of Intervaluenet and Pino Venture Partners, whose importance in the Italian and European scene is shown by their strong know how and specialization in the venture capital industry. In particular, Pino Venture Partners, provides through its funds an important link between Continental Europe and foreign companies, as its consulting services also stretch to European entry strategies, licenses and authorizations to be obtained in order to start a subsidiary. Intervaluanet, on the other side, shows a successful entry and positioning in the market place and the evolution from a simple "incubator/market intelligence provider" to an "accelerator of entrepreneurship and innovation" towards a "fully integrated incubator – accelerator – venture capital" business model, fully in line with the new market tendency to include incubators as a possible competitor to the more traditional venture capital companies.

The second section of the book (Venture Capital In The Financial System, Market Trends In Europe And The Relations With Banks And The Stock Exchange) is dedicated to the role played by venture capitalists in the broader context of the financial system (banking system and capital markets). Chapters 1 and 2 include the analysis of the special role played by venture capital in modern economies and the synergies between venture capital, bank credit and the stock exchange; particular attention is paid to the ongoing development of venture capital activities by traditional banks as a part of their corporate banking strategies. Chapter 3 highlights the crucial role played by the Stock Exchange for Venture Capital Investors: this part of the book, contributed by the General Manager of Italian Nuovo Mercato (the section of the stock exchange devoted to the quotation and trading of securities issued by small, young and high-tech firms) explores the opportunities and risks in the quotation of private equity companies. Finally, chapter 4 examines the trend of the Venture Capital market, featuring an analysis of the main characteristics of the venture capital and private equity market particularly in Continental Europe and Italy.

The final part of the book (Venture Capital In Italy: Regulatory And Legal Issues) is dedicated to the study of the main legal and regulatory issues concerning the Venture Capital business in Italy. Chapter 1 refers to the role of the Central Bank in the regulation and supervision activity: the chapter is useful in order to understand the particular point of view of Italian Central Bank in terms of need of special regulation for the venture capital business. Chapter 2 analyses the constitution of a venture capital company in the form of Italian closed-end fund. Finally, chapter 3 highlights the legal issues affecting the structuring, establishment and placement of European (and Italian) venture capital investment schemes.

At the end of this preface, the wish of the editors is that a wide audience could benefit from the results of this research work: we certainly think to students, both undergraduate and MBA and PhD's, but also to executives, professionals in the finance field in companies and intermediaries, business laywers and policymakers/regulators. After all, a fascinating topic like venture capital has something interesting to offer to very different kind of professions. The other wish, of course, is that any suggestion, comment or advice coming from the public of readers could be sent to our corresponding address in order to improve the book and its contents in the coming years.

Finally, the most challenging task for two editors: how to thank everybody that in different ways and different times helped us in completing this work without missing anyone? The first thank is for Newfin Bocconi (The Research Center for Financial Innovation of Bocconi University in Milan) and its managing directors, Prof. Paolo Mottura and Prof. Francesco Saita, whose precious support and sponsorship have been determinant for the final result of this research work. The second thank goes to our friend, Harvard Professor Josh Lerner, whose kindness is behind the foreword of this volume. Thanks a lot, Josh. Third thank you goes to Daniela Ventrone for her precious help in reviewing the final version of the manuscript together with the editors. Fourth thank to Alberto Frisiero: God only knows how many days did he spend in working with the publisher for the final release. Fifth thank, but the most important, to: mother Graziella (Stefano Gatti is writing), for her patience and love in all these years for such a bad guy and Anna and Elisa – I love you so much (Stefano Caselli is writing now) - to which the book is dedicated.

Milan, March 2003

Table of Contents

Part 1

How Does a Venture Capitalist Work and Why is Venture Capital Necessary in Modern Economies?

1 An Introduction to the Investment in New Ventures

Edmondo Tudini

1.1 Introduction

This chapter wishes to analyse, in qualitative terms, the characteristics of the demand for venture capital from the company's viewpoint. In the first paragraph of this chapter a rather broad definition of venture capital activity is given, including the financing of both newly created companies and the "development option" for small and medium sized companies already existing. In accordance with this definition, the analysis will then be directed on two different levels: one with reference to the problems of the venture creation process and the other concerning the phenomenon of the company's development. In particular, in the first part of the chapter, which analises the creation of new entrepreneurial initiatives, the reasons why the recourse to an institutional investor would be considered possible and desirable have been identified. In this sense, rather than dwelling on an analysis of the reasons why venture capital represents the only alternative to the entrepreneur's personal capital and stating the contributions in financing the venture creation process, it is preferable to focus the attention on the phenomenon of the creation of new companies and the different characteristics they may assume. This appears appropriate because if the importance of venture capital as a financial engine for the creation of new entrepreneurial initiatives has been adequately discussed in financial literature, the same cannot be considered the case in identifying the environment of its real application. Thus, in this chapter, firstly we will attempt to shed light on the extreme complexity and variety of the process of creating a new company and then identify the specific taxonomies, with reference to several explanations, such that it is possible and convenient for the entrepreneur to refer to a venture capitalist already in the start-up phases. Thus, it is shown how the intervention of an institutional investor is only possible with reference to a limited number of cases among the large number that may arise.

In the second part of the chapter, which is relative to the phenomenon of the company's development, an attempt is made to understand why existing companies, with a diversified liability and capital structure, would require the recourse of a venture capitalist in order to finance its options for growth. Therefore, analysing the main approaches of modern financial theory, with particular reference to the financial agency theory, it is highlighted how venture capital is the only financial channel capable of reducing the significant information asymmetry connected to financing options of small and medium companies. Moreover, it will be shown how venture capital, in these cases, is the only real alternative to self-financing.

1.2 The Venture Capital: Definition Aspects

The activity of risk capital investment in companies by specialised operators has age-old origins, which could be traced down, for example, to the merchant traders in England in the fifteenth century or, later on, to the role performed by the Company of Indias in developing international trade[1]. It is only since the nineteen forties however that the first true and proper venture capital market was born in the USA; the first investment company in risk capital, the American Research & Development Corp. (ARD), was in fact born in 1946[2].

Afterwards, the activity of equity financing has continued to grow in importance and has taken on varied facets, to respond in an effective manner to the different needs of the participating companies. Today, in order to describe the investment in the share capital of a non-quoted company by a specialised operator, depending on the cases, numerous definitions could be used (seed financing, start-up financing, expansion financing, turn-around financing, bridge financing etc), while the distinction between venture capital and private equity is used to describe the activity performed by the different institutional investors. However, the meaning attributed to these terms is not always clear; in fact, misunderstandings between experts are numerous. It appears, therefore, absolutely necessary to spend some words to avoid easy but just as dangerous confusion.

Over the course of time the characteristics of the institutional investment activity in risk capital have changed, depending on the entrepreneurial system in place and the degree of development of the differing markets. In fact, a common denominator is still the acquisition of significant participations in companies with a medium-long term perspective, with the objective of attaining a capital gain upon the shares sale; the additional characteristics of the investor's intervention, on the other hand, can assume

---

[1] In relation to this some authors refer to the first venture capital experiences during the times of Julius Caesar's Rome. In this sense Liaw 1999.

[2] For more detailed information see Lerner 1999.

varied connotations. The classical literature segments the different typologies of investment according to the phases of the life cycle of the participating company. The underlying logic of this segmentation is that the different development phases of the company correspond to specific requirements and interventions by operators with specialised characteristics. Therefore seed (financing of the idea) and start up financing identify the so-called early stage investments, directed at financing the first phases in the commencement of the company. Successively, when the investment is meant to support the growth and implement the development programs of companies already existing, the terms expansion financing or development capital are used, while replacement capital refers to interventions that, without increasing the share capital of the company, aims at substituting part of the shareholders no longer involved in the company's activity. All the operations focused on changing the shareholding structure of the company in a substantial manner, with the frequent use of leverage as an instrument for acquisitions (leveraged buy out), are generally regrouped in the buy outs category; just as turnaround indicates investments for restructuring companies in situations of crisis and bridge financing those interventions directed, from their realisation moment, to accompanying them to the Stock Exchange.

If the above-mentioned categories are internationally recognised and adopted for statistical purposes[3] too, the definition of Venture capital activity is not so unanimous. Some authors believe that venture capital activity is the financing of the formation of new companies, thus early stage financing; opinion shared by the A.I.F.I. too, so that all the financial interventions in risk capital in favour of existing companies are ascribable to Private Equity and not Venture Capital.

This rather restricted vision of Venture Capital is however to be considered antiquated. In fact, it could have been appropriate in defining the borders of the equity financing industry in the United States in the eighties, when the majority of the investments related to launching new companies, but today the variety of interventions realised by venture capitalists is largely superior. Moreover, the increasing complexity of the commodity sector in which the companies are involved and the distinctive related problems (e.g. the broad Information Technology sector, where often newly formed companies are ready for quotation on the Stock Exchange) mean that, in certain cases, the companies' development stage, and the related financial requirements, do not lead to a classic schematization. In

---

[3] EVCA in elaborating their own statistics use the following aggregations: seed and start up, expansion – including also bridge and turnaround financing -, replacement and buy out. See EVCA, 2000 Yearbook.

other words, the difficulty in identifying a standard categorised path for all the companies renders the boundaries between the various investment categories in risk capital even weaker. However, today more than in the past, the main elements for describing the Venture Capital activities are not the life cycle phases of the participating company or the level of risk undertaken by the investor, but the nature of the risk. In particular, one can talk of Venture Capital whenever the institutional investor takes on, other than risks of a financial and operative nature, true and real entrepreneurial ones. This happens when the investor's intervention depends on a change in the entrepreneurial formula (business model) of the company financed.

The entrepreneurial element is certainly important in the creation of a new company that by definition is founded on a new entrepreneurial formula, but it is also present in all those interventions of expansion financing that allow already existing companies, perhaps after many years, to redefine in a substantial manner their positioning, therefore causing a period of strong discontinuity in their management. Therefore, the supply of "capital for development" is an integral part of the venture capital activity.

Therefore, it appears evident that the Venture Capital demand comes from newly formed companies or from already existing small firms (Small Mature Firms, SMEs) focused on stamping a strong acceleration in their growth path. As a consequence, the Venture Capitalist's intervention can take on various facets, with reference to the technical complexity of the operation made (generally greater in expansion financing) and the managerial contribution, in terms of Know How, required for the success of the investment.

1.3 The New Companies: Why and in Which Cases Venture Capital May Represent a Convenient Strategic Opportunity

1.3.1 New Companies and Entrepreneurship: The Principal Theoretical Contributions

The theme of company creation and the new entrepreneurship has taken on a central role in the debate relating to the choice of economic policies within industrialised countries, and is today encountered with renewed interest in the context of business and organisational economic research. We are going to expose the principal approaches elaborated by economic literature on the subject, in the belief that a comprehension of the critical processes in the origins of a new company is absolutely necessary to be able to fully understand the problems that will be met in the start-up phases and the contribution that the venture capitalist can supply to overcome them.

The idea of confronting the theme of company creation necessarily requires the handling of a combination of contributions of very differing nature. On one side, there are the industrial economy studies, that attempt to explain the phenomenon of company creation in aggregated terms with reference to particular sectors and/or geographic areas. On the other, there are the most recent studies that underline the processes of originating the individual companies. Obviously, the two currents of research are not completely independent, as they often tend to overlap and in any case reciprocally influence each other, therefore it is necessary to make reference to both.

In relation to the first, economic literature allows us to trace four principal hypotheses interpreting the birth of a company:

1. the hypothesis of the market;

2. the hypothesis of the innovation;

3. the hypothesis of the incubator, in its two derivations, territorial and company;

4. the hypothesis of the self-employment.

The first interpretation refers to the particularly significant contribution offered to the formation of new companies by the positive economic climate, characterised by the expansion of the market and/or sector. In fact high returns, associated with a pro-cyclical trend, stimulate the potential entrepreneurs to start their own activities, and in the sectors characterised by the largest growth some employees may also be induced to start up on their own due to the improved opportunities in the market. In addition, an important role is also assigned to the necessity for a reorganisation of the production process: in fact, the response to the processes of enlargement and globalisation of the markets, if at times require the realisation of acquisitions and mergers, at other times fuel production decentralization operations. This leads to the development of marketing, distribution and service activities to the new companies, especially where there is a greater diffusion of smaller ones.

The second hypothesis, formulated for the first time by Schumpeter[4], sees in innovation the determining cause for the creation of new compa-

---

[4] Shumpeter 1912 identifies in particular five distinct typologies of innovation that can give origin to the creation of a new company: the introduction of a new product, the introduction of a new production method, the introduction of a new mar-

nies. With Schumpeter the entrepreneur's figure, neglected in the neo-classic economy, assumes a central role in the process of creating a new company, as he is the person who decides to break the market equilibrium, thus benefiting, for a certain period of time, of a higher level of profit. According to this model (defined as paleo-schumpeterian approach), therefore, the potential entrepreneur decides to give life to a new company based on his own innovative capacities, breaking the flow of the economic system: the innovation is the origin of the extra-profit. This approach is distinguished from a second one, defined as neo-schumpeterian (Schumpeter 1942), according to which the relationship is turned upside down and it is the profit invested in the research and development activity that allows the realisation of new ideas and projects. This second approach also links innovation to the creation of new companies in relation to the spin-off phenomena, joint venture and commercial collaboration agreements.

Under the hypothesis of the incubator, the accent is placed on the impact of environmental factors as elements assisting the process of venture creation. Within the territorial context, this approach identifies the characteristics of a determined geographical area as the most critical factors in creating new companies. One of the most important theoretical references is represented by Marshall, who defines the external economies that can be enjoyed by a company located in a highly industrial area as "economies of scale that favour all the companies of a sector and depends exclusively on its successive development" (Marshall 1920). The presence of widespread entrepreneurship, of a certain sectorial structure and a favourable socio-cultural fabric contribute in establishing fertile conditions in the local system for creating a company. Among the principal factors that favourably influence high rates of creation, reference can certainly be made to the model of industrial areas, which is particularly widespread in Italy. A second version of the incubator hypothesis refers to those companies that take on the role of nursery company for potential entrepreneurs. This approach finds its principal verification in the fact that those who create a new company tend to do so in the same sector in which they were previously an employee.

The hypothesis of self-employment finds its motivation in the choice of revenue, based on the evaluation of the conditions and opportunities of employment, due to the comparison between the possibility of an entrepreneurial activity with uncertain revenue and the alternative as an employee with secure revenue (Vivarelli 1994). According to this hypothesis, the choice to become an entrepreneur depends thus on the comparative evalua-

---

ket, the capture of a new sourcing of raw material supply and the implementation of an industrial reorganization.

tion between a salary receivable for work as an employee and the revenue realizable in the case of an autonomous activity. An important role is attributed to the current economic cycle of activity: in fact, in the recession phase of the economic cycle, the growth in the rate of unemployment represents the determinant that most significantly stimulates the processes of formation of new companies, as the necessity to find an alternative source of income represents a significant drive in giving life to a new initiative[5].

Obviously the theories discussed up to this point are not reciprocally excluded, rather each one contributes to shed some light on the process of creating a new company, that is a complex and varied phenomenon on which numerous factors have an influence and that can be analysed from several points of view. However, in more recent times there are increasingly numerous studies on entrepreneurship and the creation of new companies, where accent is placed on the individual company and its processes of origin. Their target is explaining this process in its different phases and predicting the outcome. Even though this researches relate to a relatively new field and discipline boundaries that are not yet well defined[6], it is possible to trace several common elements as reference points: the centrality generally attributed to the figure of the entrepreneur in the process of the creation of a new company; the recognised importance of environmental factors as sources of opportunity and elements capable of influencing the process of venture creation; and the awareness of the existence of a wide variety of conditions and processes that characterise the creation of a company which cannot be interpreted through comparing their variation in a typical company or entrepreneur. In relation to this, an important contribution is represented by enucleating the variables characterising the genetic process and the identification of some taxonomies capable of interpreting the variety of behavioural strategies of new companies. This approach defines the presence of common characteristics, several typologies of new entrepreneurs and companies (Vesper 1989). These taxonomies are naturally unvalid tout-court, but they make sense as they help clarify specific aspects in the process of creating new companies: therefore, depending on the objectives stated and consequentially the variables considered, there will be different categories. Up to now, the variables most frequently used refer to: the entrepreneur's personal characteristics, the environmental fac-

---

[5] For an empirical verification of this approach see Storey 1991.
[6] For a description on the state of the art of the entrepreneurship studies see Bruyat and Julien 2000.

tors and the strategies adopted by the new companies[7], i.e elements considered by the main interpretive models as the most important determinants in the success of new company initiatives.

1.3.2 The Process of Venture Creation

The process of creation of a new company is a complex phenomenon, which is influenced by many factors that can be combined in very different manners. Notwithstanding, numerous studies attempt to provide a complete interpretation: for example, the models proposed in economic literature aim at identifying the principal choices that must be accomplished in order to create a company, and therefore defining the different phases that distinguish the process of venture creation. Obviously these interpretative schemes, given the extreme variety characterising the phenomenon, do not pretend to indicate an ideal development path to be taken in each situation. Rather, they provide a key to interpreting the process of creation of a new company, that is by nature a unique event with its own distinctive characteristics.

The formation of a new company starts at the entrepreneurial idea, which is in turn affected by the abilities of the potential entrepreneur and the context in which the economic initiative is developed. In detail, four phases can be underlined in the process of venture creation:

1. the definition of the idea;

2. the development of the entrepreneurial idea;

3. the raising of the resources;

4. the commencement of the activity.

The birth of the entrepreneurial idea always derives from the identification of a potential opportunity exploitable for economic purposes, as it can be said that this is the first manifestation of the idea itself. Different factors may forster the identification of this opportunity, such as:

- the data relative to the demand may suggest a deficiency in the offer of a particular product or service;

---

[7] Gartner, in a recent work mentioned 85 explicative variables referring to four different thematic areas: 1) Individual Characteristics; 2) Entrepreneurial Behaviours; 3) Strategy; and 4) Environment. (Gartner et al. 2000).

- the presence of a unique combination of price/quality in a particular sector can reveal the existence of a broad area for the pursuit of differentiated strategies;

- the consumers' request to a company that has developed a product or service to respond to particular needs can reveal the ability of that product/service to satisfy needs other than those it was initially designed for.

Once the opportunity is identified it is necessary to decide how it will be exploited, thus relating to the definition, even with a considerable approximation, of the market it is going to address and the main competitive advantage for the success of the new initiative.

The second phase of the process of venture creation is more complex and generally requires a considerable commitment of resources from the neo-entrepreneur. In fact, the technical, operating and economic feasibility of the entrepreneurial idea must be verified. This involves the necessity of performing a series of activities such as:

- market research;

- for industrial companies, the realisation of a prototype;

- the identification of possible contractors and suppliers;

- the forecast, even if only approximate, of the financial needs connected to realising the project, and the investment's pay-back period;

- the identification of the sources of financing attainable.

Therefore, these are the passages from the first idea to performing a real and proper project, that may in some cases take on the form of a document describing all of the characteristics of the initiative to be launched. Clearly, in this phase, due to the increasing knowledge on the business and the awareness of the difficulties that can be met, there is a continuous learning process that leads the entrepreneur to redefine several times different aspects of the project as a function of the restraints, demands and responses coming from the operating environment.

The next step to be undertaken is provisioning the necessary resources in order to start the activity. In particular, the constitution of a company requires:

- human resources;

- financial resources;

- logistical resources;

- administrative resources.

The human resources are identified by the combination of those contractors that the company must be equipped with: employees, consultants, relevant professionals and in some cases shareholders. The involvement of other parties as shareholders can be necessary from a managerial and financial point of view, however it is fundamental that they share the same system of values and especially business vision. The ability to be equipped with a capable team that works well together is fundamental for the success of the entrepreneurial initiative.

The provision of financial resources is also an unavoidable factor for the creation of a new company. Normally in this phase, given the high risk of the investment, it is difficult to obtain the financial resources under the form of debt capital: therefore, the financing is provided by the personal means of the entrepreneur and in some cases through the intervention of a venture capitalist. Here, the convenience for the recourse to outside finance depends on many factors such as: the chosen sector of activity, the degree of innovation in the strategies pursued and obviously the amount of financial needs of the newly created company. In many cases however, not having considerable financial resources available rather than being a constraint can reveal itself as an opportunity for the company that is obliged to undertake innovative steps in order to reduce its needs to a minimum (Bhidè 1999).

The logistical resources are represented by the necessary infrastructures. In particular, reference is made to the physical locations where the activity is carried out (warehouses in the case of industrial companies, offices for service and floor space for trading ones, etc.) and the different machinery and equipment in order to realise the production.

Finally, the administrative resources relate to the different authorizations that must be obtained from the relevant authorities before the commencement of the activity.

The provision of these resources is a crucial moment in the creation of a new company. As a matter of fact, during this phase there often arises unforeseen difficulties whose overcoming requires large commitment and considerable abilities by the entrepreneur. It is not by chance that the inability or impossibility to recover the necessary resources resulted in the failure of numerous entrepreneurial projects.

The last step to be undertaken consists of the commencement of the activity and in the the product/service marketing. This relates to verifying the quality of the offer and testing the market response. In this phase an iterative process begins, disciplined by a feedback mechanism that allows the entrepreneur to learn from the clientele's responses. Often this process results in redefining substantially the characteristics of the offer, in a continuous adaptation to the market requirements. However, in some cases several attempts are necessary before the process of adjustment leads to the hoped-for results and thus a long period of time elapses before the start-up process can be really considered as concluded.

1.3.3 The Financing of the Venture Creation Process

The creation of a new company obviously involves the emerging of financial needs that can be achieved by referring to different sources. In economic literature, the approach that links the financial needs to the model of the company's life cycle is very widespread. However, the explicatory abilities of these models are limited, particularly with reference to those companies operating in emerging sectors, whose development process may involve peculiar trajectories, which are difficult to interpret with traditional analysis schemes.

It appears, therefore, necessary to develop new approaches that are able to explain the behaviour of these business realities. In this sense, a scheme can help understanding the variables that most influence the financial models in the process of venture creation. In particular, it is possible to identify, adopting a taxonomical approach - the cases when the launch of a new entrepreneurial initiative requires the involvement of institutional investors; i.e., when it is convenient for the entrepreneur to seek outside finance early in the development phase of the entrepreneurial project (seed financing) and start-up (start-up financing).

Evidently, at least in theory, all those wishing to commence a new activity are potentially interested in obtaining financial resources from third parties. However, the start-up of a new company is a very risky activity and as such little desirable to the suppliers of debt capital that, due to information asymmetry, would find themselves particularly exposed to phenomena of opportunistic behaviour and adverse selection[8]. Therefore, the start-up typically must be financed through risk capital that can be supplied by institutional investors. In this case as well the information asymmetry is important, but the problems that derive from it are inferior as the entrepreneur and the venture capitalist, finding themselves on the same level, pro-

---

[8] The relationship between investment choices and financing choices are shown in the second part of this chapter, in which there is also a brief review of the principal theories that consent an interpretation. In this manner also the agency financing theory is taken into consideration, which sheds light on the capacity to interpret effectively the relationship borrower lender. See paragraph 1.4.1 of this chapter.

pose the same objective (Reid 1998). Therefore, in understanding when a real demand for venture capital can arise it is indispensable to make reference to the investment logic of the venture capitalist, as in fact in order to describe the demand of a particular service it is first of all necessary to explain the characteristics of that service.

The venture capitalist finances, in general through the acquisition of participations, new entrepreneurial initiatives and small non-quoted companies, with the objective of sustaining growth in order to realise an adequate gain when exiting from the participation. The investments realised by these intermediaries are thus distinguished by several distinctive characteristics in terms of returns expected, time horizon and minimum investment size.

The expected returns are normally very high, and in fact only the prospect of attractive gains can justify the support of a considerable risk such as that associated with, for example, the financing of a start-up. With reference to the time horizon, this is generally comprised between 4 and 7 years; however the venture capitalists, although qualified as investors in the medium-long term, are not permanent partners of the company financed, and the possibility to exit easily from the participation acquired is a fundamental element. Finally, in relation to the minimum size of the investments realised, it appears appropriate to recall that the selection of the projects to invest in and the successive assistance and monitoring activity of the company financed require the commitment of significant resources, that can only be justified for investments of a certain amount. There exists, therefore, an equity gap, that is a minimum threshold under which it is not convenient for the investor to finance the company. In this sense, it may be useful to make reference to the average size of investments in start-ups during the course of 2000 in Italy by institutional investors: that is equal to Euro 1,5 million (A.I.F.I. 2001), and thus it is reasonable to consider improbable the involvement of a venture capitalist for investments inferior to Euro 0,5 million.

Given the characteristics of the capital offered by venture capitalists, as outlined above, it is important to understand in which cases it is possible and convenient for the entrepreneur to seek this source of financing already in the venture creation phases. Here, the most significant variables are: the sector in which it is intended to launch the new initiative, the strategy followed and the preparation level of the potential entrepreneur.

The typology of activity and the strategy followed influences in a determining manner the financial needs of the new company and its potential growth, while the level of preparation of the potential entrepreneur plays a fundamental role in the ability of the project to attract external financing (Fig. 1.1).

Fig. 1.1. The possible recourse to venture capital

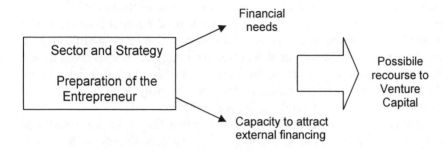

Generally speaking, in all of those cases where the launch of a new initiative occurs in traditional sectors, with an imitative strategy and by parties without any particular reputation, the financial needs must be, naturally, covered through the entrepreneur's personal resources and the art of bootstrapping financing[9]. In this sense, it is interesting to underline that the scarcity of financial resources can represent an opportunity, rather than a restraint, as it is the source of innovative behavioural strategies.

Therefore, it is not imaginable to have an involvement of institutional investors already in the phase of venture creation, because of objective limits, such as extremely large financial requirements, or when the involvement of a venture capitalist would not result in any real advantage for the success of the initiative. In fact, the added value that the institutional investor could supply, in terms of knowledge of the competitive dynamics within the (very traditional) context of the initiative and managerial ability relating to the phenomenon of rapid growth, would be very limited. Rather, the intervention of an external financier in these cases would have the single effect of complicating the management of the new company, undermining in some measure the flexibility essential in the commencement phase. In addition there could arise problems of agency with further detrimental effects on the successful outcome of the start up.

Therefore, the demand of venture capital, and more precisely seed and start-up financing, mainly relates to promising initiatives by parties equipped with technical-specialisation knowledge in innovative sectors of activity and significant growth potential, typically: biotechnology, tele-

---

[9] Financial bootstrapping refers to the use of methods for meeting the need for resources without relying on long- term external finance from debt holders and/or new owners. It is a general phenomenon, appearing in different contexts, such as R&D activities in large businesses, financings start-ups, etc. Winborg and Landstrom 2000.

communications, information technology etc. . In fact, in these cases, in the commencement phases of the company it may be already necessary to make large investments, and in addition the personal standing of the initiative promoters represents a guarantee for potential external financiers. The possible intervention of a venture capitalist may also be of significant help for the success of the initiative, as often the entrepreneurs involved do not have the managerial expertise; therefore the venture capitalist can assume the role of permanent consultant of the new company, with both reference to strategic aspects and purely operative problems.

At this point, it is necessary to make a further distinction between those cases where involvement is possible and convenient already in the definition phase of the entrepreneurial project (seed financing) and those cases where, instead, this is only considered necessary in the later phases of the process of venture creation (raising the resources and launching the activity, i.e. start up financing).

The development of the business idea requires mainly investments of an immaterial nature, typically based on activities of research and development, analysis of the market, identification of potential collaborators and employees etc. Therefore, the financial requirements related to carrying out these activities are not high, while the risk of failure of the initiative is very significant, and as a consequence the birth of the company highly uncertain. In general, although in the presence of high growth potential, it is difficult to have an involvement of the venture capitalist already in this phase. The financial needs are obtained by drawing on the personal resources of the initiative promoter and any financing obtained from state agencies operating as incubators.

An exception is represented by the medical-pharmaceutical and more specifically biotechnology sector: as a matter of fact, on the one hand, the investment in research and development can be very onerous, on the other the positive outcome of the research realised may result in enormous profits. Therefore, it may be convenient for both the potential entrepreneur and institutional investor to start the collaboration early in the seed phase.

The recourse to venture capital in financing new entrepreneurial initiatives assumes greater significance in the start-up phases than in that of the development of the entrepreneurial project, which for its characteristics is better financed by incubators.

1.4 The Capital Requirements for Developing Small and Medium Sized Companies: Venture Capital as a Possible Solution

1.4.1 The Companies' Choice of Financing: The Principal Theoretical Approaches

Venture capital is not a phenomenon confined to financing newly created entrepreneurial initiatives, but it also relates to financing the expansion projects of those SMEs who wish to impose a strong acceleration in their own development course, resulting in strong discontinuity in its revenue streams. Reference is made to the realisation of those projects that can represent a turning point in the life of a company and that can mark in a decisive manner future developments. In particular, these investments relate to: entering new markets, the internationalisation of the company, its diversification, integration processes, the re-conversion of mature businesses into more profitable activities etc. Rather than dwell on the characteristics of each of these typologies of investment, what should be noted here is the entrepreneurial risk: in fact, the capacity to reimburse and adequately remunerate the financiers mainly depends on the successful outcome of the investments realised and not on the cash flows from the company's Assets in place. Unfortunately the companies, especially those of small size, find great difficulty in recovering capital that is prepared to support this type of risk: in this sense it is possible to talk of a real and proper rationing which often compels them to give up profitable investment opportunities. The development of venture capital can represent, therefore, a significant opportunity, as the capital offered has all the characteristics that make it suitable to finance growth.

This reasoning presupposes the existence of a link between nature of investments and ways to finance them, in contrast with the classical theory of financing that is founded on the separation between investment and financing choices. A brief review of the principal theoretical contributions that have marked the passage from the classical financing theory to the modern one is appropriate, in order to understand the characteristics distinguishing development capital, the difficulties met by the companies in its sourcing through traditional channels, and the contribution that can be supplied by the venture capital industry.

The classical financing theory is based on three strong hypotheses:

1. the efficiency of the market;

2. the homogeneous distribution of information between lenders and borrowers of funds;

3. the separation between investment and financing choice choices (so-called Fisher theory).

The theory explains how, in a world characterised by stagnation, i.e. in the absence of company development, the financial choices must be interpreted on the basis of a trade off between the advantages connected to the utilisation of indebtedness - connected to the tax shelter generated by the fiscal deductibility of the financial charges - and the disadvantages related to debt, due to the costs of failure. There is, therefore, a point of optimum level of indebtedness of a company, where the marginal advantage connected to the tax shelter is equal to the marginal disadvantage of the costs of failure, where the company's value is maximum. As a consequence, the financing choices are dependent on reaching an optimum financial structure and are not influenced by the nature of the investments realised by the company. This separation between the investment decisions and the financing ones is founded on two hypotheses: first, the convenience of the investment opportunities can be unequivocally established by comparing the rate of profitability on the investments and the cost of company capital; second, it is possible for the companies to receive unlimited quantities of capital at market prices. Obviously this is not so, as the largest part of transfers of capital are characterised by a combination of requests, where the remuneration test is only one. Other important characteristics refer to the timing distribution of cash flow payments/reimbursements and their matching with the incoming cash flows of the company.

According to the classical financing theory, existing differences between the typologies of financial means mobilised by the companies, for example between self-financing and fundraising through an increase in share capital, or between debt raised respectively from shareholders, financial intermediaries, or directly on the market, are not considered.

In the successive theories, thanks to the removal of several strong hypotheses the financing choices have an important role. There would no longer be solely own capital and debt capital in the financial structure of companies, but a variety of financing forms, each characterised by particular obligations connecting the lender to the borrower.

The first hypothesis to be removed is that relative to the homogeneous distribution of information, thanks to the contribution supplied by the agency financing theory, that is founded on the asymmetrical distribution of information[10], e.i. the existence of hidden information and/or hidden action. The first can result in the phenomenon of adverse selection[11], while the second can of moral hazard[12].

In an asymmetric information approach, a characteristic element is the ability to interpret the relationships between the different subjects who will finance the company. Therefore, a distinction between inside finance and outside finance[13] and, in relation to this latter, between bank finance and

---

[10] The first systematic formulation of this theory was proposed by Jensen and Meckling 1976.

[11] The adverse selection is a phenomenon that explains a financial market failure, in particular the failure of price, that is of the interest rate, in explaining the relationship existing between borrower and lender. If the borrower has an information advantage, that is to know something about the company that the lender does not know, there is the possibility that this information will be used to the detriment of the lender, despite the fact that he wished to defend himself from that risk. In fact if the lender in defending himself, wished for example to increase the rate of interest on the financing, he would achieve only the effect of drawing himself to, or better to be drawn by, parties who find the implicit premium in the interest rate requested by the lender still low compared to the effective risk of his company. In short, for all that the lender can "think bad", there will always be a category of applicants for funds with information advantages and characterised by a risk worse than that imagined by the lender. For this reason the high level of interest rates cannot protect the lender from this risk. As the high interest rates draw higher risk parties one talks of adverse selection. This phenomenon explains why the financial markets are not able to construct contracts relative to any typology of risk in a world of asymmetric information distribution, such as the real world.

[12] The phenomenon of moral hazard is related to the liberty given to the borrower to utilise the funds raised from the lender. It is clear that any party could raise funds to realise investment projects, for example, very risky if the lender did not defend himself from this behaviour. Even a company in a stable sector and characterised by low risks could in virtue of this situation raise funds and successively decide to enter in very risky sectors or businesses. This circumstance, that is defined by opportunistic behaviour, is depicted in a situation where the lender attempts to defend himself introducing restraints in the behaviour of the borrower. When it is not possible to oblige the borrower to certain behaviour, the lender does not have any other choice but to defend himself from opportunistic behaviour by requiring guarantees. Thus if the phenomenon of the adverse selection explains the rationing of the credit, the phenomenon of moral hazard explains the reason for the request of guarantees from the borrower.

[13] The distinction between inside finance and outside finance arises from the conviction that the companies prefer to finance their own investment opportunities through internal financial resources available rather than through the different types of external sources. Beneath this hypothesis, is that, the self-finance is the normal form of financing investments, and the recourse by the companies to out-

market finance[14], is fundamental, as each financing channel is burdened in different measures by agency costs, i.e. the premium requested from the financiers to be at least partially compensated for risks that they are subject to in virtue of the distribution asymmetry of information. The distinctive element between financing forms is no longer founded on the financial instruments or on legal characteristics, but rather on the lender's possibility to have access to information and on his commitment to the borrower (Hellwig 1991).

In this sense, the successive developments of the agency theory and more precisely the signaling hypothesis (Ross 1977) and pecking order theory (Myers and Miluf 1984), have permitted the definition of a hierarchical order in the recourse to the different forms of financing in relation to agency costs. Obviously, the least costly form of financing is represented by self-financing: as there are no external lenders, the agency costs are zero. Then, in hierarchic order, we meet the funds raised from the intermediaries, due to the ability of these institutions to reduce the problems related to information asymmetry through screening and monitoring activities. Lastly, the company can turn to the emission of new shares on the market, but in this case the agency costs are very high as the subscribers do not have effective mechanisms to monitor the parties they have financed, and therefore require a premium for the high risk.

The modern theory is based on the removal of another hypothesis of the classical theory: the separation between the investment and supply decisions. This highlights a further distinction between different financing

---

side finance is motivated by specific circumstances. The hypothesis was refined gradually into the pecking order theory developed by Myers and Miluf, based on the same hypothesis of preferences for internal financing and in the signalling hypothesis, which provides further substance for the preference of the companies for inside finance, demonstrating that the recourse to external resources is accompanied by the considerable fall in the capitalisation value of the company. (Ross 1977; Meyers and Miluf 1984).

[14] The distinction between bank finance and market finance arises from the removal of the hypothesis of homogeneous distribution of information, which follows inevitably in the risk that the receiver of the funds would follow opportunustic behaviour in the realisation of the investment decisions, due to the exploitation of information and personal interests unknown to the lender. In this situation the role of the intermediaries is justified in that they are capable of reducing the costs of agency and monitoring compared to the market, through the achievement of economies of scale in the obtaining and elaboration of the information on the merit of the credit from the applicants for funds (pre- monitoring) and through the performance of successive control activity on the financing commited (pots- monitoring and commitment logic).

forms, connected to the covered or uncovered nature of the capital[15] in relation to the future cash flows generated from existing asset structures.
Therefore, it is necessary to distinguish, in the capital available to the
company, the quota destined to financing future earnings opportunities
from the ones destined to the present ability to produce earnings. The main
characteristic of first type of capital (so-called uncovered capital) is that its
remuneration and its reimbursement depend on the realisation and the success on these future opportunities. Here, it becomes important to distinguish between the self-financing quota already committed to the renewal
of assets (so-called repeat effect of the investments) or to the current remuneration of the shareholders, and the free one. This perspective permits
to understand how the distinction between renewal investment and development investment is key in order to distinguish the forms of financing capable of freeing the company's choices compared to the requirements of its
current functioning, taking into account that only the capital in excess
compared to the requirements of renewal financing can be used to development investment. This is a further enrichment prospective compared to
the vision founded on the distinction between inside and outside finance,
that allows to distinguish in the context of outside finance between the financing of the assets in place and the financing of development investments.

1.4.2 The Development Capital and Venture Capital

Development capital allows the company to use it in order to identify and
realise risky opportunities of wealth creation. It therefore finances the
company's development options, enlarging all the profitable investment
opportunities and/or the possibility to regenerate profitable investment opportunities that have reached the end of their economic life.

Another significant characteristic relates to the fact that this financing
represents the unique development potential within the resources available
to the company, it is not contended and has a so-called corporate nature.
The non-contentious nature is the exclusive character pertaining to these
opportunities by the companies promoting them: it is the condition neces-

---

[15] The introduction of the idea of covered capital in financing theory is attributable
to Myers 1984, who was the first to separate the total amount of the capital of the
company in quota supported by growth opportunities and quota supported by the
assets in place on the basis of the presupposition that the financiers tend to evaluate the companies globally and that this evaluation reflects an expectation relating
to future investments. One can therefore identify in every company which part of
the capital is supported by future investment and uncertain risky realization.

sary in order to guarantee the effective increase in the average profitability of the share capital. On the contrary, the opportunity of achieving instant profit in a competitive background, by the very fact that it is contended, is destined to have a short life span and scarce impact on the average profitability of the share capital. The corporate nature, on the other hand, highlights the fact that the development option always consists of an industrial project and cannot be identified only by the original investments (tangible and/or non-tangible, visible and non-visible)[16].

The risk that afflicts the development investments is, therefore, an industrial project risk, that may originate from incorrect valuations, or unforeseen difficulties which, striking the company at a delicate moment of transformation, can undermine the success of the project, suffocating it before it is born.

Development capital relates to capital with a slow turn around, as it finances the capacity to produce income. Moreover, it presents little possibility of asset-backing, realizable in only small percentages, as it does not finance assets in place, nor their enlargement[17], but the formation of new initiatives, or significant enlargement compared to the pre-existing organisation. For this reason the transfer of capital involving development capital is normally uncovered and can rarely be guaranteed, and the capital is characterised by a very high financial risk profile. This occurs as the development capital is normally configured as high multiples of the company's covered capital: it is in fact characteristic of the development project to induce a leap in the quantative and qualitative earnings power of the company. The main problem is that it generates financial requirements which self-financing can cover only in small measures: for this reason, the total means available plus those raised ad hoc by the entrepreneur as own

---

[16] The economic theory has identified a particular typology of investment in those actions of company policy that, although constituted by actual earning sacrifices in order to achieve greater earnings in the future, they are not recorded in the financial statements as investments and therefore risk not to be seen by the shareholders, creditors and stakeholders in general. This typically relates to reduction of revenues caused by the realisation of an aggressive pricing strategy, with the objective of increasing market share; costs for the improvement of product/processes not capitalised. The original contribution that introduced the distinction between visible investments and non-visible investments in company accounting is attributed to Stein 1989.

[17] In this there is the difference between development capital and the forms of financing dedicated, typically project financing, which on the contrary are based on the possibility to detach from the company environment the earning capacity of specific investment projects and create new financial structures supported by covered capital.

capital result in any case very modest when compared to the total investment.

The development options are characterised to supply low possibility of asset backing and to be information intensive, as they are dependent of the success of specific entrepreneurial capacities: therefore, this kind of financing is particularly sensitive to the financial tightness conditions of companies.

The point is, that given their characteristics, development capital does not lend itself to be raised through traditional financial channels (credit intermediaries, stock markets). In fact, being uncovered capital by nature the one that finances industrial projects and on which weights an entrepreneurial risk, its sourcing from third party lenders results in the emergence of important agency problems that in many cases can result in the failure of the financial system. In particular, the risks the suppliers of development capital are exposed to are twofold: the risk of transfer of wealth in favour of receivers of funds causes information advantages for the latter; and the risk of the "roll of the dice", that is the transfer of risks to the lenders that are not remunerated for their engagement. In other words, the supplier of development capital must confront two problems: the difficulty, caused by the presence of information disadvantages, to an adequate prior evaluation of the risk they are requested to finance; the possibility that retrospectively the projects effectively realised by the company financed result in being riskier than those initially programmed. The raising of development capital by companies from third party lenders is subordinated, thus, to the circumstance, that these parties are capable of adequately managing these agency problems minimising its cost.

Unfortunately, neither the traditional credit intermediaries nor the capital markets are capable of reaching this result. With reference to the credit institutions, the first problem is that often these intermediaries do not have the resources and expertise necessary to adequately evaluate the development projects of the SME and, therefore, in order to avoid the phenomenon of adverse selection, opt for capital rationing. The second problem concerns the fact that, even where a credit institute decides to finance the SME's development options, in any case the financing through debt capital only would be inadequate. In fact, this solution particularly exposes the lender to the phenomenon of the roll of the dice, as the shareholders, facing up only the favourable part of the risk, are stimulated to realise investments that increase the volatility of the cash flows generated by the company. This behaviour, therefore, has the effect of reducing the value of the credit for the lenders to the company. The problem can be resolved only in the presence of perfect systems of control on the shareholders' operations. Therefore, the monitoring activity exercisable by the credit intermediaries does not appear so effective, and the only way in which they have to defend themselves from the risk of the roll of the dice is that of requesting adequate guarantees, i.e. to transform the uncovered capital requested by the company into covered capital. Therefore, it is not possible to raise development capital through this financing channel.

Sourcing of development capital for SMEs through recourse to the market also appears to be very difficult, as other problems related to the company's size must be added to the already mentioned agency ones. In particular, reference is made to the fact that the quotation on a regulatory market has such a high cost as to render very often this way impossible for a SME. In addition, the problems deriving from information asymmetry remain, even if with different characteristics. In fact if, on the one hand, it is possible to have a partial realignment of the interests between shareholders of the company and third party lenders that can be identified as subscribers to company's new shares, on the other hand it is evident that the power of control of the investors on the company's commanding group is very limited. In other words, the market, as a mechanism for governing the relationships between lender and borrower, is not able to reduce the information asymmetry that typically weights on such relationship, often with reference to small companies, where evidently the financial community do not concentrate their interest. Therefore, the market is not the channel in which the small companies can finance their first and decisive development, but rather the place where the growth already achieved can be confirmed.

At this point it appears evident that the only intermediaries capable of satisfying the requirements of the small companies for development capital are the venture capitalists. In fact, they are the only ones who succeed in managing the relationship with the company in a way to reduce the agency costs to a minimum. The venture capitalists, acquiring a participation in the company financed, find themselves in the condition to meet both the favourable part and the unfavourable part of the risk weighting on the promised initiative. This means that they find it appropriate to be equipped with a structure, in terms of resources and expertise, that allows them to adequately evaluate the risk-return profile of the companies' development options and thus to reduce in a significant measure, compared to other typology of financiers, the risk of transferring the wealth in favour of the receivers of the funds. In addition, due to the participation acquired, they own control instruments on the company financed capable of minimising the risk of opportunistic behaviour by the old shareholders, both in terms of consumption of perks and in terms of roll of the dice with reference to that part of development capital supplied under the form of debt capital.

In conclusion, venture capitalists succeed in better governing the relationships with the companies financed[18]. In particular, these intermediaries provide the financing of the development options of the company, not in one unique occasion but through several successive transfers. In this manner, they maintain the possibility to abandon the company in case the profitability prospective should not prove satisfactory. This is fundamental as it means that the entrepreneur has an incentive to behave prudently in the use of its resources. Secondly, the venture capitalists impose on the companies financed the adoption of management remuneration schemes, in which the variable part, subordinated to reaching specific objectives, is absolutely dominant. In this case as well, the effect obtained is that of giving the company's managers necessary incentives in order to avoid opportunistic behaviour. Finally, the venture capitalists actively participate in managing the company financed, as often one of their representatives sits on the company's board of directors. This consents these intermediaries to manage a very effective monitoring system capable of timely anticipating and/or signalling the emergence of possible problems.

It therefore appears evident how the companies' only alternative to self-financing its development phase is represented by the venture capital industry, and how it is in these terms that the demand from small companies should be interpreted in dealing with venture capitalists: the small companies are interested in institutional investors entering their capital as this is often the only via through which they can seize development opportunities.

The reasons for the demand of venture capital by small companies are, thus, of a purely financial nature. In this sense, it is interesting to underline that here the management contribution of the investor, much more than being simply appreciated, is supported by the entrepreneurial group that sees it as a price to pay in order to permit the company's development.

---

[18] For a more detailed analysis of the models through which venture capital companies succeed in managing agency problems from the relationship that ties them to the company financed see the fundamental paper of Sahlman 1990.

2 A Broad Vision of the Investment Process in Venture Capital

Renato Giovannini

2.1 The Characteristics of the Managerial Process of the Venture Capitalist

The creation of an entrepreneurial project takes place through several phases of development, from birth to maturity, characterised by a series of business problems that from time to time may relate to various aspects of its entire structure. The venture capitalist proves to be, much more than the most immediate vision of simple supplier of risk capital, a financial intermediary capable of supporting and strengthen the growth of a company and assisting it "with strategic planning, management recruiting, operations planning or introductions to potential customers and suppliers" (Gorman and Sahlman 1989). It is, therefore, of fundamental importance from the commencement and take-off phases that the entrepreneur, often gifted with great creative and innovative potential, is supported by this type of consultant, capable of supplying solutions to possible qualitative problems, by the choice of optimal managerial resources for the requirements of the organisation. One can thus understand how the venture capitalist invests in an initiative still in the embryonic stage, or an idea or research programme that must be completely or a large part of it still developed. From here one understands the importance and problematic nature of the analysis that we wish to make.

The characteristics of the managerial process of venture capitalists represents one of the key aspects for the comprehension of the activity of investment in risk capital. In the first part of the chapter we will deal with the institutional activity of the venture capitalists, the intervention phases, the key role of assistance to the managerial entrepreneur, the methodology and the process of selection, of investment and of exiting. The underlying aspects to the investment process in risk capital can be associated with the considerations, still valid today, made by George Doriot in 1940:"always consider investing in a grade A man with a grade B idea. Never invest in a grade B man with a grade A idea" (Bygrave and Timmons 1992).

The second part reflects on the problems relating to agency between venture capital and entrepreneur, on the setting up phases and on the overall structure of a venture capital initiative. In addition we examine the critical factors that have historically influenced the development, the formation and the success of institutional investment activity in risk capital in Italy. Finally, we focus on the principal aspect of the managerial process, namely on the importance of the managerial resources in the development of the business.

2.1.1 The Activity of Institutional Investment in Risk Capital

In the classical meaning for venture capital it is commonly intended the activity of investment in risk capital of emerging entrepreneurial initiatives, of small sizes, at the commencement of their development or the creation of new companies. The parties (specialised companies, closed-end funds, regional development financiers, professional persons, industrialists etc., diversified for the modus operandi) invest significant financial and managerial resources in innovative projects that show high potential, often with minority interests and in order to make capital gains. These investors supply capital for initiatives differentiated by risk, diversifying the resources on a sufficiently high number of initiatives («portfolio approach») that allows them to balance the successful cases with those not having the hoped for developments. In order to realise the expected capital gain, the venture capitalist after a period of time, usually 3-5 years, proceeds to the exit.

It is therefore appropriate to underline that the venture capitalist:

– intervenes with established time limits;

– acquires only minority interests, due to the entrepreneurial risk and to the difficulty in selling the controlling interests;

– places particular emphasis on returns (capital gain/goodwill) as the participation in risk capital is only partially remunerated during the period of ownership, through the receipt of dividends or compensation for the consulting activity performed;

– they may supply some services contrary to the closed-end funds that cannot perform these services, due to the exclusion clause in their statutes.

2.1.2 The Intervention Phases in Risk Capital

The intervention in risk capital, relating to the various stages in the development of a company (origination - implementation - exit closing) each characterised by diverse sizes, prospective and requirements, realizes the necessity of an institutional intervention different depending on the case and especially in relation to the possible combination of the two key aspects of this activity: the capital and the know how.

The most classic and widespread segmentation of the risk capital market, whose categories, though with minimum geographic adaptation, are internationally adopted by the operators, by the associations and by the research centres, even for statistical purposes (Dessy and Vender 2001), classify the types of investment, substantially, according to the different phases of the life cycle of the target company.

The possible types of venture capital intervention are normally classified and characterised by the participation in the initial stage (Early-stage financing) that is sub-divided in:

- seed financing (experimentation phase). The investor in risk capital intervenes already in the experimentation phase, when the technical validity of the product/service still has to be demonstrated. The investor intervenes with modest financial contributions in the initial development phases of a project and in the preparation of the relative commercial feasibility plan. The risks are very high;

- start-up financing (commencement of activity phase). In this stage the production activity is financed, but the commercial validity of the product/service is not yet known. The financial contributions and risks are high;

- first-stage financing (first development phase). In this stage the commencement of the production activity has already been completed but the commercial validity of the product/service must still be fully evaluated. The intervention provides for high financial contributions and lower risks.

The increasing complexity of the analysis of the sector and the particular problems relating to each of these stages[1] means that, in certain cases, the stage of development of different companies, and the financial re-

---

[1] One thinks in the context of the Information Technology sector and what is happened in the more advanced economic systems, where "newly born» companies are ready for quotation on the Stock Exchange.

quirements connected to them, are little suited to a classical schematisation.

In addition the investors of risk capital have developed advanced instruments of financial engineering, increasingly more complex and sophisticated, through which simultaneously they can turn to different possible leverage, resulting for this reason difficult to classify.

It would therefore be coherent to define a more analytical classification in relation to the possible strategic requirements of the company, to the considerations of the threats/opportunities facing the sector and to the final objectives considered satisfactory for the investor. In this environment it is possible to group and classify the transfer of risk capital, by the institutional investors, in three principal types:

- financing of the commencement;

- financing of the development;

- financing of the change / reflection.

2.1.3 The Role of Managerial Assistance in Managing the Commencement by the Institutional Investments

The management of venture capitalists in the commencement of new or not yet consolidated entrepreneurial activities (start-up financing) results in a contribution of financial resources in the initial phase that is limited if compared with the managerial support required for the development of the company's project. The strategies adopted in this phase, in fact, require large amounts of managerial expertise to create, in support of the underlying entrepreneurial idea, a position of competitive advantage, which can be maintained over time. There is therefore created a true and real synergetic interaction between the capacities of the entrepreneur, who contributes all of his relationships with economic agents (customers, suppliers, industry associations ...), and the venture capitalist, which makes available his network of relationships as tutor and successively sponsor of the initiative.

The aim, therefore, is to create a management capable of supporting the entrepreneur and his ideas with business strategies, with the purpose of avoiding the phenomenon of one-man one show.

In the commencement projects (be that seed financing or start-up) it proves to be determining, with a view to limiting to a minimum the risks of failure, the analysis of the feasibility of the underlying entrepreneurial idea of the project and the capabilities of the entrepreneur (Millan and Zeman 1987) to achieve it.

2.1.4 The Importance of the Pre-Selection of the Investments

In the selection of the investments the venture capitalists in order to choose in relation to sufficient economic rationality, operating in an environment of high information uncertainty, relies on the interaction between impartial components, analyses with strong methodological rigour, and subjective, experience and intuition. The first, results as determining in relation to the management of the main "scarce resource" for a venture capitalist, the time necessary for the support and the consulting to the entrepreneur. It follows that in order to optimise the problem of time consuming, the aim is to limit the possible initiatives with a strong pre-selection, working towards a selection by quality and not quantity. This occurs in virtue of the key role of venture capital in its strongly participatory function of assistance, professional advice and intervention towards the companies. From here results the fact that a good number of the investors in risk capital do not place more than 5 initiatives in their portfolio in a year.

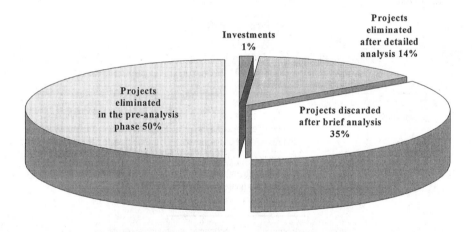

Fig. 2.1. The selection of the investments by the venture capitalist

As it can be observed from the above table (see Fig. 2.1) more than 85% of the initiatives are discarded before the selection process, and just 1-2 % reaches the financing and managerial (Collitti 1990) contribution phase.

2.1.5 The Methodology of Pre-Selection Adopted by the Venture Capitalist

The first evaluative step is represented by the pre-investment phase, where the process is made analysing by importance a series of critical factors that may condition the interest of the investor. This first screening is strongly influenced by the strategic orientation of the investor in risk capital: the geographic location, the sector and the type of product (technology used, trademarks, leadership in differentiation or of cost). The proposals received by the venture capitalist in this phase that do not overcome this initial analysis represent 50% of the total.

The proposals that reach the second step are examined with greater analytical detail in relation to the following key elements: analysis of the depth of the chosen market and its development, economic-financial results expected, amount of financing required. This results in the elimination of a further 35% of the requests for investment received by the venture capitalist.

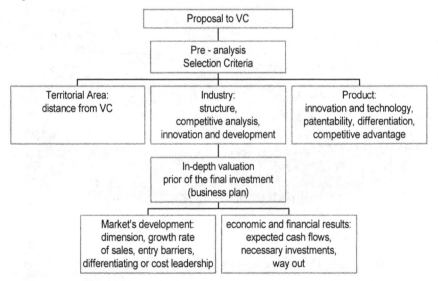

Fig. 2.2. From the analysis of the proposals to the pre-investment phase

It follows that the successive more detailed selection process is characterised by approximately 15% of the total proposals.

2.1.6 The Process of Selection of the Venture Capitalists

The selection process, as already mentioned successive to a pre-analysis of the validity of the proposal from the entrepreneur, is concentrated on several key steps that can be identified in the following aspects:

– the business plan (the detailed analysis of the pre-investment phases)

1. the business

2. the market

3. the entrepreneur and the management team

4. the competitive advantage of the initiative

5. the strategy

6. the economic-financial equilibrium

7. the timing

– the investment process:

1. the capital budget

2. the pricing and the structure of the investment

3. the exit closing

The Business Plan (the Detailed Analysis of the Pre-Investment Phases)

The business plan of financial investments has always favoured industrial sectors where the prospective returns in financial projects are very high. In Italy the success of recourse to venture capital, as a form of financing entrepreneurs' ideas, has contributed to the growth in the sector in an exponential manner, that today invests almost Lire 3,000 billion yearly in newly created or recently formation[2] companies.

The notoriety even amongst the general public of this form of financial support has made the search for funds an increasingly more difficult process. The determining variable in the choice of the investments to finance becomes therefore the time available to the venture capitalist in evaluating the various proposals.

At a company economic level, the projection plans thus developed become defined the business plan. This becomes the base for the request for

---

[2] See Fig. 2.2; it has already been noted that of one hundred proposals presented not more than five succeed in obtaining financing.

risk capital, and the first instrument of contact for the creation of the relationship between entrepreneur and institutional investor. It is therefore easy intuition the importance of the elaboration of this financial plan, in that it is a visiting card for the entrepreneur, the company and the project.

The objective in preparing an adequate business plan, therefore, is to achieve the double objective of transmitting in a clear and effective manner all of the information genuinely relevant that renders the project unique and interesting.

Reaching this goal, in fact, is not easy but without doubt it is the most delicate and important phase in the processes of searching for financing.

A valid financial plan must be accompanied by another document called the executive summary which substantially summaries in a space not greater than four pages the fundamental elements of the project: opportunities, risks, expertise of the management team and development time. The executive summary is the only instrument that can result in the venture capitalist dedicating attention also to a detailed business plan. It is essential that this document is written in an effective manner and captures the attention of the potential investors. The objective of the entrepreneur must be to communicate the concept of having an absolutely feasible business idea that is only missing (or prevalently) sufficient capital.

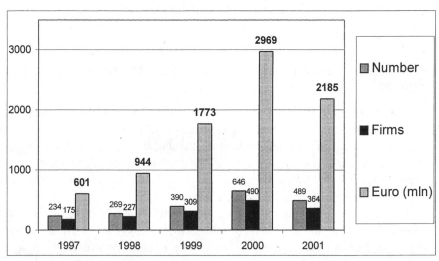

Fig. 2.3 Annual investments of venture capitalists in Italy (Source : A.I.F.I., Italian Private Equity and Venture Capital Association, 2002)

The executive summary must be written placing oneself in the perspective of the investor, highlighting therefore that, where existing, all the risk situations have been considered and evaluated. It must be demonstrated that the strategy contained in the business plan can be implemented with success.

To obtain a business plan prepared in an exhaustive and complete manner, recourse should be made to an analytical and detailed definition of a series of key aspects that we will examine:

1. The business

It is important for the venture capitalist to understand if it relates to a product or a service, a combination of the two or the creation of an original and more complex model. The description of the product or trademark must be focused on the principal attributes making it unique, which may be taken for granted by the entrepreneur but not so for the investor. Based on this it is necessary to specify what are the strong and weak points, the innovative validity of the idea, the potential problems and the developments planned. The investor is interested in knowing the limits of the product and the service offered. In addition, in the identification of the weak points of the product or service it is necessary to analysis the role that could be played by the introduction of new production technology, or a new consolidating advertising campaign or re-launch of the trademark.

2. The market

The analysis of the targeted market (or markets) must be very detailed not only in the definition of the boundary but also in terms of the impact that the project can have in relation to the forecast market share. The depth of the market is just as important: a venture capitalist is not interested in financing a project that has as an objective a 50% share in a potential market of ten billion lire but may be much more attracted by a market share of 10% in a market of a thousand billion lire. In detail the investor needs to know:

– the reference market in terms of the size of the demand, location and target (meant as product-market combination, which segment of the sector);

– the market share of the company and its defence;

– the competitors of the company;

– the coherence of the marketing plan (the overall actions relating to publicity and promotion, the distribution, the pricing policy,…) with the three preceding points.

3. The entrepreneur and the management team

The investor, prior to investing in the business, invests in the management, in its expertise and experience, in its cohesion and motivation. It is not always necessary to define a compete organisational structure but at least the group of founders of the project must be presented highlighting their capabilities and limits, interest and dedication to the project, essential for the achievement of entrepreneurial success. The investor wants, in fact, to understand if the entrepreneur, or its management, has a history of success to their name; if they were capable, thus, of reaching the pre-fixed objectives in the past and in what way they were achieved.

4. The competitive advantage of the business

A project is successful especially if it has the potential to maintain this status over time. It is necessary, thus, to explain what are the possibilities to defend the competitive advantages of the project, if it is possible to construct entry barriers to potential competitors and how it is intended to defend the idea of an attack from competitors already operating in the market. For this purpose the analysis is based on the competitive model[3], making reference to the "five forces" that interact in the sector and influence the performances:

- suppliers: contractual powers

- substitute products: threats of substitute products or services

- consumers: contractual powers

- potential entrants: treat of new entrants

- competitors in the sector: rivalry between existing companies

It is also necessary to demonstrate knowledge and awareness of the life cycle stages through which the sector passes and the critical factors for success (for example high entry barriers, legislative environment, unfavourable commercial and technological environment, …).

5. The strategy

The analysis of the strategy is the most important part of the whole business plan, especially as it permits the investor to evaluate what targets the company has established and how they intend to reach them. The strategy, however, does not offer a response to the individual problems that are inevitably encountered over time, as this task must be resolved in the "tactical plans", better known as "operating plans", and relate to the daily operations and not the medium/long term, which is expressed in the strategy.

---

[3] For more details on the competitive model of Porter see porter 1985.

It is of fundamental important that the strategy is also coherent and co-ordinated with the economic-financial projections that will be elaborated within the business plan. Often the investor utilises the collaboration of external consultants specialised in due diligence in order to verify the validity of the strategy and to monitor the risks (for example of the market) not considered within the business plan.

6. The financial-economic equilibrium

The central nucleus of every project is represented by the financial projections: costs and revenues, investments and cash flows are the basis for the evaluation of the feasibility. The projections must be interpreted as the means through which the strategy expresses itself. Although the concept of the strategy and the estimate of the economic-financial size can be made in different occasions, between the two aspects there should exist a perfect co-ordination and careful cohesion. In the estimate of the results an excessive optimism or, on the contrary, an unjustified prudence, must be avoided. The time horizon must be equal to at least a three year period, even if would be preferable to cover at least a five year period. For each year an income statement, balance sheet, and cash flow statement must be prepared and the principal assumptions upon which the estimates were made must be always stated.

7. The timing

In relation to the reference time horizon to be taken for the business plan, this varies depending on whether the plan relates to a project or a company. While in the first case the reference time period can cover the whole life of the project, in the second case it generally varies between three and five years, with a very detailed degree of analysis in the first year and a more generalised approach for the successive years.

In the drafting of the business plan, the management must consider that the potential financiers judge the work performed with different criteria compared to that utilised by the management of the company. The emphasis, in fact, is on the capacity of the business plan to create value for the investor. Finally, a successful business plan must also be capable of transmitting the enthusiasm and the determination of those who propose the investment, giving the investor the "vision" that was the basis of the decision to commence an entrepreneurial adventure.

The Capital Budget (the Investment Process)

1. The investment process phases

The investment decision by the investor, other than the evaluation of the entrepreneur, is made based on diverse factors, among which the current and potential market relative to the company, to the technology held, and to the presence of the conditions to create "value" in the exit phase of the participation in the company.

In the case where negotiations begin, generally a period of three to six months pass from the preparation of the business plan to the decision by the investor to finance it. This time is shortened depending on the clearness and completeness of the information supplied by the entrepreneur. Of particular importance are the terms negotiated not just in relation to the price, but also to the timing and method of payment.

The agreement of the preliminary contract provides for the signing of a confidentiality clause that commits the parties not to release company information except to external consultants directly involved in some of the aspects in the valuation of the investment. The commitment undertaken refers strictly to company information, but not that relating to the market or the sector in which the company operates. In setting up the negotiations the entrepreneur must impose it in a manner that will maximise the chances of success, as in the eventuality of failure, the investor can by right decide to activate a research on competitor companies operating in the same sector.

If there emerges a concrete possibility of agreement following the in-depth study by the investor of the key points of the operation, letters of intent are signed in which the salient economic, legal and company structure aspects are defined that will then be taken and re-elaborated in detail in the investment contract. In this letter essential aspects to both parties are explicitly included for a satisfactory conclusion of the agreement, among which: the value of the company, the presence of the investor on the Board of Directors, the informative obligations the entrepreneur commits to in case of investment and the eventual clauses provided for exiting.

The signing of the final contract occurs only successively, that is once a series of verifications of both a formal and substantial nature have been made.

Thus, a further phase commences, commonly defined as "due diligence". The term, taken from Anglo-Saxon terminology, refers to all of the activities, performed by the investor, necessary to reach a final valuation, analysis the current state of the company and its future potential. The activity of due diligence contributes to the protection of the capital of third parties invested by institutional investors. This activity, although present during all the phases of negotiation, intensifies in the phase following the signing of the letter of intent, up to the signing of the final contract. The institutional investor can directly follow this activity or entrust this to selected professionals, experts in specific industry and business areas.

Good progress of the due diligence guarantees a quicker closing of the negotiations and consents the investor to acquire all the necessary information to make the investment in a professional manner, avoiding the subsequent emergence of challenges and contentious issues.

The principal aspects on which the due diligence activity focuses on are classified as follows.

a. Market due diligence

Allows the investor to fully understand the potentiality and risks of the specific market in which the company operates and positioning within the market, in order to compare the results with the prospective plans presented in the business plan of the entrepreneur. Greater emphasis can be placed on the current market and on the future potentiality, as well as new market opportunities (internal and external) or new products/methods of distribution. To guarantee consistency of the company data presented, historical or prospective, with those of the market and in order to understand any differences, the activity of market due diligence must be co-ordinated with the financial due diligence.

b. Environmental due diligence

Provides for a comparison of the company profitability with legislation and environmental regulations, to the internal organisation of environmental control and pollution, to identification of the impact on the environment of current activities and to verification of environmental problems not yet resolved. If all of these analyses made by the investor give a positive outcome, the signing of the contract can proceed which defines in detail the terms of the agreement between the company and its shareholders on the one hand and the investor on the other. This contract provides for the normal guarantee clauses by the acquirer on the correctness and completeness of data, of the facts represented and on the inexistence of hidden liabilities, as well as the procedures for the resolution of disputes.

The investment contract also performs the function of shareholder agreements, with agreements governing the relationships between shareholders, that may relate to corporate governance[4], or exit plans[5].

---

[4] Instrument that governs the rules for the nomination and the functioning of the supervisory boards, for the functioning of the control of the business, and greater control required relating to particular acts.

[5] In relation to the agreement on the existing rules and on the exercise of option rights.

Once the agreement on pricing, on the quota of the participation and on the administration aspects regulating the final contract are reached, the operation is finalised with:

- the transfer of the shares;

- the payment of the agreed price;

- the issue of the guarantees (possible surety ships);

- the re-formation of the Board of Directors and the substitution of the directors;

- and the signing of any collateral contracts.

c. Financial due diligence

This is performed with the objective of evaluating the economic-financial aspects of the company's plans and in the definition of the necessary financing. This analysis highlights in particular the factors of success, and in addition pays attention to the historical economic trend in the past number of years, in relation to the sales, margins, trend of production costs and on fixed costs.

The financial due diligence develops, other than the trend of the historical and future financial structure, the following analysis:

- cash flow and working capital;

- budget and business plan for 3-5 years, considering the results of the market due diligence;

- the organisation structure of the company and its coherence with the market and economic-financial objectives;

- highlighting any liabilities and risks connected to the company's activity.

d. Legal due diligence

Brings to light the problems of a legal nature in relation to, either the valuation of the company or the drawing up of the final contract. In particular examines: any lawsuit in course or threatened; identification of particular commitments undertaken with third parties and the relative risks; the existence of legal requisites for the carrying out of the activity; the necessity of contractual guarantees; the situation of employee labour agreements; the employee agreements of the principal directors; the stock option plans and similar.

e. Tax due diligence

Analyses fiscal aspects related to the company, such as for example:

- potential liabilities and contentious matters (contingent liabilities);

- structuring of the acquisition operation;

- identifying future fiscal benefits (tax assets);

- preliminary identification of the fiscal effects of an exit strategy.

At this point the investor and entrepreneur are for all intents and purposes shareholders of the same initiative and must begin to work together in order to maximise the creation of value.

2. The definition of the price

The final valuation (Gervasoni and Sattin 2000) has as its principal objective the determination of a price, whose result does not depend as much on the valuation methods characterised by the general parameters, but on calculations based on data as near as possible to reality and testable, leaving to the market and the negotiations the identification of the correct parameter. For this, at an international level, the most widespread calculation method, despite the validity of exemplary techniques from a theoretical point of view, in the venture capital market is that of multiples.

3. The exit closing

The exit phase constitutes the final part of the investment process, an extremely delicate phase as it is in this stage that the gains in terms of goodwill must be realised, representing the ultimate goal of the institutional investor in risk capital[6].

The exit moment of the investor in risk capital of the company is almost never pre-determined, but is dependent on the development of the company. Increasingly, however, the investors attempt to anticipate, at the acquisition moment of the participation, the eventual exit channels and the time frame, in order to better plan the conclusive phase of the operation.

In the case of success, the exit is at the moment in which the company has reached its expected development and the value of the company, and thus the participation, has consequentially increased. In the event that the initiative fails, due to, for example, the new product or the new technology

---

[6] This operator, in fact, does not remain by his very nature connected to the company financed for too long a period, in that he is a temporary partner and with objective of realizing a capital gain in the medium-long term. If it was not such, it would be transformed into a holding of participation.

is not capable of establishing itself in the market, the exit is made when there is complete conviction that it is no longer possible to resolve the crisis situation created.

In both situations, the time and manner of the exit is normally defined with the agreement of all the shareholders.

Wishing to schematise, the manner of exit can be distinguished between the following methods:

- the quotation, after having made a public offer (public offering);
- sale through private agreement with varying methods:
- sale of the participation to new shareholders, industrial or financial,
- merger or incorporation with other companies;
- the sale of the participation to the majority shareholder or to the management (buy-back, leverage buy-out) or the acquisition of own shares (treasury shares);
- the zeroing of the participation following liquidation (write-off)[7].

Among the above mentioned methodologies of exit, that realised through the quotation of shares of the company on a regulatory market represents, in the majority of cases, the most sough after. From the moment it is possible to place even a minority of the company's capital on the Stock Exchange, this road permits the investor to sell with a profit the shares held, and the entrepreneur to maintain control of the company. In addition, having an institutional investor as a shareholder means having already taken important strides towards transparency and thus be largely ready to confront the quotation. At the same time, the admission to an official listing on the Stock Exchange is not a simple process for smaller companies and, thus, this channel can be inserted within a medium/long term viewpoint, as a method having a reasonable degree of certainty, only for companies that have reached a certain development and maturity.

The Stock Exchange quotation presents advantages and disadvantages. They, in fact, if on the one hand increase the liquidity of the shares, consent the making of new emissions and to sell off further amounts of capital, on the other hand, subject the company to controls that may not be appreciated and, not least, result in a certain cost.

---

[7] The methodology used by the EVCA subdivides the tipology of exits into: initial public offering, trade sales, write-off and other means, in which the modalities of the third point are included.

The presence of the institutional investor, anticipating many of the necessary passages, contributes in mitigating the risks and disadvantages deriving from the quotation process.

The significance and extent of the specific advantages related to the quotation and the relative charges depend on the parties involved.

2.2 The Venture Capital Operations and the Problems of Agency

The venture capital operations provide for the presence of three key figures: the financier of the venture capital operations (pension funds or other types of institutional or private investors); the venture capitalist for managerial and financial support to the entrepreneur; and finally the entrepreneur who developed the idea considered to have a high innovative content.

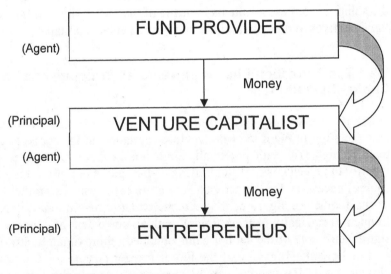

Fig. 2.4 The structure of a venture capital operation

The typical structure of venture capital contracts, in the relationships between institutional investor and company financed, may suffer the negative influence of asymmetric information situations, principally once the investment decision has been taken. Thus situations of moral hazard, see figure 4, may arise either due to the entrepreneur (Principal) or due to the venture capitalist (Agent). The critical factors that are added once the financing has been paid, firstly relates to the entrepreneur upon whom depends principally the positive outcome and thus the success of the company's project. In the absence of valid control mechanisms, this latter could have opportunistic behaviour following personal interests; for example, given that the entrepreneur is not capable of financing the idea with his own capital, he could motivate the investor in risk capital to continue financing a project even when there are no longer the conditions for its valid and effective development. On the other hand the venture capitalist could have opportunistic behaviour in retrospect, exploiting the ideas of the entrepreneur and financing other competitive companies resulting in the exportation of competitive advantages related to the idea of the initiative already financed.

At this point it is clear the importance related to the agreement of binding contracts concerned at resolving possible conflicting interests between the entrepreneur, wishing to limit possible leakage of information relating to competitive advantages, and the institutional investor, interested in maximising the motivation of the company's management for the most efficient development of the plan. The form of the financing chosen consents the solutions to these problems in virtue of the possibility to influence both the incentives of the entrepreneur and of the venture capitalist.

2.2.1 The Formation Phases and Overall Structure of a Venture Capital Operation

Other than the analysis carried out up to this point, in relation to the importance relating to the role of the venture capitalists in the support of initiatives at high risk with potentially high remuneration, it is necessary to identify the mechanism of fundraising and employment of the financial resources necessary for the activity of venture capitalists. From the sourcing of the funds, see figure 5, the recourse to large institutional investors is frequent (principally pension funds) with an agreement of concession of a significant quota of the capital gains obtained (approximately 80%) of the eventual positive outcome of the ipo or merger operations. From the employment side the venture capital companies provide the financing of a portfolio of companies, technically granting managerial support and financial resources to the entrepreneur, with the counterpart the transfer to the investor in risk capital a minority shareholding (equity between 5% and 15%). The primary objective of the operation made by a venture capitalist is, from the viewpoint of the performance of the company, turned to the achievement of a high profit (capital gains equal to least 20%), being the difference between the cost price of the participation on acquisition (at the moment of subscription) and the sales price to third parties (with operations of: IPO, LBO or Mergers) retrospectively.

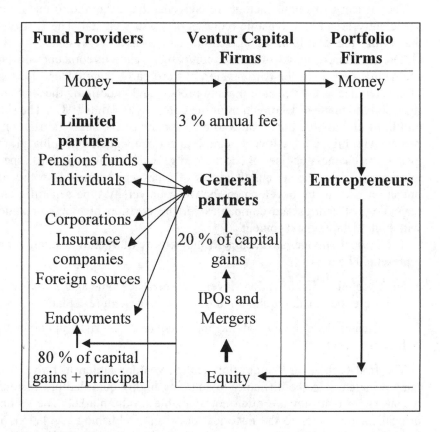

Fig. 2.5 The overall structure of a Venture Capital operationSource: Osnabrugge and Robinson 2000

2.3 The Critical Areas of the Management of Venture Capital Operations

The principal problems relating to critical factors that have historically conditioned the development and the formation in Italy of venture capitalists, are related to the following three aspects:

1. the closed ownership model of companies;

2. the growth process of family companies;

3. the innovative processes.

The critical exogenous factors are elements that delay the formation of sufficiently broad and dynamic requests of venture capital. The first type that limits the development of venture capitalists is represented by the closed ownership model of Italian companies. The entrepreneurs do not see with a good light the participation of third party shareholders, for fear of interference in the decision-making process and in defining the strategic lines and the managerial direction of the company (Collitti 1987). The second limit highlighted here relates to the capacity of the small Italian companies, with high competitive potential, to adopt adequate operating strategies to the business choices of accelerated growth. Finally, the third aspect is represented by the serious delay, as of today not corrected, in the cooperation and in the development between university type scientific and technological centres and companies for the creation of new companies with high technological content.

The critical endogenous factors in the activity of venture capitalists are represented by a:

1. High technical qualification, level of experience, managerial expertise and professionalism required in the activities of venture capital;

2. Insufficient development of the intermediation for risk capital in the Italian market.

The first problem in the development of venture capital in Italy, is related principally to the difficulty in sourcing highly qualified professional resources, in that they are often not available on the market. The second critical factor relates to the historical delay in establishing regulations in Italy, which up until 1987 did not provide for financial intermediaries having the possibility to operate as merchant banks or as venture capitalist companies.

Therefore, the observation, of how these critical factors in an international context with such high competitive content, typical of the Italian companies, often interact as negative discriminating factors for the development of venture capital activities.

2.4 The Importance of the Managerial Resources for a Valid Company Development

The realisation of the venture capital is a rather demanding task in the actual implementation, and as such requires an activity with a high content of consulting, an awareness of the high potential risk and an intense participation in all the aspects of definition and organisation of the final objectives of the operation (commencing with participation on the Board of Directors).

At a managerial structure level dedicated to the individual operations, the technical operations resembles more the activity of a professional practice than that of a branch of a traditional financial Intermediary, inspired by the overlapping interdisciplinary typical of the structure of investors in risk capital.

The principal critical factor of success of a venture capitalist, is represented without doubt by the professional quality of the human resources dedicated to performing managerial activity with a highly innovative function, as well as demanding. The wealth of knowledge and expertise, in fact, is one of the principal requisites to guarantee concreteness and feasibility of projects of this type.

The characteristics required, other than the above-mentioned expertise, are numerous: confidentiality, business aptitude, flexibility, critical mind, and capacity to decide based on a precise calculation of the risks.

It is of fundamental importance, thus, that the resources utilised by the venture capitalist supporting these entrepreneurial initiatives have proven experience in the field of evaluating companies. This know-how is represented by the capacity to perform a business analysis not exclusively limited to economic-financial aspects, but widened also to the strategic analysis of the sector (opportunities and threats of the possible product-market combinations), and to the study of the strengths and weaknesses (at competitive level) of the future development programmes of the company.

The role of a venture capitalist in operations with the purpose of transferring risk capital, is different from that of mere substitute to the activity performed by the entrepreneur. It consists principally of a much more complex role than support and assistance, aimed at providing a specific professional contribution, characterised by human resources that guarantee a service with a proven image of "active neutrality". From here the strong need, for all those type of intermediaries, to undertake a radical change compared to the traditional behaviour models, each starting from different levels, be that in relation to the diversity of experience, the capacity to maintain their market segment, or of their capacity-propensity to change.

The participation in the shareholding of small and medium sized companies by the institutional investors of risk capital, must therefore be based on a responsible entry in the best entrepreneurial initiatives.

It is thus possible to combine the expansion of the company while maintaining the family character (Dessy and Vender 2001). In fact, the recourse to venture capital has the significant advantage of allowing the entrepreneur to place outside the family shares that are under his control, to issue preference shares, to place on the Stock Exchange or outside the quota of minority shares in companies controlled by the holding company, creating real and proper groups of companies capable of carrying out autonomously recourse to the capital markets.

2.4.1 The Company Profile at the Foundation of an Ideal Venture Capital Operation

We can now outline the ideal profile of a company that a venture capitalist intends to support and participate in:

1. Company gifted with a management possessing undisputed entrepreneurial expertise and experience, in relation to the typical role of the venture capitalist in working alongside and assisting in a professional manner but without having to substitute the entrepreneur;

2. Company gifted with an analytical business plan and innovative market strategies, capable of guaranteeing a potential level of development compatible with the dynamics of the market and with the situation of the company;

3. Opening of the company to the entry of new financial shareholders: such as the breakdown of the typical psychological and cultural barriers that prevents Italian entrepreneurs offering share capital to external partners;

4. High company performance: current and future profitability considered coherent with level of risk-return, with a good economic-financial equilibrium in order to guarantee a "secure" return of the investment both in terms of dividends received and capital gains;

5. Company management characterised by absolute transparency, obtainable by certification of company financial statements.

The intervention of venture capitalists, therefore, excludes a significant amount of companies based on their strategic development plans, economic-financial conditions, and finally the level of managerial expertise held.

We will attempt to identify in a clear manner what are the real and potential advantages of this type of intervention. Wishing to list the potential advantages for the company, we can recall:

- Entry of new financial resources, for the development of the company's initiatives;

- New opportunity to access financing, for the family operated companies;

- Entry of a prestigious shareholder, permitting greater contractual power in relation to both suppliers and competitors, and from customers in the greater guarantees offered;

- Optimisation of the company image with the markets and traditional financiers, thanks to the presence of solid financial partners that assure the validity of the company and its programmes;

- Strong distinction in the corporate governance between corporate and personal interests, for greater weight towards market policy and the business strategies adopted by management;

- Possible synergies (strategic choices, organisation, selection of personnel), between the expertise of the company's management and those not only financial of the minority partner.

Wishing to summarise all of the advantages outlined above, it is possible to affirm that the greatest value is constituted by the change in mentality that consents the small-medium sized entrepreneur to widen his financial horizons, which often result in having limited space to manoeuvre.

2.4.2 The Possible Factors of an Unsuccessful Financial Participation: the Principal Threats for the Venture Capitalists

The analysis of the profile determining the possible failure of an investment in risk capital, shows that the principal cause is related to the activities performed by management. Several empirical verifications permitted the testing (Gorman and Sahlman 1989) whereby in 95% of the cases the actions of the senior managers was the most significant factor in the failure of the venture capital operations[8]. It was also noted how the managerial

---

[8] The analysis was performed on a scale of 1 (factor that contributed only partially to the failure) and 5 (determining factor in the failure).

functions, relating to the financial, marketing, distribution and production, were further crucial elements for the final success. In relation to the final product, it is noted that in half of the cases the delay in its development was fatal for the positive outcome of the operation. In figure 6 a further series of possible critical factors are highlighted: from the market entry to the distribution channels to finish with the level of competition in the sector.

In relation to what has been said the conclusion is reached that the more frequent a problem arises, the greater is its gravity and thus its negative impact on the final result.

It can be concluded that the management, as well as in concurrence with another series of problems, results in being the principal critical factor in the development of the venture capital initiative. In relation to this the theory maintains that "the managers are the dynamic element and the source of life of any business. Without this factor the production resources would be only resources and would never become productive " (Drucker 1989). It follows that the role of management is without doubt very complex, in that other than having to manager the key company functions, it is fundamental in the strategic choices in relation to the determination of the business in which it operates, or in confronting possible external threats.

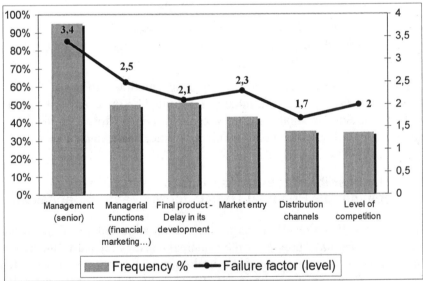

Fig. 2.6 The problems determining the possible failure of venture capital operations

2.4.3 The Problems of the Participation of the Venture Capitalist on the Board of Directors of the Company

The last aspect that will be analysed relates to the problems concerning the control of a venture capital operation. The effective participation in the capital of the company results in the requirement to monitor information asymmetry due to the activities performed by the entrepreneur and the management. The principal instrument safeguarding the capital transferred is represented by an active involvement by the venture capitalist in the management of the company. The possibility have own people to exercise a control on the Board of Directors, permits the institutional investors to participate in the key decisions in relation to suppliers, market policies, extraordinary financial operations, and in the ordinary management of the company. The theory of the financial intermediary leads this type of monitoring to two forms: the first, defined "soft facet" (Hellmann and Puri 2000), where the interaction with management in relation to the principal operating choices and decisions is of mere support; the other typology, "hard facet", provides for a form based on the control of the activities of the entrepreneur, with the purpose in limiting the creation of possible conflicts of interest between the objectives of the entrepreneur and those of the venture capitalists, and to the creation of company value. What has just been said, is clearly aimed at establishing the best conditions for the obtaining of a final return that permits the institutional investor to consider the operation in line with his own expectations of success.

For what has been discussed up to this point it is clear how the salient aspect of the activity performed by a venture capital company is much more different than a simple support to the entrepreneur and of consultation to his projects. As previously established, in fact, at the foundation to the positive outcome of an investment initiative in risk capital there is the irreplaceable role covered by the venture capitalist in the support, assistance and participation in the development of the various phases of the entrepreneurial project, supplying his specific professional contribution, characterised by human resources which permit the increase in the chances of success of the operation undertaken.

3 Funding Processes

Manuela Geranio

3.1 General Features of Venture Capital Funds

In the past five years the activity of venture capital and private equity has seen unprecedented development in the European market (see Fig. 3.1).

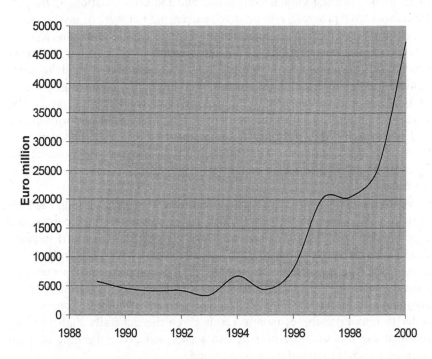

Fig. 3.1. Venture Capital funds raised in Europe in 2000 (Source: EVCA 2001)

From the demand side, greater numbers of entrepreneurs were looking for funds in order to develop and realise their business idea, particularly in those sectors related to the so-called new economy (software, telecommunications, biotechnology, etc). At the same time, the development of stock markets and in particular the creation of new markets dedicated to companies with high growth potential increased the possibilities to play exit strategies, thus facilitating the realisation of interesting capital gains for the venture capital investors and the raising of further financial resources for companies.

From the offer side, the investors have searched for new investment opportunities, in order to increase return on investments following the reduction in interest rates that has characterised the European financial markets in the last five years. An increasing number of investors, both professional and not, have thus come in contact with the venture capital activity, deciding to invest their capital in the industry. At the same time the presence of a venture capitalist as a shareholder is often identified as a good sign of quality in the selection of companies that are either ready for entry on stock exchange listings or already quoted. At a sector level, the development of the venture capital companies and the consolidation of its performances have progressively attracted increasing numbers of investors.

A common definition accepted for the venture capital term is as follows "..share capital contributions or the subscription of convertible shares by specialised operators, over a medium-long term time period, made in companies, not already listed in the market, with high growth potential in terms of products and services, new technology, new market concepts …The venture capitalist considers the holding as temporary, of a minority nature and performs its role not only through financial resources but also with know-how contributions …The final aims is to increase the value of the investment and realise a high capital gain on exiting." (Gervasoni and Sattin 2000).

Venture capital funds are characterised by a long-term although not indefinite perspective. In fact, there does not exist a pre-defined contractual time horizon with the companies that received the funds (even if in practice this period varies between three and five years). Vice versa there exists a maturity date by which the venture capital company must return the funds raised to the investors. Hence the need for a careful selection of the companies to be financed and a close management of these investments, in order to maximise the economic result given the generally high level of risk that the investors are prepared to accept and given the time horizon limit by which the capital must be returned.

The stability of the funds is essential in order to guarantee substantial financial resources to the companies in the start-up phase of their business idea, where cash flow is typically negative and thus not capable of securing the interest payment on any loan financing obtained. On the other hand loan financing from banks would be difficult to obtain given the absence, in a start-up company, of tangible activities as a guarantee or trustable forecasts on future cash flows capable of assuring the servicing of the loan. Thus the preference for financing through equity financial instruments rather than loan financing.

A second important contribution deriving from the presence of a venture capitalist in the share capital of a company is related to its non-financial resources, e.g. managerial expertise which is particularly needed in the corporate governance phase of the start up. A recently formed company, which often operates in an innovative market, does not possess a track record or an historical past sufficient to present its credentials and therefore it derives its value from the future growth potential. These characteristics require significant evaluation, monitoring and management capabilities, not common to all investors. Only the so-called "active investors" are typically capable of valuing these types of investments, as the maintenance of a direct shareholding in the company allows them a direct access to monitoring and if necessary to the management. On the other hand, not all institutional investors are accorded the role of active investor, because either not all of them have the necessary capabilities to perform this task or the entrepreneur would prefer to communicate with one single interlocutor only, in order to maintain the maximum flexibility from the interference by external parties which is crucial in the start-up phase of an enterprise.

The relationship that is created in a venture capital company is twofold (see Fig. 3.2): on the one hand, the relationship of financial support between entrepreneur and venture capital company; on the other, between this latter and the investors from whom the funds have been received. In both cases contracts are drawn up in order to avoid financial asymmetry problems and moral hazard emerging between the parties. The present chapter will focus on the contracts and relationships between the venture capital company and investors, attempting to identify the morphology of fund raising for the venture capital sector in Italy.

Fig. 3.2. The twofold relationship managed by the venture capitalist

3.1.1 Information Theory and Fundraising

The fundraising originates a principal-agent relationship between venture capitalist (general partner) and the individual investors, therefore coming financial asymmetry and moral hazard. The first series of problems are connected to the difficulty the investors have in monitoring the venture capitalist, since the investments made in start ups are difficult to value and by definition do not have a comparable value in the market before their conclusion or a possible sale to third parties. In relation to moral hazard, it can be observed that given the limited undertaking of risk made by the venture capitalist on its own behalf (generally equal to 1% of the capital of the company financed) the general partner could have an interest to employ funds in excessively risky activity, in order to obtain greater personal income in case of success (and a contained loss in case of failure).

To reduce these information problems the contract for fund raising provides for various clauses that define the raising and utilisation methods for funds. The information asymmetry is typically solved by imposing an obligation of constant and periodic communication on the trend of the investments made.

In relation to moral hazard, there are usually three contractual characteristics designed to contain it. First of all, a maximum time horizon is determined by which the funds must be returned to the financiers. This time horizon is on average approximately ten years. Therefore the investors acquire the possibility to condition, even with long time horizons, the venture capitalists actions, limiting the availability of the funds invested to a determined period. Secondly, it is usually provided that the revenues from exiting are distributed to the investors and cannot be freely reinvested by the venture capitalists. As a third recurring contractual characteristic it is usually provided that the initial raising of capital from investors is realised through a series of payments over time, or conditioned to the good operations of the venture capitalist (although limited to the early stages of its activity). These clauses have the objective of controlling the venture capitalists actions in the early stages of the initiative. In particular the takedown plan generally provides for a payment between 10% and 33% of the total agreed amount upon signing the contract. The subsequent payments are made in two or three successive occasions, at the general partner request and in relation to the investment requirements (particularly in funds of greater dimensions) or according to a precise schedule. In the majority of cases 100% of the funds are raised in a period varying between two and four years from the beginning of the contract.

3.1.2 Micro and Macro Economics in Fundraising

After having briefly mentioned the problems of information theory in the venture capital sector, in order to analyse the fundraising market it is necessary to mention two other categories of factors, of a micro and macro economic nature, that can influence the methodologies and parties intervening in the fundraising.

Firstly the business factors (reputation of the venture capital company, age, track record, etc.) and those related to the contractual structure used. Just as the venture capital company must select appropriately the most profitable companies, the investors must choose the company in which they entrust their money. Typically investors screening between good and bad venture capitalists is performed on the basis of the experience that the venture capitalist can show as well as on the quality and detail of the information supplied on the due diligence and governance rules that the general partner is prepared to be subjected to. Other contractual variables can influence the facility in fund raising. For example, the size of the fund is certainly an important element as an excessively high number of investors entail increased administrative costs and disclosure. At the same time too few investors could discourage the subscription as this is a symptom of an unsuccessful fundraising caused by asymmetric information among the investors themselves (one could for example suppose the presence of informed and not informed investors, where the former decided not to subscribe to the operation due to information in their possession and the latter observed such behaviour as a negative signal). In such a situation it would be better for the few adhering investors to abandon the initiative. For this reason, the fundraising contracts generally include a minimum and maximum amount both in terms of capital value raised and number of investors[1]. In order to assure equal treatment among the investors new entries after the beginning of the investment activity are generally not permitted, except where there is a prior authorisation and the contribution paid by the new entrants includes all the costs incurred by the existing investors.

An empirical verification carried out in the English market (Robbie et al. 1999) attributes factors such as past performance, the tendency of concentrating the investments in a small number of funds of large dimension and the necessity of a diversification of the offer by venture capital funds as the determining factors for the success of the fundraising by the venture capital companies.

Parallel to this it is observed that the venture capital funding is similarly influenced by a multiplicity of macroeconomic variables (for example the GDP trend and the interest rates) and context (degree of development of the IPO and stock market, legislation, etc.). In the latter case empirical evidence is greater.

---

[1] The maximum value is in some cases surpassed with the agreement of the investors already present up to a maximum of 10-20%.

A research focused in the United Stated market (Gompers and Lerner 1999) outlined that the development of venture capital activity in the period 1972-1994 was largely attributable to changes in the pension funds regulation (now qualified to freely invest in venture capital funds), the renewed taxation regime of capital gains, the economic development and the growth in research and development expenditure.

A second research (Jeng and Wells 2000) extends the theme of the critical factors in the venture capital fund raising to an international comparison between 21 countries. The results show the importance of the initial public offering market as the principal driver of the investment choice. The development of pension funds assumes grater importance over time but not in a country comparison. Surprisingly the GDP level and capitalisation of the financial markets do not seem to exercise a particular effect, while government policies exercise a strong influence. A final result underlines the negative relationship between the rigidity of the labour market and the degree of development of venture capital activities in the so-called early stage. This is an interesting starting point in explaining several particularities employed by venture capital activity in Europe and in Italy, such as the lower importance given to early stage investments compared to those concentrated in the later phases of a company's life.

At present for the Italian market there does not exist a detailed empirical research on the micro and macro economic factors that influence the fundraising, due partially to the relatively recent beginning of venture capital activity. However the results of the principal foreign experiences offer several useful pointers and are extendable to the analysis in the Italian context.

3.2 Origins of Funds

The organisational forms of venture capital fundraising and management are rather heterogeneous. In theory, it is possible to trace them to the following categories:

1. Business Angels

   They are private investors, who finance individually new entrepreneurial initiatives, characterised by a high availability of personal finance, generally together with a detailed knowledge of the sector in which they wish to invest. The risks of the projects undertaken are particularly high and superior to that potentially sustainable by the institutional venture capital companies (restricted by procedures and rules of risk management). Consequently also the returns, in case of success, are particularly high. The principal limit of this form of intervention is related to the limited, although considerable in absolute terms, dimensions of the assets that the individuals are prepared to invest.

2. Private pools of funds

This relates to a partnership in which several shareholders (limited partners) decide to invest part of their own assets. Typically these funds originate from entrepreneurs or holders of substantial assets interested in jointly investing part of their wealth in young enterprises with potentially high returns. The first historical experiences were realised by entrepreneurs who, having consolidated their own company's success, intended to use their own know-how and available financial resources in new initiatives with excellent development prospects. Among the first experiences were those undertaken by family companies such as Rockfeller and Phipps and Whitney. Also in more recent times private single entrepreneurs have given life to private organisational forms meant to invest in innovative initiatives. The existence of experience matured in the field and a solid financial background allows them to sustain projects with high risk in sectors that are similar to those known (thus exploiting the experience matured) or in different ones (in order to diversify their portfolio). The transfers made between shareholders can have very differing dimensions, from 25,000 to 10,000,000 dollars individually. The investments are typically concentrated on deals of small dimensions.

3. Corporate funds

This relates to funds and financial resources belonging to companies and not private individuals, destined to be managed by a venture capitalist and thus focused on financing companies in the development stage. Typically in this case the corporation aims at both financial and strategical objectives, such as information gathering in the target sector with relation to research, marketing, development of new products or production processes, etc. If the presence of consolidated companies represents an important channel also for the later financing of the start-up phase, in certain cases the culture and interests of the former can refrain the development of new initiatives due to the lack in defining precise objectives and to a poor capability in managing rapidly evolving situations. Normally a corporate sponsor will be less likely to abandon a non-profitable initiative due to the loss of image that could result for the parent company.

4. Mutual investment funds

Mutual investment funds represent a very important financial channel for venture capital, given the size of the capital raised from the public and the consequential requirements to diversify. It relates to a financial vehicle that provides for the raising of capital through the emission and placement of quotas with investors (both institutional and private) to be invested in participations. In particular, the closed-end funds represent an instrument capable of meeting the requirements of the start-up companies, given the stability of the capital for a medium-long term horizon and the high risk profile that can characterise the single investments. In the Anglo-Saxon markets the raising of closed-end funds is prevalently realised through pension funds, both public and private, while in the European case the largest part comes from the banking system and other categories of institutional investors. The closed-end fundraising is realised during the initial formation phase through financial intermediaries, till the capital requirement is reached. The fund will therefore be managed by a specialised intermediary, who will invest in development projects whose returns will be distributed to the investors on exit.

5. Public venture capital companies

This relates to venture capital companies who have gone public in order to obtain greater financial resources compared to those obtainable by a private placement. The greater inflow of capital is a consequence of the greater notoriety and transparency that the company derives from the quotation as well as the possibility for the investors to exit from the operation due to the existence of a secondary market where the shares can be sold. The principal experiences relate to American companies, while in the European case only 3% of the companies are quoted (although many are currently considering the opportunity of a stock market quotation)[2]. In Europe the founders and families connected with the general partner are usually the larger shareholders of venture capital companies.

6. Banks

The banks and financial intermediaries (insurance companies, finance companies, and investment banks) normally have the necessary expertise to value industrial projects and to raise funds for their realisation. In particular the merchant banks are the institutions that are most oriented to long-term investments and prepared to sustain risk levels that may be considerably high in the hope of achieving greater returns than the av-

---

[2] Data from Coopers & Lybrand Corporate Finance 1999.

erage. This incentive acquires growing importance even among the traditional financial intermediaries, given the necessity to get back through the stock market activities the profitability compromised from the progressive reduction of intermediary revenues. The operations of the traditional banks in the venture capital and the private equity sectors are expanding also due to the removal and reorganisation of numerous vehicles that in the past prevented the investment in non-financial holdings. However, there still remain regulatory elements that discourage investment in venture capital (intended as seed and early stage phases, favouring the banks to acquire holdings in companies already in the growth stage, i.e. real private equity)[3].

7. Public Funds

The intervention of public operators in the venture capital sector generally aims at promoting objectives of a collective nature through the development of new company innovations, such as research, creation of new employment, growth of specific geographical areas. The size of public funds paid is generally however limited and their payment are subject to the satisfaction of rather long requisites and bureaucratic procedures. Some states have jointly accumulated wide-ranging funds, between 5 and 10 million dollars, reaching up to hundreds of millions. The size remains limited if compared to the largest venture capital private companies and the average investments vary from 100,000 to 500,000 dollars.

8. Others

Among the other institutions that may make investments in venture capital, not directly but generally through the acquisition of holdings in closed-end funds, the academic foundations and institutions focus their contributions to rather limited risk levels and generally through the acquisition of holdings in closed-end funds.

The proportion undertaken by each of the above-mentioned operators is different in each country (see also chapter 4). In the Anglo-Saxon context the majority of venture capitalists are structured as a single company that manages simultaneously several funds, each legally distinct from the other in limited partnership. The success of this organisational structure is essen-

---

[3] In particular with reference to the rules established by BIS in terms of the adequacy of the bank's capital it is noted that holding shares in companies less than two years old and not yet able to produce profits imposes on the bank a capital requirement equal to 2, thus absorbing the double of the capital requested for other investments.

tially attributable to the fiscal advantages that can be obtained. In each limited partnership the venture capitalist assumes unlimited responsibility, when subscribing for a share in the investment (generally 1%), while the investors are liable only for the amount of capital subscribed. In each fund the capital is invested in new activities in the first 3-5 years. Following this period new investments are rare and vice versa the exiting phase commences. The profits realised are thus distributed to the initial investors rather than reinvested. Normally before the exit of a fund is completed a new fund is launched in order to invest in new initiatives. Therefore the exiting period for old funds and the formation of new ones overlap one another, motivating the previous investors to continue financing new initiatives (it is estimated that approximately 72% of the increase in private venture capital in the USA between 1977 and 1998 was attributable to follow-on funds).

In the European context the birth of venture capital companies is to be found in the financial institutions, although successively some of them acquired their own independence. In Italy the prevalent structure is that of closed-end funds, similarly to the Anglo-Saxon limited partnership. The principal peculiarity depends on the fact that the largest operators in venture capital sector are of a captive nature, or rather held by a bank, alongside several private companies. A second peculiarity, common to the majority of European venture capital companies, relates to the type of investments chosen, concentrated for a 90% in buyout operations. A third characteristic relates to the lesser specialisation by sector of European funds, where only in recent years have "niche" products been developed, compared to the available offer in the United States.

3.3 How to Raise Funds

The fundraising represents a fundamental step for the concrete realisation of the successive investment strategy. Moreover, on average the fundraising phase lasts not less than one year and absorbs considerable resources in terms of time and costs. Therefore the fundraising strategy should be planned carefully, identifying the category of investors potentially interested in the initiative and establishing the kind of approach to be taken.

The principal channel for venture capital fundraising is by direct contact between company and investors. These contacts are often established using previous direct contacts with the investors or thanks to the creation of a network of contacts. For a recently formed company this latter is initially developed through common acquaintances and the intervention of agents.

Only in a second phase, when the venture capital company has established its own reputation in the market, if not a true brand name, the network of contacts tends to widen due to the investors initiatives and the venture capitalist encounters less difficulty in gaining credibility. In recent years alternative channels have arisen for the fundraising, in particular via Internet. The phenomenon, however limited to a few examples, appears to be characterized more by a promotional and marketing aim rather than being an effective distributive channel.

The fundraising strategy will be profoundly influenced by the circumstances whether the fund is encountering its first experience or represents the continuation of a previous initiative. In fact in the first case the absence of a track record and the necessity to develop a network of contacts makes the exercise rather complex and burdensome. Further variables that can affect the fundraising methodology (technique, strategy, …) are the size of the fund, the number of reports already in existence and the availability of managerial resources for the fundraising.

In the latter case an important decision must be made in relation to the possible use of the services of a placement agent, i.e. a specialised operator that can count on numerous relationships with potentially useful clientele for the raising of capital and, at the same time, in virtue of its experience can contribute to the definition of the characteristics of the fund and the marketing strategy to follow for the placement. The choice to take advantage of a placement agent is thus considered at the outset, given the necessity for its involvement from the very initial stages. On the other hand, the agents do not appreciate being called for a placement to resolve difficult situations in the fundraising. Therefore, when choosing just for the placement strategy, the general partner or manager of the fund must consider the advantages in terms of enlarging and diversifying the relationships with the investors. This acquires value in the short term as it can facilitate a quicker conclusion to the fundraising, also due to the greater experience of the agent, who often is capable of identifying more effectively the target clientele potentially interested in the initiative. In the long term, the enlargement of the potential investor base represents an important element for the consolidation of the manager's success, as it enlarges the reputation and reduces the dependence on a narrow group of initial investors. A further advantage deriving from the employment of an agent for the fundraising activity is the saving of time and energy for the manager, who can therefore concentrate his efforts on other critical activities such as the selection of investments. Meanwhile it is necessary to consider the cost of employing a placement agent. Normally the remuneration consists of a commission, which is generally high, applied only in the case of success. The most convincing argument is probably the cost of a failure in the funding phase, which consists in the fixed costs already sustained by the venture capital company (e.g. legal expenses) and in the loss of image for the company itself.

Once chosen the channel and parties to be employed in the fundraising, the next step is to identify the target market and develop the fundraising strategy. For a representation of the process see Fig. 3.3.

1. Identification of the target market

The definition of the potential clientele interested in investing in the fund represents the starting point for the fundraising. Attention must first be given to potential domestic clients, as the confidence shown by domestic investors is an important signal in attracting the confidence of foreign capital. On the other hand this latter is often essential in completing the fundraising, given that few non-United States funds succeed in completing the process in a totally domestic environment. In addition to this the foreign investors take into account other information, such as the economic prospects of the country where the fund is originated, the characteristics of its capital markets, the presence of interesting entrepreneurial initiatives, etc.

The size of the fund becomes particularly significant when there is an interest to involve large institutional investors in the initiative. For example in the United States these investors are generally prepared to invest a minimum quota varying between 10,000 and 50,000 dollars, on condition that this amount does not represent more than 10% of the fund. This means that directing the offer to United States investors only makes sense for initiatives of a significant size (200-300 million of dollars). Moreover, this conclusion is reached also in consideration of the marketing costs a placement in the United States would have for a European operator.

In the selection of potential clients it is also necessary to note the increasing role played by gatekeepers, i.e. e. institutional investors that offer consulting or management services for pension funds and fund of funds. Originated in the United States, but now also relevant in Europe, the gatekeepers raise funds from small or medium sized institutions, large institutions with no experts in the private equity sector or high net worth individuals who wish to invest in private equity initiatives. Given the attention and severity with which the gatekeepers select and follow their own investments, their presence in a venture capital fund represents a useful positive signal in attracting further potential clients. Thus the particular attention given to the involvement of gatekeepers from the early phases in the fundraising.

In the European context it is frequent for banks or consulting companies with a well-known and widespread reputation to sponsor the fund. The purpose of the sponsorship is to offer a form of reassurance to the investors on the validity of the venture capital company, which is often small and little known, in new and under-developed markets. Nevertheless the sponsorship obviously involves costs, such as profit sharing with the venture capital company in terms of fees or as a consequence of an effective holding in the share capital. Moreover, if the sponsor is a bank or a financial intermediary the sponsor are likely to take part in the investment decisions, with potential advantages in terms of capability in selecting and evaluating the investments but also with possible conflicts of interest.

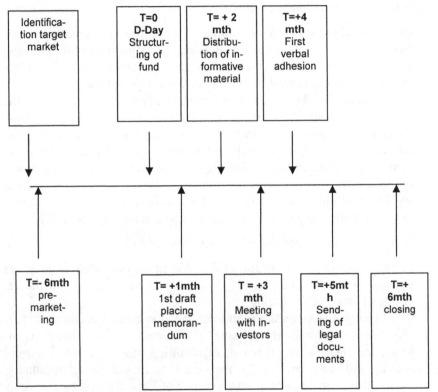

Fig. 3.3. The process of fundraising in venture capital

The sponsor must however be ready to show his track record (for example towards potential foreign investors), without making exclusive reliance on his local reputation. When selecting the sponsor it is also necessary to consider that the professional specialisation of the sponsor contributes to the quality of the message sent to the potential investors: that is, a commercial bank or consulting company specialising in M&A deals does not always represent the most suitable sponsor in strengthening and enhancing the competences of a venture capital company.

2. Pre-marketing

This phase focuses on sounding the potential market for the project, in order to evaluate the interest and gather useful information for the final structure of the new investment proposal. Normally this occurs through meetings with existing investors, in order to update them on the new possible initiatives. Moreover, meetings of a purely informative nature will be made with new potential investors.

3. Definition of the fund structure

Once gained the approval of the market through the pre-marketing contacts, the fund structure must be defined, in cooperation with legal and fiscal consultants. This phase ends up when the first version of the placing memorandum is established and successively verified.

The definition of the structure of the project and the drawing up of the relative documents must aim at eliminating as far as possible the legal, fiscal and technical factors that could discourage the investors. They often receive numerous proposals and the first rapid selection is based on simplicity and transparency. Usually the investors appreciate comparable structures with previous investments undertaken, that clearly specify the fiscal impact and states possible conflicts of interest (e.g. the investment policy of a new fund compared to that already existent).

4. Preparation and distribution of informative material

In this phase the final version of the placing memorandum is prepared together with any other useful documentation in response to the requests of potential investors.

The placing memorandum represents the first crucial occasion to present the venture capital company. The document must therefore enhance the investment proposed, highlighting past experience, current activity and gains realised for the investors, as well as differentiation factors from competitors. In the presence of legal difficulties that delay the publication of the document some companies shows this information, not yet finalised to the sale of a quota in the fund, in an intermediate informative publication, to promote the manager's image.

The placing memorandum comprehends essentially three parts:

~ the investment proposal and track record (attractiveness of the geographical area and sector, timing of the initiative, competitive advantages of the fund, etc)

~ the terms and conditions of subscribing (size of the fund, timing of the payments, remuneration, reporting, management costs, etc.)

~ the legal structure, the fiscal and regulatory aspects (legal structure, place of jurisdiction, restrictions to investments, etc).

To the memorandum further documents are often added in response to specific requests from potential investors. Among these in particular the side letters, which confirm the absence of privileged investors.

5. Frequent meetings with potential investors

Meetings are useful to present the investment to the previous investors and raising of the first indicative domestic adherences. At the same time meetings are organized to propose the project to potential investors already contacted in the pre-marketing phase. International road shows can are also be organized to present the project to potential foreign investors.

The intensity and frequency of the meetings vary depending on the level of previous acquaintance already existing between the parties. In order to obtain the confidence of a new investor, for example, often at least four individual meetings are necessary, while with previous investors two meetings may be enough.

6. Due diligence

Investors and placement agents usually request numerous additional information in order to evaluate the proposed initiative. A rather significant value is attributed to the references of other managers of closed-end funds, or professionals involved in the investments under realisation, directors of the companies financed, previous investors, banks and partners.

The managers must however be prepared to offer precise information, if not contained in the memorandum, relating to the track record of past initiatives, specifying for each one the details relative to the structure of the operation, cash flows, growth of the investments and values and timing of exiting. At the same time it is useful to offer a list of references, previously advised, about the possible request for information from potential investors.

7. Preparation of the legal documentation

Sending of the legal documentation to the probable adhering investors (partnership agreements, copy of contract, fiscal and legal matters, etc.)

8. Conclusion of the operation

Gathering of the definitive adherents (possibly in two rounds, one for domestic and previous investors, the other for new/foreign investors).

A final observation relates to the consolidation of the relationships with the investors after the beginning of the project. Maintaining an active and transparent relationship during the life of the fund is fundamental to guarantee the success of new initiatives in the future (according to some experts a good manager must obtain approximately a 70% retention rate of investors). The most common techniques adopted are the sending of periodic information (usually on a quarterly basis) and the organisation of direct meetings with the most important investors.

As already mentioned the fundraising phase requires a considerable absorption of resources. The management must dedicate time to the realisation of the placing memorandum, to the definition and realisation of the marketing presentations, to telephone contacts and meetings with potential investors (three or four steps with the same institution are not rare).

The costs of the legal and fiscal consultants vary between 300,000 and 500,000 Euro, depending on the complexity of the operation (i.e. costs that in case of failure during the fundraising phase represent a net loss for the venture capital company). Moreover a further compensation for marketing consultants should be added. If the company decides to employ a placement agent there is also a commission of about 2% of the capital raised, as well as the reimbursement of the costs sustained by the agent during the promotional activity. However, it is necessary to remember that the time and resources employed in the development of relations with the investors is itself a long-term investment. As a matter of facts, if it is not rare that at the launch of the first initiative a fund succeeds in obtaining capital from one investor for every ten contacts, it is likewise true that if a fund achieves good performance a large part of the initial investors will probably show greater interest in future investments.

3.4 The Current Funding Policies in Italy

The Italian offer of venture capital funds is influenced by the rather recent development of this sector in the country. As shown in Table 3.1 and Fig. 3.4, in the two-year period 1999-2000 the fundraising more than doubled compared to the previous two years; the investments made have increased even more evidently, as for the first time in 2000 they exceeded the capital raised. Previously, in fact, the balance between fundraising and investments made by Italian companies, technically defined as overhang, was always positive, due to the presence of companies who raised funds without completely using them for the investments. The growth process of venture capital funds however suffered a sudden halt in the first half of 2001, during which both the fundraising and the investments substantially halved compared to the first half of 2000 (when the capital raised amounted to 1,320 million of Euro). Moreover the overhang in the last six-month period was negative, thus showing that some companies financed part of the investments with funds raised in previous periods.

Table 3.1. Capital raised and invested in Venture Capital in Italy (millions of Euro)

ITALIAN MARKET	1997	1998	1999	2000	2001 (First six months)
Capital Raised	1,068	1,052	2,207	2,925	513
Capital Invested	602	944	1,779	2,969	674
Overhang	466	108	428	-44	-161

Source: AIFI, 2001

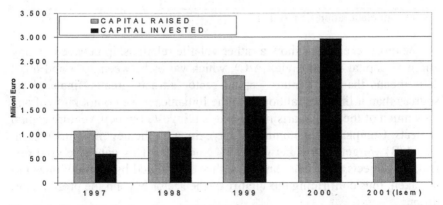

Fig. 3.4. Capital raised and invested in Italy (Source: our elaboration AIFI data 2001)

The slowdown is certainly due to the worsening economic climate and the crisis arising particularly in the new economy sector, rather than specific to problems in the Italian venture capital industry. In fact at the European level[4] there has been a reduction in the volumes invested in the first six-months of 2001 equal to 17.4% compared to the first six-months of 2000. Even more significant was the reduction in the amounts raised (-78%) and investments made (-73%) in the United States market, where over 81% of the funds were invested in follow-on projects already existing, thus underlining the tendency of consolidating existing portfolios in expectation of an improvement in the general economic context[5].

A further signal of the attractiveness of the venture capital sector for the lenders of funds in Italy is the proportion between the capital freed from exiting and the new resources raised and employed, as shown in Tab. 3.2.

Table 3.2. Exit, investments and funds raised for Venture Capital sector in Italy

	1999		2000		2001 (First six months)	
	Number	Million Euro	Number	Million Euro	Number	Million Euro
A – Exit	170	1,119	186	900	88	302
B – New Investment	390	1,779	646	2,969	295	674
C – Raised		2,207		2,925		513
A/B	44%	63%	29%	30%	30%	45%
A/C		51%		31%		59%

Source: our elaboration EVCA data

The most recent data show a rather volatile relationship between exiting and new capital raised (index A/C), which varies between 30% and 60%. This means that the venture capital sector attracts greater financial resources than it liberates, although in the Italian case we do not know if and how much of the capital arising on exit is reinvested in new venture capital projects. Comparing the number of the exits with the new investments (index A/B) we get a ratio between 44% and 30%. This data also confirms that the projects concluded are more than substituted by the new ones undertaken, thus confirming the vitality of the sector beyond the present conditions.

---

[4] EVCA data October 2001 based on a sample composed of 58% of the European Venture capital companies.
[5] EVCA data October 2001.

In order to identify the most active investor categories in the Italian market, reference is made in Fig. 3.5, which relates to the type of fundraising sources. In Italy over 70% of the funds are raised from independent companies, against 25% from captive companies. The number of independent companies is slightly above the European average.

In greater detail we can examine the Italian fundraising by type of investor, whose data are currently available only for the three-year period 1998-2000 (see Table 3.3). Most of the funds come from holding companies, although there is not an unequivocal trend in the period.

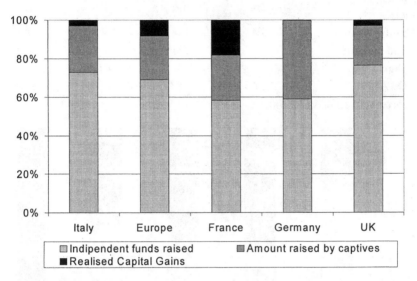

Fig. 3.5. Funds raised by source

However, if we add the quota relative to industrial groups to the first category, a more stable proportion between 40% and 50% is obtained. Banks increased their presence during 2000, taking their quota of funds raised to approximately one quarter. Vice versa, the individual investors quickly reduced their share. On the other hand, the growth of capital received from funds of funds and, although in a reduced manner, from insurance companies and pension funds, shows how the individual savings previously channelled directly into venture capital are now probably involved in more or less innovative forms of managed savings. Finally, the re-employment of the capital gains realised in new initiatives contributes rather modestly, due to the limited size of exits and the young age of the average initiatives undertaken in Italy.

Table 3.3. Capital raised by type of investor

CAPITAL RAISED	1998	1999	2000	2001 (Europe)
Holding Companies	32%	49%	24%	10%
Individual Investors	26%	6%	9%	7%
Banks	16%	11%	24%	20%
Industrial Groups	8%	6%	19%	-
Pension Funds	6%	5%	3%	22%
Insurance Companies	2%	5%	3%	12%
Capital Gain	5%	6%	3%	8%
Fund of Fund		9%	10%	10%
Other	5%	3%	4%	11%
TOTAL	100%	100%	100%	100%

Source: Aifi and EVCA 2001

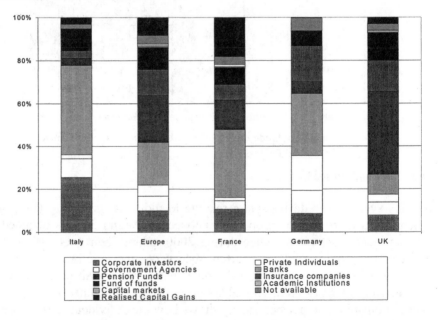

Fig. 3.6. Funds raised by type of investor

In terms of international comparison Fig. 3.6 shows a representation of the different types of investors in Italy and in the principal European countries (France, Italy, Germany and Great Britain representing jointly over 70% of the European market by volume). In Italy the presence of corporate and banking investors is above the international average; moreover, the individual investor's involvement is among the highest, while the presence of pension funds and insurance companies is considerably lower.

In relation to the geographic source of the capital raised (see Fig. 3.7) the domestic fundraising share is approximately 50%, while the remaining capital raised by Italian operators comes for a 35% from the European continent and for a 15% from other countries. Compared to the European average, the Italian venture capitalists attract less capital from non-European countries, although this does not represent an exception (e. g. the Germany situation). The most internationalised operators from the fundraising viewpoint are the British ones, who raise approximately 50% of the funds for venture capital activities from abroad.

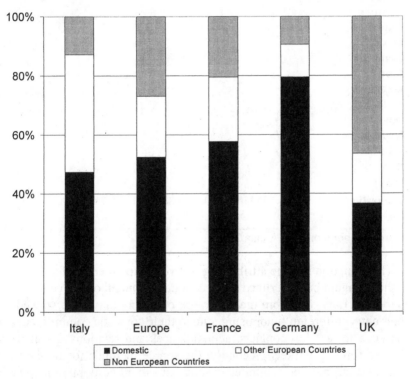

Fig. 3.7. Geographic source of funds raised

The Italian venture capitalists manage almost (93%) of the resources raised in Italy. Table 3.4 shows the total funds raised in each country divided by the nationality of the fund manager. The Italian proportion is largely above the European average (72%) and that of the main countries analysed (85%). The high presence of domestic management is mainly due to the fact that Italy shows strong barriers to international fundraising competition.

Table 3.4. Management of the domestic funds by domestic venture capitalists (millions of euro and %)

	Italy		Europe		France		Germany		UK	
	€	%	€	%	€	%	€	%	€	%
Domestic funds managed by domestic Venture Capitalists	1,387	93	25,185	72	4,317	85	4,863	85	6,479	85
Domestic funds managed by foreign Venture Capitalists	101	7	9,931	28	770	15	879	15	1,100	15
Total Domestic Funds	1,488	100	35,116	100	5,087	100	5,742	100	7,579	100

Source: elaboration EVCA data 2000

This situation may be attributable to two complementary factors. Firstly, it should again be underlined the recent development of the venture capital market in Italy: therefore many foreign operators may have slowed down their entry in foreign fundraising, due to their domestic commitment and to the need to develop contacts, acquaintances and relationships in recently formed foreign markets. Secondly, the source of funds for Italian venture capital is strongly concentrated among banks and corporate operators and this may create obstacles to the entry of foreign venture capitalists. In particular in the case of banks it is evident that funds normally entrust the management within the bank or a management company belonging to the same banking group (i.e. for closed-end funds), leaving little space to third parties. The sector which is currently most permeable to the entry of new foreign operators for fundraising is probably that of pension funds, given the greater experience of other countries in this sector. The timing for the growth of this sector however does not appear to be a short time horizon.

In completing the analysis of the fundraising it is useful to examine the division by type of investment target. Table 3.5 shows in the Italian market a progressive increase in the early stage investments, due to the diffusion of knowledge and presence of new operators capable of evaluating and operating in riskier activities. The last six-month period shows a greater interest for expansion investments, given the necessity to consolidate the initiatives in portfolio and the preference for investments with contained risks against the uncertain economic backdrop. Finally, it should be noted that buyout operations have historically shown constant growth, however abruptly interrupted in the last six-month period.

Table 3.5. Funds raised by type of investment target in Italy

	1998	1999	2000	2001 (First six months)
Early Stage	7%	21%	33%	54%
Expansion	16%	42%	25%	39%
Buy Out	27%	36%	40%	4%
Other	50%	1%	2%	3%
TOTAL	1005	100%	100%	100%

Source: AIFI 2001

The international comparison (see Fig. 3.8) signals that the Italian prevalence for early stage projects in 2000 is above the European average and very similar to the German case (although in Germany these early stage investments did not relate solely to the high tech sector, as was the case of Italy).

Finally it may be interesting to analyze the operations by category of operators, i.e. the intermediaries who perform the functional fundraising after the investment is made. Fundraising and investment are two phases particularly interconnected in the venture capital sector, given the necessity to explain clearly in the funding phase the types of investment to which the funds will be destined.

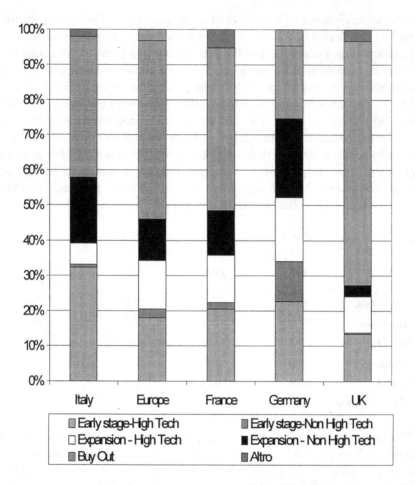

Fig. 3.8. Expected allocation of funds raised (Source: our elaboration on EVCA data 2000)

Table 3.6. Number of investments by category of operators

INVESTORS	Number of Investments				Data in %			
	1997	1998	1999	2000	1997	1998	1999	2000
Italian banks	53	93	93	119	23%	35%	24%	18%
Public operators	59	60	62	82	25%	22%	16%	13%
Regional financiers	57	42	55	54	24%	16%	14%	8%
Private/industrial operators	15	15	23	42	6%	6%	6%	7%

Table 3.6 (cont.)

INVESTORS	Number of Investments				Data in %			
	1997	1998	1999	2000	1997	1998	1999	2000
Italian closed funds	12	19	50	41	5%	13%	13%	6%
International operators	38	40	107		16%	27%	27%	
Advisor of VC funds				233				36%
Advisor of PE funds				69				11%
Emission of foreign bank				6				1%
TOTAL	234	269	390	646	100%	100%	100%	100%

Source: AIFI 2001

It is therefore evident that the total number of investments made has increased from 15% in 1997 and 1998, to 45% in 1998 and 1999, to 65% in 1999 and 2000 (see Table 3.6). The growing trend of amounts paid is even more evident, as it was respectively 57%, 88%, 65% (see Table 3.7). Consequently, as illustrated in Table 3.8, the average amount of the investment has increased too.

Table 3.7. Amount invested by category of operators

Investors	Amounts in million euro				Data in %			
	1997	1998	1999	2000	1997	1998	1999	2000
Italian banks	168	279	547	647	28%	30%	31%	22%
Public operators	69	75	100	138	12%	8%	6%	5%
Regional financiers	44	17	31	30	7%	2%	2%	1%
Private/industrial operators	98	189	132	81	16%	20%	7%	3%
Italian closed funds	39	84	159	120	7%	9%	9%	4%
International operators	182	301	811		30%	32%	46%	
Advisor of VC funds				369				12%
Advisor of PE funds				1,395				47%
Emission by foreign bank				189				6%
TOTAL	602	944	1,780	2,968	100%	100%	100%	100%

Source: AIFI 2001

The most active operators in Italy, both in terms of amount invested and number of investments undertaken are the managers of venture capital and private equity, who realised 47% of the total operations by number and 59% by amount. These operators are largely international (in the past they were identified by the item "international operators"). This evidences a greater internationalisation of the Italian venture capital market from the investment perspective compared to that of the fundraising (or rather the foreign investors are more prepared to invest funds previously raised elsewhere that finance Italian projects rather than raise funds from Italian investors). In the second position we find Italian banks, which represent 18% of the investments and 22% of the volumes. In previous years, when the volume of activity was lower compared to the present day, the banks represented an even greater share of the market. This is explained by the major role played by the banks in financing Italian companies, traditionally by debt capital and more recently also by risk capital, following the reduction of constraints in undertaking industrial participations[6]. The share of banks in the venture capital investments has therefore reduced in the previous two years not due to a reduction in volumes, that on the contrary have increased, but following the growing presence of other operators in the sector, principally the Italian and foreign fund managers.

The public operators (public companies and cooperative associations) have a significant share both in terms of number of operations and volumes. In most cases the initiatives relate to promoting new entrepreneurs and economic growth in slow developing areas. Moreover, in some areas the initiatives undertaken by regional financiers are rapidly multiplying. This involves companies partly held by Regions that accept to operate in venture capital. At present, the most significant examples are in the North of Italy and in particular in those Regions with higher autonomy from the state government (i.e. Friuli, Trentino). That is, the most interesting entrepreneurial initiatives and the greater freedom of action deriving from the regional autonomy obviously favour the operator. As for the role of private investors and industrial companies, we notice a basically stable but contained share in terms of number of operations; while in terms of volumes the share has decreased due to the entry of new operators. Finally, the contribution of closed-end Italian funds is limited although constantly growing, due to the recent diffusion of these instruments and to the still poor development of pension funds in the Italian context.

In terms of the average amount of investments (see Table 3.8) the largest shares are associated with private equity funds and especially if originated by foreign banks, that prefer to finance a few selected operations of

---

[6] Introduced by the new banking law in 1993.

rather considerable amount. This may be due to the greater expertise of larger operators, who are capable of evaluating more complex operations, and at the same time do not focus on smaller size investments as they are insufficient to cover the operation's structural costs. The small size investments refer to financing public entities, for the development of new small companies.

Table 3.8. Average amount of the investment in Venture Capital in Italy

AMOUNTS IN MILLION EURO	1997	1998	1999	2000
Italian banks	3.2	3.0	5.9	5.4
Public operators	1.2	1.2	1.6	1.6
Regional financiers	0.8	0.4	0.6	0.6
Private/industrial operators	6.5	12.6	5.7	1.9
Italian closed funds	3.3	4.4	3.2	2.9
International operators	4.8	7.5	7.6	
Advisor of VC funds				1.6
Advisor of PC funds				20.2
Emission by foreign bank				31.4
TOTAL	2.6	3.5	4.6	4.6

Source: AIFI 2001

3.5 Conclusions

The analysis performed in the present chapter allows us to assert that the fundraising in the Italian market still appears as a rather concentrated process in terms of type of operators and method of operating.

The most active operators are represented by banks and other financial captive operators (managers of closed-end funds). The fundraising is thus prevalently realised in a semi-direct manner (often between financial intermediaries) or in case of private investors through a network of pre-existing relationships. The fundraising method in Italy is thus strongly influenced by the institutional characteristic of the financial system, founded on the centrality of the credit intermediation. Unfortunately the lack of explicit information on fundraising strategies from venture capital companies and the fact that the few data available on this topic come from an unique source limits at the moment a more detailed description of the specific Italian reality.

Domestic operators almost exclusively manage the resources raised in Italy. Funding Italian venture capital companies however does not limit itself to the national boundaries. More frequently Italian companies raise funds from foreign investors, due to the growing interest of foreign institutional investors for a greater international diversification of their portfolio. At present the greatest efforts appear to focus on the main American investors, though the Euro will facilitate a growth in European fundraising. Therefore this context represents an ideal ground for the growth of Italian companies, as long as there is still a constant growth in terms of volumes exchanged. A more severe competitive threat could emerge with the entrance of the sector in a mature phase, first abroad than in Italy, and the worsening of the economic climate. These factors could in fact encourage greater fundraising activity in our country by foreign operators with consolidated reputations. Hence a warning for Italian companies to consolidate without delay their own reputations in the internal and international markets, in preparation for the greater competition challenge in the near future.

4 The Valuation of the Target Company[1]

Stefano Gatti

4.1 Introduction

The topic of company valuation is crucial when talking of investment policies put in place by venture capitalists. In the past few months we have experienced both here in Europe and in USA a hype on technology stocks that has fevered the stock markets of both continents. Prices have surged to unexpected levels in a short period of time and at the end they have lost contact with the fundamentals of the issuing firms.[2]

What has happened after march 2000 has shown that many stocks had been overvalued, that new metrics of value had been developed only to justify the extremely high prices of stocks, that too much money had been routed to venture capitalists without sound investment policies, that finally many investors had not exactly understood where the value of firms laid. Now that prices have come back to more reasonable levels, the participants to the market of high tech firms are turning back to an assessment of company value rooted on more traditional methods of valuation; market multiples and the more complex real options models have progressively disappeared from the scene.

---

[1] I would like to thank Claudio Giacomiello for his precious help in writing paragraphs from 4.4.1.1 to 4.4.1.3. The final responsibility for the chapter contents is of course mine.

[2] As some authors suggest (Lewis 1999), the phenomenon of IPO's of technology companies without profits began with Netscape. Six months after Jim Clark founded Netscape , he agitated for the company to go public. The company had few revenues, no profits, and a lot of employees. No one else inside the company thought it should do anything but keep its head down and try to become a viable enterprise. But Clark said there was a reason for this: he needed more money to buy a big new boat. The first day of trading the price of Netscape shares rose from $12 to 48$, and three months later it was $140. The IPO gave credibility to anarchy and anything could be quoted. The example of Netscape is representative of what happened during 1999 and 2000. According to IBES consensus estimates in March 2000, in terms of price to 1999 sales, Nasdaq 100 was trading at 11x, S&P 500 at 2.1x and DJIA at 1.7x. Amazon was trading at 14.2x, Ariba at 400x and Siebel Systems at 530x. Everyone can see the end of the story reflected in current stock prices.

The cruciality of company valuation, at the very end, is evident considering that, after the signing of the confidentiality agreement, the due diligence phase and the bargaining between the equity holders of the target company and the investment company, the closing of the deal lays on the agreement about the amount of money that is required by the firm and about the number of shares that the shareholders are open to give up in favor of the venture capitalist.

The importance of the topic is more evident when the valuation concerns a high tech company[3], with no history (or a limited track record), highly risky and often with poor economic and financial performance. In this situation, the funding of development cannot come from traditional bank loans and a large part of its financial needs must be matched with equity capital from investment companies. As a consequence, the determination of the correct price of an equity stake becomes one of the most important key success factors in the venture capital industry.

Moreover, apart from the acquisition of an equity stake by an investment company, the theme of valuation is also important for high tech firms in other situations, mainly due to the rapid and constant evolution of the sectors in which they operate:

1. the structuring of a partnership or joint ventures with other firms;

2. mergers with other companies;

3. an initial public offering of stocks on the capital market.

In this chapter, we will address some of the questions outlined above. More precisely, we will focus on how to choose a proper method of valuation for high tech companies and the main cautions that the use of each one requires. In the following paragraph we will explain the determinants affecting the choice of a specific valuation method among the different op-

---

[3] The precise definition of high tech company is not agreed among pratictioners and academics. In this chapter, our position is to consider the concept of high tech in a broad sense, including ICT, computer industry (hardware, software and semiconductors), internet and web, electronics, engineering, robotics, advanced chemistry, pharmaceutical, biotechnology/bioengineering, aerospace/aeronautics and telecom. Although in the paper we refer generally to high tech firms, it's easy to notice that some of the industries are still based on tangible capital (see for example the computer industry, electronics or telecom) while others are strongly linked to unbooked assets as marketing or R&D (take for example the internet related business or biotechnology); for the former traditional methods can be adapted in an easier way, for the latter old methods remain useful but more attention has to be used in order to apply them in a correct way. For definitions of high tech see Web-Based high tech company directory available at www.corptech.com.

tions available. In paragraph 4.3 we will show the special features of a new economy firm and the consequences for the valuation of the equity capital. In paragraph 4.4 we will briefly review the basic concepts underpinning the methods traditionally used for estimating the value of the firms and will explain why these methods cannot be straightforwardly applied to high tech companies. Finally, paragraph 4.5 concludes and indicates some further developments for future research on this topic.

The aim of this chapter is neither an in-depth analysis of all the technicalities of the different evaluation methods nor the discussion on the more recent real options evaluation models. Interested readers can address to the books cited in the references list for a thorough analysis of these aspects. On the contrary, we will give as known the basic principles of company valuation and will rather focus the attention on the solutions aimed at adapting the methods to high tech firms.

4.2 The Determinants in the Choice of a Valuation Method

The first problem addressed by someone valuing a company is how to choose a proper method of valuation. Put differently, given that different methods are available in practice, which is the most suitable for the company to be estimated?

This is a common problem in valuation, for old economy firms too. For this reason, we start considering the main factors that could affect the choice: this is particularly useful in the field of high tech firms because not all of them share the same value drivers.

At the very end, the determinants in the choice can be summed up in the following four:

1. the country in which the company that has to be estimated is set up;

2. the industry in which it operates;

3. the availability and the reliability of data on which the valuation is based;

4. the status of quoted company.

4.2.1 The Country in Which the Target Company is Set up

This factor is important considering that in Europe and in USA most part of venture capitalists are interested in assembling a portfolio of companies operating in the same industry but not necessarily in the same country. When co-investment agreements are set up between domestic and international investment companies, this situation becomes more frequent.

When a company has to be appraised, the parts of the transaction must be aware that the traditional best practices in different countries aren't necessarily the same. This is an important point for a successful closing of the deal. If a venture capitalist estimates the value through a method that is not commonly used in the country or, worse, not accepted by the counterpart and its advisors, the probability of the deal closing is strongly reduced. The differences in the best practices of valuation among the countries – especially between Continental Europe and Anglosaxon countries – are becoming lighter[4]; anyway, they are still in effect.

Continental Europe has traditionally seen academics and pratictioners opting for a valuation based on data taken from the balance sheets or income statements of a firm. Income-based methods or balance-sheet based methods – sometimes mixed together – are still used in Italy, Germany, France in deals of companies. Decidedly less used are cash flow based methods and market based (or multiple) methods; not the former, considering the high level of subjectivity surrounding the final estimation of value; not the latter, given the reduced importance of stock exchanges and capital markets in the financial systems of Continental Europe countries.[5]

The position of anglosaxon colleagues is opposite. They argue that the value of the firm lays only on the attitude of the firm to generate positive free cash flows to debt and equity holders. Moreover, if a company is going public, their assumption is that the value could not be too far from the prices the markets are open to pay for similar companies. That's why the use of multiples of figures like net earnings, EBIT or EBITDA are commonly used by most investment banks.

---

[4] Americans for example are using the concept of EVA, very close to the "goodwill" of the European tradition; European venture capitalists and investment companies are increasingly using free cash flow or multiples for valuing not only high tech firms but also old economy ones. This is not surprising given the pace of the globalisation of financial markets and relative practices.

[5] In Italy, the use of net worth based methods or earning methods has been ratified also at a statutory level by the Bank of Italy which recommend these metrics for the valuation of companies included in closed funds portfolios. Cash flow methods are recalled but not openly recommended (AIFI 1996).

4.2.2 The Company Industry

It's common sense that different industries have different value drivers. These value drivers tell the professional where the value of a firm comes from.

Common sense tells also that industries in which the tangible capital (fixed assets and working capital) is a large part of the capital employed by a firm – think for example to the manufacturing of metals, banking and insurance, chemistry or the real estate business – the value of such capital should count in the final valuation, maybe mixed with the results given by other methods.[6]

Other sectors in the old and the new economy – take the example of fashion or consulting or, in the new economy, biotechnology or internet-related firms – do not show large amounts of capital invested because large part of their competitive strength is due to intangible assets that do not find place in the balance sheet (they are unbooked assets). Brand equity, R&D or marketing expenses aren't accounted as assets because the legislations of many countries do not allow this practice. Anyway, there's no doubt that these costs will have a beneficial effect on future performance of a company.

It can be easily understood that in such sectors, only cash flow based methods or income based methods can give a correct estimate of the company value. As we will show in next paragraph and paragraph 4.4, however, some adjustments made to the traditional balance sheet based methods (especially a cost based approach using the reproduction value of intangible assets) allow an application also to high tech firms, especially to biotech firms.

4.2.3 Data Availability and Reliability

The choice among the methods available for the valuation is strongly influenced by the availability of data on the firm, its markets, the future evolution of the sectors, the competitors. More important, the availability

---

[6] Actually, balance sheet based methods and income based or cash flow based methods are strictly linked one another. The tangible capital (and the intangible capital, indeed) is what the company can use today in order to obtain in the future a stream of net income or cash flow. That's why the estimation of a company with a simple balance sheet method is not correct: it considers only the wealth of the company today but fails to recognize that the value is generated only through a correct use of present wealth and, as a consequence, by the revenues produced by the capital.

should encompass data regarding the past, the present and, above all, the future value of all the relevant variables affecting the performance of the firm.

Together with the availability, the other key point is the reliability of the available data. On one side, we could experience a situation in which data are available and reliable; on the other side, data could lack or they would be available but not reliable. In this situation, the valuation carried out with cash flow or income based methods can be strongly subjective and sensitivities could not be of help. An estimation based on data from the past is probably more sound.

On the contrary, the availability of reliable data on the firm and the competitors allow the professional to carry out the valuation with forward-looking methods (again, cash flow and income based methods).

4.2.4 The Status of Quoted Company

The fourth determinant in the choice of the proper valuation method is the availability of market prices referred to the stocks of the company. If a firm has publicly traded stocks, the valuation can be based – at least as a form of control of the consistency of other estimates – on the judgement of the market mirrored in the prices at which the stocks have been bought and sold during a not too short period of time.

As we will discuss in depth in paragraph 4.4.3, the use of market prices can allow the analysts to calculate multiples derived from stock quotes of comparable firms. This can be a useful solution to the problem of valuation of high tech firms, especially for those with no history or with poor set of data available. Multiples derived from average values of ratios of similar companies can be easily obtainable for most part of high tech stock markets; however, the main problem lays in finding companies sufficiently similar to the one to be valued.

4.3 The Special Features of a High Tech Company and Their Effects on Valuation

In the preceding paragraph, we have presented the factors affecting the choice of a valuation method. Net worth based methods, earnings based methods, cash flows based methods or market multiples are definitely the four main classes of methods used by professionals in order to cope with the problem of estimating the value of a functioning firm. Put differently, "old valuation metrics" tend to estimate the value of a firm by considering net worth, net earnings, cash flows or multiples as good proxies of the fair value of the firm.

Unfortunately, these four parameters are not viable for a high tech company for the reasons discussed in the following pages.

A high tech company has a low value if we look at its balance sheet: a large part of high tech firms (take for example the case of internet firms or biotech R&D laboratories) have a small amount of tangible assets in their balance sheet; their book value is not high, much part of the company value lays in "unbooked" assets represented by the present value of opportunities (or "options") the firm will be able to reap in the future. As we will discuss in the next paragraph, this feature of high tech firms strongly limits the use of net worth based methods.

Figure 4.1 shows the difference between the balance sheet of a generic "old economy" and "new economy" firm. Although not referred to any specific industry, the figure is useful because it compares the different sources of value in the two firms: in the old economy a good part of the value can be referred to the difference between the market value of assets and liabilities and only a small part is generated by future growth opportunities (this part is smaller in mature industries in which the growth is not far from the growth rate of the whole economy). On the contrary, "new economy" firms have value because a large part of equity value is linked to the present value of future results in terms of income or cash flows. Unfortunately, this part does not find place in the balance sheet and book value for these companies is very low.

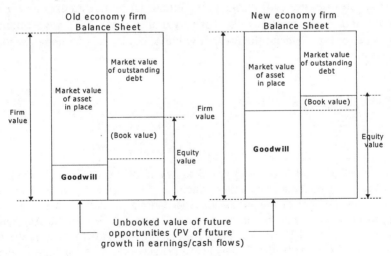

Fig. 4.1. Balance sheet and company value in the new and old economy

A high tech company, especially if recently started up, has a poor earnings track record. This is a common feature also for old economy firms with few years of life but it becomes a crucial point for a venture capitalist whose business is exactly the investment in firms always at the beginning of their life. Newly born firms cannot be estimated by discounting their net income to present because if the period in which the firms have negative results is sufficiently long the result could be negative. The result could be far more negative if the efforts made to establish a good position/brand/market quote generate large amounts of costs not paralleled by a corresponding increase in sales. This is particularly true for high tech firms whose R&D or marketing expenses cannot be – in most part of European countries – deferred or capitalized in the balance sheet.[7]

A high tech company shows poor free cash flows. We will discuss about the free cash flow (that is the cash generated by the company in excess of the investments in fixed and working capital). Here, we merely note that if a firm is at the beginning of its life, the building up of a stable business requires a large amount of investments (the larger the growth of the industry, the larger the amount of the investments needed) and, of course, of cash.

Just to give the reader an illustrative example of the fast pace of growth in the internet-related business, in table 4.1 we have calculated the compound average growth rates of some business areas referred to the business to consumer (B2C) market in North America and Europe. The magnitude of the growth is impressive when considering that a "normal" growth rate of revenues in traditional industries equals 2/3% per year. Such impressive growth rates requires, all things being equal, large investments in working capital and fixed assets; if the cash available after these investments is negative for many years, the present value of equity will be negative too.

---

[7] IASC (International Accounting Standard Committee) is the body devoted to the issuance of international accounting principles. Although every european country is not obliged to include the international accounting standards in national legislation, in many countries the national principles of Certified Public Accountants heavily rely on the IAS (International Accounting Standards) issued by the committee. IAS 38, in particular, is devoted to the accounting treatment of intangible assets and allow capitalization of internally generated intangibles (but not of marketing and advertising costs) only when a defined set of criteria is fulfilled. Italian legislation is based on such a framework.

Table 4.1. Business to consumer e-commerce forecasts: breakdown of the value chain and cumulative annual growth rates

	Europe			
	2001/00	2002/01	2003/02	CAGR 2000/03
B2C E-commerce	140,27%	86,31%	n.a.	111,58%
Sell-side Software	35,65%	23,69%	12,77%	23,68%
Buy-side Software (E-procurement solutions)	135,80%	48,28%	40,05%	69,81%
E-commerce related IT consulting services	-6,53%	22,71%	n.a.	7,10%
E-commerce related IT development and integration services	25,77%	41,81%	n.a.	33,55%
	North America			
	2001/00	2002/01	2003/02	CAGR 2000/03
B2C E-commerce	72,49%	44,22%	n.a.	57,73%
Sell-side Software	18,00%	16,97%	7,23%	13,96%
Buy-side Software (E-procurement solutions)	63,64%	25,91%	14,72%	33,21%
E-commerce related IT consulting services	n.a.	n.a.	n.a.	n.a.
E-commerce related IT development and integration services	n.a.	n.a.	n.a.	n.a.

Source: our calculation based on European Information Technology Observatory, 2000.
Note: when data are not available, the Cagr is referred to the period 2000-2002; Europe include Western and Eastern Europe except for IT consulting services and IT development and integration services.

A high tech company lacks of a sufficient number of market comparables. The exercise of valuation is quite easy for a professional if a company is unique but has a sufficiently long set of accounting and market data or, on the contrary, has been recently set up but can be compared to other similar firms that have been in place for a sufficiently long time. In both cases, in fact, the professional can rely upon a complete set of information to derive the value from. Unfortunately, high tech firms are both news and unique. The use of comparables becomes harder and the estimate of the value cannot be considered as a definite valuation of the firm especially if the comparable firm is not exactly operating in the same business/market segment, has not the same scale or has not been set up at the same time.[8]

After this rapid review of the features of high tech firms, one could wonder if new methods of valuation are necessary in order to quantify a proper equity value and if old methods have to be straighforwardly dismissed. This mindset has been actually dominant during the past few years during the boom of high tech capital markets both in the USA and in Europe when analysts tried to develop "new metrics" in order to find positive equity values for otherwise negative evaluation estimates. In particular, "real options" and "operating multiples" were devised as new tools widely used by venture capitalists and market analysts.

Our position with reference to "new metrics" is quite skeptical. Apart from the fact that "real options" are basically an extension of basic NPV models based on earnings or cash flow discounting and that, as we will discuss later, empirical multiples can have a useful application only when the company has sound perspectives of earnings and cash flows increase in the future, we are firmly convinced that the problem does not lay in the methods (that remain valid also for high tech firms) but on the different business model that the high tech firms have with reference to the old economy counterparts. Different business models mean different asset structure, different earnings and cash flow patterns; first of all, they mean different factors on which the competitive advantage can be built. At the end, in our opinion, the problem of valuation of high tech equity refers not to the metrics but to the different strategy.

Mauboussin et al. 2001 talk about "business category" dividing firms into three groups:

1. Physical: it is constituted by firms that strongly rely upon tangible assets; their growth in terms of earnings and cash flows is determined by the investments in capacity and physical assets. All traditional "metrics", net worth based methods included, can be used for the value appraisal.

---

[8] This last factor is particularly relevant in that an older firm could have built up a brand image sufficiently strong in comparison to the firm that has to be valued. If the brand image is strong, the potential of scalability and the "loyalty effect" further explained in this paragraph could be already in place and market values could reflect such situation. Using multiples could overprice the equity of the firm under valuation.

2. Service: this group is composed by firms operating in the business of services, often delivered on a one-to-one basis (see, for example, banking and insurance). They found their business success on people: their growth depends on people increase and productivity gains. That's why the choice of valuation methods starts shrinking: net worth based methods do not catch the source of value because the people value – a typical intangible asset – is not accounted for in the balance sheet.

3. Knowledge: this is the group formed by firms whose success depends from people but that, in contrast with the service group, build up intellectual contents that can be reproduced horizontally and vertically in different businesses without losing value.[9] The growth of these firms is based on the ability to use the present capacity and cost structure to leverage the growth in sales (the so-called "scalability effect"). Scalability of business is very high when intangibles count the most. An intangible can be used many times without losing its value; moreover, the larger the number of people that use the good or service, the larger the value of the good or service itself (network effect). In terms of valuation methods, the situation is similar to the service group with the additional necessity to consider the potential of scalability of the firm in terms of growth rates in revenues and costs. Put differently, a 10% of sales growth in the industry could become a higher figure for the single firm if the scalability of its business allow the use of its intellectual property horizontally or vertically in other sectors.

Most part of high tech firms falls in the third business category. It's important then to consider the most relevant value drivers that can contribute to value creation in these companies. They are also the features that a venture capitalist should look at when determining the correct price for the equity stake:

---

[9] Examples are many in different sectors. Take for example the B2C business in which an e-tailer has developed a good brand image and customer loyalty. The e-tailer can widen the mix of products sold without paying the costs for the increase in the customer base and, if the technology platform allows it, without making further investments in technology. Another example is the group of science-based sectors. In the biotech industry the R&D process can lead to unexpected results and the final outcome of a research program can benefit other programs leading to new developments in other research fields. In this case the sales can be increased by using the results of a research program for other uses without a corresponding increase in costs. Scalability and network effects are common also in the old economy: take for example the luxury industry in which the scalability of the brand is crucial for stylists and their maisons (Gatti 1997).

- good management and good team. Knowledge-based firms rely upon people and their value is strongly dependent from that of people working in them. For this reason, it's necessary to assess the ability of the management in terms of technical/industry experience and in terms of people management. Teamworking, strong coordination capabilities, decentralized decision making are all catalysts for innovation.

- scalability of business. As we said earlier, business scalability is the condition that allow a start up high tech firm to increase revenues at a faster pace than its costs. This situation is particularly important in order to overcome the problems of excessive cash absorption (the so-called burn rate)[10] occurring during the first years of operation. High scalability means that the financial equilibrium of the firm can be reached in an easier and faster way with positive effects on shareholders and venture capitalists. Scalability depends mainly on the ability of the company to replicate its business in different sectors, it's strongly linked to intangible assets and requires investments in marketing and R&D.

- good investment policy in the critical value drivers. A high tech firm faces high financial needs in order to reach a credible and sustainable business. These investments must be oriented towards the key success factors of the firm. For dotcoms, for example, investments should be dedicated to the consolidation of the customer base and to the building up of a strong brand image, not to pure and simple extension of the number of clients (especially if their churn rate is high). A stable customer base and strong loyalty can allow the firm to reduce costs (the experience in use increases and assistance to customers becomes less expensive), to increase sales (through cross selling), to increase the switching costs of the client from the company to another alternative supplier, to ask for a convenient premium price. Investments aimed at simply increasing the number of customers do not necessarily lead to these results especially in the internet domain in which loyalty is not a common virtue among customers. Another example is the one of biotechnology firms. Here investments must be constantly addressed to the R&D field in order to foster the growth of the research project pipeline. Considering that in the USA on average only 1 over 5.000 drugs experienced during pre-clinical test can be tested on human beings and that

---

[10] Analysts talk about burn rate to indicate, even not explicitly, the value of the negative free cash flow (or sometimes of the net loss taken from the last available annual report) necessary to sustain the business development. The concept of free cash flow is explained in paragraph 4.4.1 and the burn rate is further analysed in box 1 discussing about multiples in the biotech industry.

only 1 over 5 medical drugs in the "human testing" phase can obtain the approval of the Food and Drug Administration, it's easy to understand why venture capitalists give attention to the number of projects in the pipeline when valuing the equity. In any case, investments in the critical value drivers should award flexibility and time-to-market: innovators in the new economy are the first movers and can benefit from this in terms of margins and future investments needed to sustain the growth.

4.4. Adapting Old Methods to the New Economy: Cautions and Suggestions

4.4.1 Discounted Cash Flows (DCF) and "New Metrics": General Remarks

Financial theory argues that the value of any investment, from the simplest one to the more complex, depends directly from the present value of the stream of cash flows that the investment will generate in the future. Regardless of how difficult the estimation process of all relevant variables can be and regardless of the possible opportunities (options) an investment can benefit from, this assumption holds true also for company valuation. A company, in fact, is a complex investment at the very end.

We are firmly convinced that the value appraisal has to be rooted in a free cash flow based method because only cash matters, for both old and new economy firms. Although European continental countries have started using financial methods only recently, one can legitimately wonder why in the past few years also US analysts have been rapidly dismissed their beloved DCF models as long as dotcoms were concerned even considering their satisfying results for old economy firms valuations.[11]

---

[11] This is true especially for dotcoms and not for other high tech firms. Empirical evidence shows that in investment banking dotcoms have been valued, during the past few years, through DCF models integrated and corrected with comparables based valuations, in particular in the IPO's market. Evidence from Italy, on the contrary, shows that the first venture capitalists have based their valuation on standard DCF models. For biotechnology, things are a bit different: a standard DCF model can always be used without major adjustments because in these firms the processes of investment and of results generation follow a pattern similar to the one experienced by old economy firms (of course, apart from the scalability of business and the importance of flexibility that can suggest the implementation of an extended NPV model taking into consideration the value of real options). See for example the valuation model proposed by Kellogg and Charnes (2000).

We will analyze first Free Cash Flow to Equity (FCFE) Method (also known as levered free cash flow method), then Free Cash Flow to the Firm (FCFF) (or unlevered free cash flow method) Method. Then, we will turn to some remarks concerning the application of DCF models to new economy firms.

FCFE Method (Levered Free Cash Flow)

We define Free Cash Flow to Equity (FCFE) as the residual cash flow left over after:

1. meeting all financial obligations;

2. covering capital expenditures and working capital needs.

If we define the debt ratio as:

$$d = \frac{Debt}{Debt + Equity} \tag{4.1}$$

and assuming a stable leverage for the firm and absence of extraordinary costs and revenues in the P&L, FCFE is simply given by:

$$FCFE = NI - (1-d) \times [\pm \Delta \text{ Working Capital} + (Capex - Depreciation)] \tag{4.2}$$

where NI indicates the Net Income of the firm in a given year.

Now we can build up a two-stages Discounted Cash Flow Model, based on FCFE.

$$ValueOfEquity = \sum_{t=1}^{n} \frac{FCFE_t}{(1+k_e)^t} + \frac{\frac{FCFE_{n+1}}{k_{en} - g_n}}{(1+k_e)^n} \qquad \text{for t=1 to n} \tag{4.3}$$

k_e is the required rate of return to equity investor (cost of equity) in the high growth period, k_{en} is the required rate of return to equity investor (cost of equity) in the stable growth period and g_n is the growth rate in FCFE for the firm forever in the stable growth period.[12]

---

[12] The theory and the empirical observation of real markets show that a firm passes through a period of sustained growth (the CAP, competitive advantage period) and then enters in a steady state of business. In the first period, a firm invests large

If a firm is stable and it pays out all its FCFE as dividends, the value obtained from FCFE Model will be the same as the one obtained from a Gordon Dividend Discount Model (DDM). If this is not the case, FCFE Model will give different results from a DDM and usually the value from FCFE Model will be higher than the value derived from DDM.

The use of the FCFE Model is not recommended as long as high tech companies are considered. One can correctly use it when:

1. he/she is a minority investor in a quoted company: minority shareholders, in fact, are particularly interested in receiving a stable flow of dividends by the company in which they have put their money. For this reason, dividends make the difference in the value appraisal of the stocks of the firm;

2. he/she is evaluating a firm for a takeover: when deciding to takeover a company, a raider usually looks at the poor performance of the company in terms of dividends paid to the shareholders. A long track record of low levels of dividends suggests that the stock price will start a downward pattern, making it convenient for a raider to buy stocks on the market, obtain the control over the firm and substitute the old management for a new one.

3. firm has an almost stable leverage: if this condition is not fulfilled, the equity capital needed for the funding of new investments in fixed and working capital can change over the time. This is particularly true for newly set up companies whose high level of risk is badly matched by debt financing; strong capitalization in the earlier stages of development is needed in order to increase the confidence of the banking system, so it is realistic to suppose that in newly born companies d ratio of equation (4.1) is increasing as long as time passes and company success strengthens.

---

amount of money because the return on capital invested is higher than the cost of capital. Competition, as long as time passes, reduces this competitive advantage and the growth rate of the company as well. Opportunities of investment fall and the investments in fixed capital become a simple replacement of installed capacity. As this situation occurs, the firm enters in the so-called steady state. In a two stages model, then, in the stable growth period capital expenditures will be offset by depreciation, beta of the firm won't be significantly different from 1, g_n will reasonably be less or equal to the nominal growth rate of the economy. For further analysis on the CAP see Copeland, Koller and Murrin (2000).

FCFF Method (Unlevered Free Cash Flow)

We define Free Cash Flow to the Firm (FCFF) as the amount of cash flow available to all claimants to the firm, including holders of common stocks, preferred stocks, bonds and lenders as well.

FCFF is given by:

$$FCFF = Ebit \times (1-t) + Depreciation - Capex \pm \Delta \text{ Working Capital} \qquad (4.4)$$

where Ebit is earning before interest and taxes and t is the tax rate for the firm.

The Value of the firm based on FCFF in a two stages model can be calculated by discounting the future unlevered free cash flows with an appropriate discount rate:

$$ValueOfFirm = \sum_{t=1}^{n} \frac{FCFF_t}{(1+WACC)^t} + \frac{\dfrac{FCFF_{n+1}}{WACC_n - g_n}}{(1+WACC)^n} \quad \begin{array}{l} \text{for } t=1 \\ \text{to } n \end{array} \qquad (4.5)$$

The assumptions are that the firm reaches a steady state after n years and starts growing at a stable growth rate g_n after that. WACC is the Weighted Average Cost of Capital and $WACC_n$ is WACC in steady state.

FCFF is obviously pre-debt, so its value is much less likely negative than FCFE. Starting from the value of the firm, the value of equity can be obtained by the difference between the value of the firm and the market value of debt outstanding at the moment of valuation.

$$\text{Value of Equity} = \text{Value of Firm} - \text{Market Value of Oustanding Debt} \qquad (4.6)$$

The value of equity derived by the discount of future FCFF, in theory, should be the same of that calculated through the implementation of a FCFE model.[13] However, this identity is verified only if:

---

[13] One of the primary problems using FCFE is the frequent occurrence of negative FCFE in highly leveraged or highly cyclical firms because of compulsory debt payments: in general FCFF is higher than FCFE for the same firm. Another important consequence of FCFF is that its growth will not be affected by increased leverage: in general growth rate in FCFF is lower than growth rate in FCFE.

1. Consistent assumptions are made about growth in the two approaches about the growth rate. Growth rate in earnings should be adjusted for leverage in FCFE, in particular for the terminal value.

2. Firm's debt is correctly valued: if firm's debt is overvalued, with FCFF you will subtract overvalued debt from value of the firm and you will get lower value of equity than with a FCFE.

3. The firm has no debt: in this case FCFF equals FCFE (d ratio of eq. (4.1) is zero) and WACC is k_e. Although this should appear at first only a theoretical case, many newly born firms are close to this situation because their credibility on the debt market has still to be set up.

FCFF Model is particularly suitable for:

1. the valuation of a private company: if a firm's stocks are not publicly traded, the shareholders are less interested in a stable and continuous flow of dividends. In these firms there's often a coincidence between the role of shareholder and the role of manager. For the latter, it's much more important to know the amount of cash still available after the investments in working capital and fixed assets rather than the amount of dividends at least in the short run.

2. the valuation of a firm in which you are partner or controlling shareholder, not a minority investor: this argument is related to the previous one. A controlling shareholder or a financial partner (a venture capitalist for example) must always consider the amount of cash needed to fund new investments. If free cash flow is negative, shareholders have to decide the way to fill the gap and this is often a strategic choice for the future success of the company.

3. the valuation of highly leveraged firms or that are in process to change leverage over time: as we noticed in the previous paragraph, high tech firms at the very beginning are likely to be low leveraged due to their high level of business risk. Assuming a constant d ratio is not correct. For highly leveraged firms (take for example the case of some European Telecom Companies that have recently issued jumbo loans on the euro-market) FCFF is important because the high debt can become a constraint for the development strategy of the firm if the cash available after the needed investments is not sufficient to repay the old debt.

4. the valuation of turnaround plans: when companies decline or face business/financial crisis, the turnaround is viable and should be carried on only if the turnaround plan is able to create favorable conditions to the

generation of sufficient unlevered free cash flow to reduce the high leverage. The FCFF figure can properly highlight whether the plan is sustainable or not.

The Problem of Terminal Value (TV)

Two important computing factors are the two second members on the right side of equation (4.3) and (4.5), the so called terminal value (hereafter indicated as TV)[14]:

$$\frac{FCFE_{n+1}}{k_{en} - g_n} \qquad \text{for the FCFE model}$$

$$\frac{FCFF_{n+1}}{WACC_n - g_n} \qquad \text{for the FCFF model}$$

In traditional two-stage DCF models, typically 70-80% of value is inherent in the terminal value calculation. For growth stocks already quoted on capital markets this is usually around 90%. For start-ups or new economy stocks it is not uncommon to have over 100% of total value to be in the terminal value. Since the terminal value is based upon the cash flow in the final year of the esplicit forecast period, this becomes the critical figure in the valuation.

In order to show the impact of the TV value on company valuation, we consider in Figure 4.2, the potential outcomes of the valuation of Amazon.com, assuming 4 scenarios of stable growth for the calculation of TV.

---

[14] The reader should note that in the unlevered free cash flow model, FCFFn+1 is just equal to EBIT (1-t) - ΔWorking Capital. In fact, in steady state it is assumed that capital expenditures are not significantly greater than depreciation; otherwise we should see extraordinary growth again.

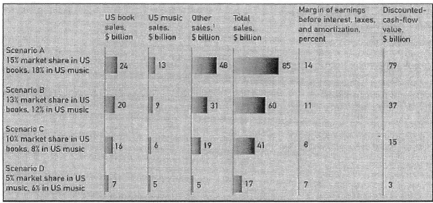

	US book sales, $ billion	US music sales, $ billion	Other sales,¹ $ billion	Total sales, $ billion	Margin of earnings before interest, taxes, and amortization, percent	Discounted-cash-flow value, $ billion
Scenario A 15% market share in US books, 18% in US music	24	13	48	85	14	79
Scenario B 13% market share in US books, 12% in US music	20	9	31	60	11	37
Scenario C 10% market share in US books, 8% in US music	16	6	19	41	8	15
Scenario D 5% market share in US music, 6% in US music	7	5	5	17	7	3

Fig. 4.2. The impact of different scenarios on Amazon.com's value (Source: Desmet, Francis, Hu, Koller, Riedel 2000)

Scenario A is the optimistic one, it is assumed that Amazon can become the next Wal-Mart, the US retailer that has radically changed its industry and has taken a significant market share in all its target markets: we imagine that by 2010, Amazon continues to be the leading on-line retailer and in certain markets it is leader both on- and off-line. So total sales are $85 billion, EBITDA margin is 14% and DCF value is $79 billion.

Scenario B has Amazon capturing revenues almost as large as it happens in Scenario A, but with EBITDA margin, just lower, of 11%. DCF value is $37 billion, less than half of the previous scenario. Please note that $37 billion was the value Amazon had as of the fourth quarter of 1999.

Scenario C assumes sales and margins closer to those of more traditional retailers and a valuation of $15 billion, that is half of scenario B and just 20% of the valuation of scenario A. In scenario D sales are less than half of scenario C with an EBITDA margin of 7%: DCF value is $3 billion, that is 3.8% of the valuation of scenario A and one fifth of scenario C.

Even assuming that the input variables needed by the DCF Model have been correctly estimated, one cannot be sure to be protected from making big mistakes because in technology stocks more than 90% of the value come from the terminal value, and the terminal value is based on the assumptions of sustainable margins and sustainable growth.

When you evaluate a common old economy company you start from the present, the current performance, and year by year you assume a rate of growth. With a high tech company you cannot do this. You must perform a quantum leap in the future, imagine what the company could look like when it develops from today's very high growth, unstable condition to a sustainable, moderate-growth state in the future; and then extrapolate back to current performance.

DCF Metrics and Their Use for High Tech Firms

Summing up some of the concepts discussed above, we can conclude that a high tech company shows, at least, three main features:

– it is a private company, not quoted: shares are in the hands of founders or venture capital/private equity partners;
– it is lowly leveraged in the beginning, becomes highly leveraged as long as credibility increases together with the confidence (overconfidence?) of the banking sector or the bond market. The use of a constant d ratio is then incorrect in principle;
– it produces negative cash flows and negative earnings for the first years of operation.

Considering what we told in the previous sections of this chapter, the model that best fit this kind of company is the FCFF Model.

After a period in which financial methods seemed disappeared from the scene of valuation practice also in the US investment banking and corporate finance consulting in favour of other methods more able to explain the irrational level of market prices reached before march 2000, the rehabilitation of free cash flow estimates in the new economy sectors – especially for internet-related businesses – has started with some empirical studies that have reviewed the economic and financial performance of dotcoms and of comparable companies working in the old economy in order to highlight the differences between the two samples. Unfortunately the empirical evidence refers to companies quoted on US capital markets but, given the rapid process of globalisation of markets and strategies, the result should be valid also for european firms.

Mauboussin and Hiler (1999), using accounting and market data, build up a framework analysis in which a dotcom is compared with a traditional old economy firm operating in a comparable industry. The authors' conclusions are that DCF methods are still able to quantify exactly the value of a firm but that in the past analysts haven't exactly understood that the evolution patterns of free cash flows of dotcoms (in the unlevered version) are heavily different from the ones of old economy firms (see figure 4.3[15]).

---

[15] Note that the diagonal going from the upper left corner to the lower right corner of the square represents in any point a situation of financial equilibrium (a positive

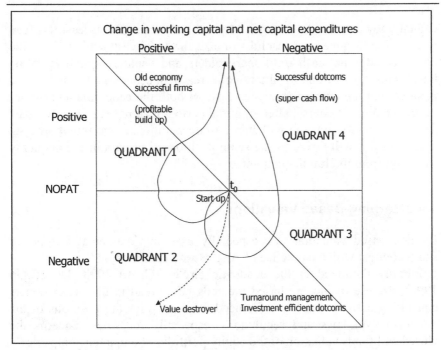

Fig. 4.3. Cash flow patterns in old and new economy firms (Source: Adapted from Mauboussin and Hiler 1999)

Traditionally, a start up company in an old economy context passes through an initial phase of intensive investment (both in fixed assets and in working capital) in which NOPAT is negative. If it is successful in establishing a good brand image and a good market position (if not, it is the classic lemon), economic results turn from negative to positive. Finally, when the industry slows the growth rate (the industry enters the steady state), investments fall and free cash flow becomes high and positive.

Dotcoms seems to behave differently. In the start up phase the pattern of free cash flow is similar to old economy firms. However the evolution to cash flow generation follows a different way. Apart from value destroyers, whose managers will continue to put money in the firm without shifting from negative to positive NOPAT, successful firms will invest in fixed assets but will manage efficiently their working capital needs. Dotcoms, in fact, can manage efficiently stocks, accounts payable and receivables thanks to the clever use of technology in a way that is unavailable to old economy counterparts. If they succeed in reducing the level of working

(negative) NOPAT is offset by an equivalent value of positive (negative) change in working capital and net capex (gross capex less depreciation).

capital, they can also be unprofitable in the short run but at least they can count on cash generation useful to make further investments without the need of additional cash from shareholders and venture capitalists. If the business is scalable and loyalty effect is reached, economic results become positive and the dotcom generates cash on both the economic and the investment side. Of course, after an initial period of activity as pure "virtual" firm, a successful dotcom will start a more traditional investment process. That's why it will arrive to the same position of successful old economy firms but from the fourth quadrant.

4.4.2 Earning-Based Valuations

Earning based valuations are especially used in Continental Europe, in sharp contrast with the practice of Anglosaxon countries.

Initially theorized in the academic circles (Guatri 1998; Dallocchio 1996), the use of income based methods has spread in the practictioners consulting activity and has influenced in many ways the positions of the Authorities in laws and regulations, especially for what concerns the closed end funds valuation of the equity portfolio of unquoted companies.

Actually, there is no difference in principle between the anglosaxon approach, based on free cash flow discounting, and the continental Europe's earning-based one. In fact, taking the whole life of a firm, earnings and cash flow will assume the same values because distortions due to the application of the accounting principles are completely crowded out in the long run. Only if shorter periods of time are considered, the two methods differ: cash flows and earnings haven't the same value and, under certain circumstances, the two can be deeply different. In any case, it's not wrong to consider earnings as a proxie for the true value generator, cash flow.

In the following pages we will briefly introduce the fundamentals of earning based methods and then discuss about their application to high tech or newly set up firms. As usual, attention will be primarily devoted to show the difficulties arising from the use of the methods in the context of the new economy.

Basically, earnings based methods derive the equity value of the firm by discounting the net income of the firm during a defined period of time (corresponding to the years necessary to reach a steady state in the industry and to the cancellation of the competitive advantage of the firm). The net income must reflect "normal operating conditions", so usually it excludes extraordinary items; in practice, if earnings are based on historical data recorded extraordinary items aren't considered in the estimates (and fiscal burden is recalculated accordingly) and if they are forecasted, the forecast does not consider the effects arising from extraordinary profit and losses. Equity value can then be determined as follows:

$$EquityValue = \sum_{t=1}^{n} \frac{NNI_t}{(1+k_e)^t}$$
(4.7)

where NNI indicates Net Normalized Income. The other symbols have the same meaning of those used for discounted cash flow methods.

Please, note that no terminal value is considered: for Europeans, in fact, the firm is considered as an investment that pays out to the shareholders a limited-in-time flow of earnings.

A position similar to the Continental Europe view on company valuation is the EVA™ approach developed in the USA and centered around the concept of "residual income" (Stewart 1991). In this case, the value of the firm is derived from the discounting of a flow of earnings net of a fair reward of capital (total capital invested or equity capital according to the unlevered or levered version used) employed in the firm.

EVA can be calculated using the following formulas:

$$EVA_L = NNI - k_e \times EC$$
(4.8)

in the levered version (where EC is equity capital) and

$$EVA_U = NOPAT - WACC \times CI$$
(4.9)

in the unlevered version (where CI means capital invested (debt and equity)).

Discounting the values of EVA to present returns the value of the equity of the firm[16]:

$$EquityValue = \sum_{t=1}^{n} \frac{EVA^L{}_t}{(1+r_f)^t}$$
(4.10)

---

[16] In eq. (4.10) and (4.11) we have used a risk free rate (r_f) in order to discount to present the future stream of residual incomes. There is a strong academic debate about this point in that some argue that the discounting should consider a risk premium embedded in the rate of interest (so they recommend the use of k_e or WACC for the levered or unlevered version respectively). Our position lays on the assumption that the use of risk-adjusted rate would depress the value of the firm twice: the first time when EVA is calculated for each year, the second time when the same rate is used for discounting (Guatri 1998; Dallocchio 1996).

when the levered version is used, and

$$EquityValue = \sum_{t=1}^{n} \frac{EVA^{U}_{t}}{(1+r_{f})^{t}} - outst.debt \qquad (4.11)$$

when the unlevered version is used.

Earning Based Methods and New Economy: Some Remarks

There are two main problems in the application of earnings based methods to new economy firms, provided that we agree on the fact that earnings can be a proxie of cash flows:

1. overestimate of firm equity value;

2. negative economic results.

The first problem is due to the fact that during the first years of operation a new economy firms invests large amount of money in fixed and working capital. Investments are not completely reflected in P&L accounts (only their economic effects are, take for example the case of depreciation related to fixed assets, tangible or accounted intangibles or financial expenses incurred for the financing of stocks or accounts receivables).

This means that a fast growing company can also become profitable in a reasonable period of time but if investments go on at a sustained pace, the firm records financial deficits that have to be financed. Using an income based method would lead to the discount of positive flows, cash flow based methods would on the contrary consider a negative flow both in the unlevered and levered version. EVA methods can cope with this overestimate but the problem can still remain.

Less significant is the accounting practice of capitalized costs. Capitalization can take place in most countries when a cost paid for during a given year has utility that exceeds the 12-months period. Extraordinary R&D programs or advertising costs aimed at building a strong brand image, for example, can be carved out from the P&L, recorded in the balance sheet and then amortized. The distortion introduced by the accounting practice of capitalization lays in the difference between the amount of annual capitalized costs (that are financially irrelevant and then deducted from NOPAT in the calculation of free cash flow) and their depreciation recorded in the P&L in the same year (that is summed to NOPAT in the free cash flow calculation). If the two values are quite similar, the difference between NOPAT and operating cash flow wouldn't be that high. That's why in our opinion the use of earning based methods why in our opinion the use of earning based methods can be acceptable only in new economy firms in which investments in fixed and working capital aren't big and that capitalize costs. Put differently, the use of earning based methods is correct if investments (the increase of the left-hand side of the balance sheet) are represented for a large part by capitalized costs.

Consider the following example of a biotech firm that is investing in important research programs for the discovery of new drugs. Suppose that the average R&D biotech firm spends 5% of annual sales for discovery programs and that our firm invest 20% of its annual sales. 15% are considered strategic programs with potential fallout on a 3-years period and then capitalized. The situation of the firm over the 3-years period is shown in Table 4.2 and Table 4.3.

Table 4.2. The effect of R&D capitalization on company value appraisal: profit and loss accounts

Years	1	2	3
Sales	1000	1000	1000
Operating Costs	-550	-650	-700
(of which for R&D programs)	-200	-200	-200
Depreciation	-150	-200	-250
(of which for R&D programs)	-50	-100	-150
Capitalizad R&D	150	150	150
Operating Profit	*450*	*300*	*200*
(less) 50% taxes	-225	-150	-100
NOPAT	*225*	*150*	*100*
(less) net interest exoenses	-50	-50	-50
Net Income	*175*	*100*	*50*

Table 4.3. The effect of R&D capitalization on company value appraisal: cash flow calculation

Years	1	2	3
NOPAT	225	150	100
(less) capitalized R&D	-150	-150	-150
Depreciation	150	200	250
Operating Cash Flow	*225*	*200*	*200*
+/- Change in working capital	xxx	xxx	xxx
- Capital expenditures	xxx	xxx	xxx
FCFF (unlevered FCF)	*xxx*	*xxx*	*xxx*

The reader can note that – putting aside the problem of change in working capital and capex – the difference between operating cash flow and NOPAT is due to the difference between the value of capitalized R&D and depreciation. The faster the period of amortization the lesser the difference and the lesser the distortion introduced by the use of earning based methods (for example an unlevered EVA method).

Of course, if the final part of cash flow calculation is not zero but shows a big amount of investments in fixed and working capital, the use of earning based methods is incorrect and should be avoided in favor of more accurate DCF estimates.

The second problem indicated at the beginning of this paragraph refers to negative economic results. As we will discuss in the coming paragraph, it's exactly the same problem affecting the use of multiples methods.

If a firm passes through a period of expected negative earnings, the application of eq. (4.7) to (4.11) will obviously return a negative equity value. If this situation is expected to be momentary (if not, the firm has no value now and will have no value in the future and is doomed to be cancelled from the market) some adaptations must be worked out.

Damodaran (2000) suggests 3 solutions:

- use of normalized values of historical earnings instead of present economic results;

- use of forecasted margins;

- use of reduced value of leverage and forecast of future results with lower levels of debt.

Actually, only one solution (the second one) is really viable. Point 3 is often meaningless in very young firms that have low debt due to the high business risk (it is more important for mobile telecom, for example, when firms are highly leveraged and their P&L are depressed by the high value of interest expenses). Point 1 and 2 must actually be considered jointly because normalized historical earnings for new economy firms are often negative so they cannot be used to forecast future earnings.

The use of normalized economic results instead of actual earnings is anyway necessary in order to evaluate how much a firm will be able to gain in the future when present market and firm conditions (considered in some ways not normal) will reach a "normal state".

If a firm has negative margins today but successful competitors in the same industry, perhaps born years before, are enjoying positive ROI or ROE we could use these margins in order to derive the "normalized value" of EBIT or Net Income for the years to come by multiplying ROI or ROE for the expected value of future capital employed (CI) or equity capital

(EC) respectively. Put differently, we will derive the value of future streams of earnings avoiding direct estimates of sales and arriving to economic results in a straightforward way.[17]

If we use data taken from reports of competitors that have achieved positive earnings only after some years since the setting up of the business, we could expect that the normalization of results will take place approximately in the same time span. As a consequence, the margins calculated with comparables data have to be considered future margins so they must be discounted to present by using an appropriate discount rate (k_e or WACC according to the use of ROE or ROI). For example, if we use normalized ROE (ROE_n) estimates from competitors that have enjoyed positive net earnings after t years from starting operation, using a simple discount of net incomes we will obtain the equity value through the following formula:

$$EquityValue = \frac{\sum_{k=1}^{n} \frac{ROE_n \times EC_k}{(1+k_e)^k}}{(1+k_e)^t} \qquad (4.12)$$

Apparently simple, this method doesn't solve the major problem: which comparables must be used for the valuation and how to select them? That's the problem we will address in the following paragraph.

4.4.3 Valuation Through Comparables

Valuing a firm through a comparison with similar firms is a practice that has been used for a long time by american and UK merchant and investment banks for old economy firms. This widespread use of comparables lays on the assumption that any analytical estimate of company value is affected by strong subjectivity. Estimation of earnings, cash flows, growth rates, discounting factors is a hard exercise that can be overcome by addressing to market valuations. Being the market the final judge of a firm and assuming that the markets behave efficiently, there are sufficient rea-

---

[17] Using ratios instead of absolute values of comparables firms' earnings allow to avoid the problem of firm size (earnings of a 1 million euro-sales firms cannot be compared to a 100 million euro sales company) even if learning effects, scale economies or loyalty and network effects must be taken in great consideration. Larger firms usually benefit from these favourable factors and we can't assume that a smaller, younger firm be in the same situation.

sons to refer to market valuations in order to derive the valuation of equity of a given firm.

Continental Europe is somewhat different in that capital markets aren't sufficiently developed, market capitalisation is far below the levels reached by US and UK markets and the number of quoted companies is not high. In this context, finding a good comparable (or a good sample of comparables) is not an easy exercise. That's why the method has never been considered as a basis for the valuation but only as a "control value" for analytical estimates obtained with the methods discussed above.

Despite their limited use in the valuation of unquoted firms, market multiples have nevertheless become a common practice also in Europe because they could justify a positive value for the equity of a high tech firm even when other methods indicated a negative or very low value. For this reason, we dedicate this paragraph to a short review of the multiples methods. Our goal is not a precise description of each multiple but rather to point out the main critical traps hidden in their acritical application.

First of all, some basic concepts. If a professional has to estimate a company by comparison he has to choose an adequate benchmark. It can be a company whose stocks are publicly traded or a private company that has been recently bought or sold. Given the difficulty of obtaining information concerning private deals, publicly traded companies are the preferred solution.

The second problem to be addressed is the choice of the multiple: two families are available, the former being represented by multiples of equity value (the most common being the Price/Earning Ratio), the latter being represented by multiples of the value of the firm (or EV, enterprise value). In the first case, the application of the multiple gives directly the value of equity, in the second case one has to deduct the value of outstanding debt in order to calculate the value of equity. Anyway, if the firm has few or no debts, and this is quite common in high tech firms, the results shouldn't be radically different.

Multiples can be calculated with different sets of data. The simplest way is to use historical accounting data derived by the last available annual report (historical multiples); another option can be the use of the data referred to the 12 months prior to the estimate date (trailing multiples). Finally, multiples can be calculated by using expected data (leading multiples). This last choice is the most common when a high tech or a start up company is considered.

Which Comparables and Which Multiples to Use?

The valuation through comparable in itself is a very simple task. Far more difficult is the choice of a correct set of comparable companies from which to derive the necessary data for the estimate. The largest the sample, the more objective the estimate even if too a large sample reduces the homogeneity of the companies included.

Good comparables should be identified on the basis of the following criteria:

1. industry: the comparables must operate in the same industry and in the same business segment. If the comparable operates different business, a break up approach is necessary in order to insulate the data referred to the business under valuation.

2. Size: comparables should show similar size in order to avoid the effects due to cost economies, bargaining power toward suppliers, banks and other counterparts. If size is different earnings or cash flows based multiples could distort the estimate (overpricing).

3. Capital structure: the sample should include firms without any major difference in their capital structure choice. A highly leveraged firm can behave differently in terms of earnings (especially net income) when compared with a strongly capitalized firm; the former shows also a higher level of financial risk. Luckily enough, as long as high tech or start ups are considered, they do not have too much debt at the very beginning of their business activity, so this problem is less stringent.

4. Brand image/loyalty effect: firms shouldn't differ heavily on this. A firm with strong brand image and strong loyalty effect performs better, has higher margins, is less risky. Using such data to estimate a firm with poor or no brand image leads to overvaluation.

5. Expected growth. Market prices always incorporate the expectations of investors regarding the growth. High growth firms are rewarded with higher prices and their multiples usually are high. If a company under valuation hasn't the same growth potential, the use of the calculated multiples must be done with caution.

Once the sample has been formed, the second problem is the correct choice of the multiple. Actually, this shouldn't be a problem at all because the value of the firm is a function of the cash generated in the future, so multiples should always be linked to this value driver.

In practice, the problem exists because during its life cycle the company passes through different phases and in the first one (start up and beginning of development phase) it doesn't show positive economic results and financial flows as previously shown in figure 4.3. Using multiples based on negative figures would lead to negative results.

Generalizing our discussion, during the first phases of the life cycle, when economic performance is still negative and the investment efforts are still high we can use only multiples based on operating variables. During the development phase (consolidation of market position) we can pass to multiples based on economic figures, for example sales (especially if the economic performance is still negative). Only after the consolidation phase the economic results and financial flows turn from negative to positive as shown in paragraph 4.4.1.4 and then traditional multiples based on earnings or cash flows can be applied successfully (see figure 4.4).

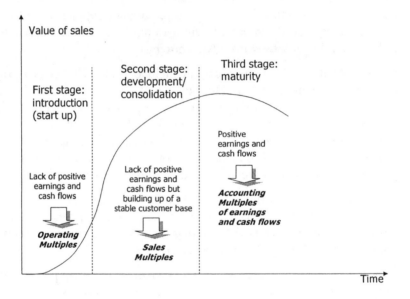

Fig. 4.4. Life cycle of the firm and use of different families of multiples

If this scheme is an effort to rationalize analysts behavior, we reaffirm strongly the distortion underlying the reasoning. Only cash flow or earning based multiples, as long as we want to use a multiple method of valuation and not analytical estimates, should be used in valuation because only cash and earnings are the key value generators.

Multiples Based on Operating Variables

Every discussion about operating multiples is doomed to be partial; every sector, in fact, has its own value drivers that can differ sharply from the ones of another industry. Accounting multiples, on the contrary, are much more similar for all the sectors being based on the same accounting variables. That's why we will discuss the example of dotcoms and illustrate the multiples used in this context for internet related businesses showing the major shortcomings of each one. The reader can instead turn to box 1 for a brief overview of operating and accounting multiples in a totally different industry as biotechnology.

In internet related businesses, the key value driver is the capacity to generate the highest level of traffic. High traffic, however, is only a part of the problem because if the high traffic doesn't turn in earnings, traffic alone is not sufficient. For example, for a horizontal portal traffic is important because of both the revenues from the interconnections paid by telecom companies and the revenues from advertising. But if the portal has no special features that distinguish it from the competitors, its customers are likely to change supplier with ease and are less likely to click on a banner or a pop-up window and generate advertising revenues.

The same holds true for e-commerce (or e-banking and trading on line). Capturing a large customer base is not sufficient to establish a sound company if the base isn't loyal. Only a strong loyalty effect can assure profitability.

Taking the previous notes as a reminder, we can now look at the most used operating multiples in the dotcoms industry:

1. EV/unique user;

2. EV/subscriber;

3. EV/customer;

4. EV/reach.

EV/unique user: unique user is a internet user (a "surfer") that has entered a site at least once during a certain period of time, regardless of the number of times he has visited a site. Particular attention must be deserved to this multiple: in fact, unique user is a synonym for "transit" on a site or, if you want, of "traffic" but it doesn't say anything about the frequency of visits and, above all, business generation. The multiple is widely use and has some importance only for horizontal portals valuation: the underlying rationale is that the higher the number of visitors, the higher the likelihood to generate advertising revenues or e-commerce fees.

EV/subscriber: this multiple is an attempt to refine the previous one. In horizontal portals, part of the services (chat, forum, sms free messages, e-mail boxes etc.) can be enjoyed only by registered users. Registration is a sort of "tie break" for a surfer in that he has to disclose personal information. Using the number of subscribers instead of the number of unique users should capture a restricted part of "loyal" customers. Secondly, if a portal has a large base of subscribers, it can use the profiles to customize its service and enhance customer loyalty. The shortcoming of this ratio is again a problem of loyalty: nothing assures that a customer is registered only in that portal. Being the registration totally free, one could register in different portals without being loyal to any of them.[18]

EV/customer: this is a multiple used for the e-business ventures. Customer is the person that has registered his/her profile and has been authorized by the site manager to make business transactions. The higher the number of customers, the higher the value of the e-commerce firm. Again, attention has to be paid: for B2B business, considering that the customers are firms and that the transactions take place regularly among a sufficiently stable set of economic agents, the relation number of customers-value of firm can hold. For B2C, on the contrary, the loyalty effect is still the critical factor that the simple number of customers can't catch. An e-commerce site could have a large customer base but if the customer is inactive or makes only a small amount of transaction per unit of time the revenue stream is weak.[19]

EV/reach: this multiple tries to overcome the problems shown by the previous three focusing on a proxie of the customer loyalty. The reach rate, in fact, is the number of visits to the site over the total number of contacts for all the sites in a given period of time (usually monthly). If the reach rate is high, the visitors transit on the site more often than on others with all the consequences in terms of traffic, advertising and e-commerce revenues.

---

[18] In order to better discern the reliability of the EV/subscriber, statistics concerning the average time spent by subscribed customers on the site are particularly useful. A high permanence on the site means, all things being equal, higher loyalty than a site with low minutes permanence. A high permanence, moreover, enhances the likelihood of clicking on advertising banners or e-commerce sites.

[19] Take the example of e-banking, a special subsegment of e-business: the multiple EV/accounts is useless if the accounts are inactive or show only a limited amount and volume of transactions. More useful is trying to split the accounts in inactive and active accounts and calculate a multiple as EV/active accounts. Better again is to link value to the number of transactions (especially if every transaction is rewarded with a fee payment) through a EV/transactions ratio.

BOX 1: *Multiples and ratios used in the valuation of biotech companies*

Given the special features of the biotech industry it is not surprising that analysts and professionals have developed a unique set of ratios and multiples in order to evaluate a firm even if measures 2,3 and 5 explained below can be applied to other high tech sectors as well. Part of them are based on operating variables, the others can be considered a subset of accounting multiples.

The most common used indicators are:

1. Cash (and marketable securities)/market capitalisation;

2. Burn Rate;

3. Survival Index;

4. Implied Technology/staff;

5. Market to book (or equity multiple).

Cash (and marketable securities)/market capitalization: the rationale underlying this ratio can be understood by referring to figure 4.1. The analysts use to determine the value of the Implied Technology of a biotech firm (that is the unbooked value of future opportunities) by calculating the difference between the market capitalization and the sum of cash and other marketable securities. The highest the value of cash for a given market capitalization, the highest the ratio, the lower the Implied Technology value. The ratio is anyway very rough. First of all it does not consider the value of outstanding debt and the value of asset in place: in a biotech firm the two values are not particularly important but if they are not zero, they must be taken into account in order to avoid an overestimate of the implied technology value. Secondly, the simple formula (market capitalisation + debt − cash/marketable securities) is unsuitable to explain the reasons of a positive a strong Implied Technology value. In a bullish market, multiples could be high simply because investors are paying high prices (require a low return for risk) and not because of the soundness of the future opportunities of the firm.

Burn Rate and Survival Index: in its simplest version, burn rate is the synonym for net loss incurred by a biotech lab. Analysts look at the burn rate in that a loss "burns" rapidly the cash available to the firm. In the biotech industry, especially in the first stages of development, a lab invests heavily in R&D in order to sustain the business. Dividing the value of cash and other marketable securities by the figure of the burn rate gives the value of the Survival Index, that is the number of years (or months) of fi

BOX 1: *continued*

nancial self-sustainability. A low Survival Index – determined by the acceleration in the marketing/R&D spending due to an increased competition and a subsequent increase of the burn rate – means lower levels of financial sustainability of the strategy. In the US, for example, 35,5% of the competitors in 1999 shown a survival rate of less than 1 year. The same figure was 25,1% only two years before (Ernst & Young 2000). Attention must be paid in the use of the net loss for the calculation of the Survival Index. Net loss is a proxie of cash burning but do not coincide with it: cash burning is more properly measured by free cash flow as shown in paragraph 4.4.1. As a consequence, the Survival Index should be determined by using free cash flow in place of the value of net loss. We could in fact experience the case of a laboratory with positive earnings (in this case, of course, the survival index would be not meaningful) but with a large amount of investments in fixed and working capital not accounted in the P&L. Obviously, no one would tell that this lab is not burning cash.

Implied Technology/staff or Implied Technology/Number of Engineers: this is an example of operating multiple for biotech firms. The rationale here is to link the value of the technology (the value of future opportunities) to the dimension of the staff of researchers. If a lab is successful, somebody argues, it will develop and increase the scale of activity so it will need a larger number of researchers. The highest the number of reaserchers, the highest the probability to be in front of a sound and well established firm, the highest the value given to the implied technology. Summing this value to the cash available and deducting the outstanding debt will give the value of the firm. The multiple has to be used carefully in that it doesn't consider the productivity of the staff but only the number of resources employed; it should lead to overvaluation of firms with large staffs but poor productivity performance.

Equity multiple: nothing to say about this ratio also known as market to book ratio (the ratio between the market capitalisation and the accounting value of net capital). It is another way to estimate the value of unbooked assets: other things being equal, the highest the ratio, the highest the value of intangible assets.

As to may 2001, the situation of a large sample of biotech firms classified into different subsegments of the industry is shown as an example in the following table (Source: Signalsmag, 5 september 2001).

BOX 1: *continued*

SEGMENTS IN THE BIOTECHNOLOGY INDUSTRY		MARKET CAP ($M)	CASH & MKT. SEC. ($M)	TECH VALUE ($M)	TECH VALUE/STAFF ($M)	EST. BURN RATE ($M)	SURVIVAL INDEX (YRS)	EQUITY MULTIPLE
1st Generation Genomics								
	Average	2,758	545	2,213	3.4	-51.5	17.05	3.00
	Median	430	138	292	1.4	-29.5	13.03	1.04
2nd Generation Genomics								
	Average	574	147	420	1.7	-50.8	0.01	1.07
	Median	261	103	211	1.2	-23.4	4.02	1.02
Agbio/Enviro								
	Average	222	17	205	0.3	-2.8	1.07	1.02
	Median	5	3	4	0.3	-1.1	1.00	1.00
Autoimmune								
	Average	7.16	131	639	8.1	-92.1	5.04	2.01
	Median	182	35	127	2.4	-14.8	1.08	1.04
Cancer								
	Average	556	102	454	4.5	-27.4	11.06	1.09
	Median	293	48	238	2.6	-17.5	2.09	1.03
Cardiovascular								
	Average	574	119	455	3.3	-22.1	7.04	2.06
	Median	257	51	134	1.4	-12.3	0.01	1.06
Chemistry								
	Average	520	104	416	1.9	-30.2	4.02	1.09
	Median	228	66	192	1.00	-17.00	2.00	1.03
CNS								
	Average	826	49	577	5.1	-76.3	2.01	2.02
	Median	224	23	204	2.2	-16.3	1.04	1.07
Delivery								
	Average	1,589	224	1,462	5.5	-54.4	5.07	5.02
	Median	286	66	244	2.1	-19.8	1.00	2.03
Diagnostic/Imaging								
	Average	262	17	245	1.4	-23.6	2.01	3.02
	Median	100	12	92	0.6	-12.1	0.09	1.07
Gene Theraphy								
	Average	211	125	98	1.1	-23.1	4.05	1.01
	Median	125	38	77	1.1	-13.7	2.06	1.00
Infection								
	Average	930	106	824	3.2	-59.6	2.04	1.09
	Median	233	44	174	2.4	-23.2	1.01	1.02
Metabolic								
	Average	430	101	383	1.9	-24.6	9.09	1.03
	Median	162	55	271	1.3	-19.5	5.01	1.01
Other								
	Average	582	64	425	1.6	-26.1	2.01	3.02
	Median	192	19	150	1.1	-14.5	1.05	1.03
Revenue driven								
	Average	20,411	1,032	19,379	8.8	n.a.	n.a.	13.05
	Median	9,643	669	8,974	5,4	n.a.	n.a.	8.01
Screening								
	Average	527	101	426	0.9	-25.1	8.05	2.02
	Median	446	97	362	1.4	-10.7	5.01	1.08
Wound								
	Average	136	9	127	1.1	-16.4	0.06	1.01
	Median	85	6	72	1.1	-13.00	0.06	1.01
Grand								
	Average	1,178	137	1,051	2.9	-38.5	5.09	2.06
	Median	257	54	209	1.7	-18.5	2.04	1.05

Accounting Multiples

Once a company has reached a sufficiently stable customer base, traditional multiples can be used also for dotcoms. In the development phase they are based on sales, in a later phase they can be determined also by using the value of net income or of EBIT/EBITDA. Some examples are shown in box 2 when the reader can also notice the strong change that has occured in the value of accounting multiples for ISP/Portals after the bubble smashing in march 2000.

The most commonly used accounting multiples are:

EV/sales: in a dotcom, we recommend the use of this multiple on a "break up basis". In other words, we suggest to split the sales of the dotcom under valuation in different brackets (for a e-commerce portal, for example, sales can be broken up in revenues from traffic, advertising and business transactions), calculate the EV referred to any bracket and then sum up all the results. It's useful to avoid a simple calculation based on single EV/sales given the possible diversity of business models of the comparables chosen for comparison.

EV/EBIT and EV/EBTDA: these multiples can be used only when the dotcom shows positive margins and are preferred by analysts to the most common EV/Net income used for old economy companies for two reasons: 1. EBIT and EBITDA avoid the distortions due to accounting policies mirrored in the net income; 2. EBITDA is a good measure of the cash generation from operating business of the firm. EBIT and EBITDA can be the actual values at the moment of valuation or prospective values. In this case, the value obtained by the analyst will be a prospective figure that has to be discounted to present for the time corresponding to the number of years considered in the estimation of future margins.

Price/earning: this is the most common used multiple for old economy firms. It's seldom used for dotcoms simply because, as discussed in paragraph 4.4.2, dotcoms are at present without positive net income. So, if one wants to use it for internet related business, he/she has to use leading price/earning based on prospective values of net income. The method we've observed in some investment banks, especially for e-business ventures, is based upon 3 steps: 1. forecasting the value of sales that the company will be able to reach in a period of 3 to 5 years (S_5)[20]; 2. estimating the "normalized" profit margin the company will enjoy based on the mar-

---

[20] This calculation can be done by using different growth rates for the coming years or, in a more roughly way, by applying the most recent growth in sales experienced by the firm.

gins reaped today in similar old economy industry (retail distribution, for example, if we have to estimate an e-tailer) (π); 3. calculating the equity price by multiplying the normalized profit margin by the price/earning ratio of the comparable industry; 4. finally, discounting the result for the corresponding time period used for step 1 by using an appropriate cost of equity k_e. The equity value, finally, is calculated as follows:

$$EquityValue(t_0) = \frac{S_5 \times \pi \times \dfrac{P}{E}}{(1 + k_e)^5}$$ (4.13)

BOX 2: *Accounting multiples for ISP/Portals in Europe and North America*

In August 2000, market expectations for the most important competitors in the ISP/Horizontal Portals business area were shown in the following table (Source: IBES, CSFB and internal research):

| | EUROPE | | | | |
	Seat	Terra	Tiscali	T-Online	Wanadoo
EV/Sales					
2001E	5,9	1,9	3,1	5,3	4,1
2002E	4,8	1,3	2,4	4,4	3,1
2003E	4,1	0,9	2	3,5	2,4
EV/EBITDA					
2001E	28,9	nm	nm	nm	nm
2002E	21,2	67,2	49	nm	nm
2003E	15,3	7,4	14,9	137,1	28,4

| | NORTHAMERICA | | | | |
	AOL	Earthlink	Excite	Yahoo	Viacom
EV/Sales					
2001E	5,7	1	2,1	11,8	4
2002E	4,9	0,9	1,5	9,9	3,7
2003E	4,2	nm	0,5	8	3,4
EV/EBITDA					
2001E	21,2	nm	nm	1600	16,8
2002E	16,9	21,6	nm	70	14,4
2003E	14,3	6,6	nm	33,1	13,2

In mid 1999, expectations in a bullish market for AOL, Terra Network and Tiscali were much higher (Source: IBES, CSFB and internal research):

BOX 2: *continued*

	AOL	Terra	Tiscali
EV/Sales			
2001E	25,8	61,8	21,1
2002E	21,6	35,1	11,8
2003E	19,7	21,8	8,7
EV/EBITDA			
2001E	185,8	nm	286,3
2002E	137,6	nm	91,7
2003E	105,8	nm	60,6

Taking the example of AOL, today EV/Sales for 2001 expected for AOL is 5,7 and in mid 1999 it was 21,6: so we can measure a loss in value of 74% in the multiple, meaning that the multiple is now 26% of the original value. Had an analyst estimated the value of a US portal using AOL sales multiple for 2001 as a comparable in mid 1999, he would have strongly overvalued his/her portal.

The same holds true for EV/EBITDA calculation for AOL; the value goes from 137,6 to 21,2 with a loss in the enterprise value of 85%. The EV/Sales analysis for Terra and Tiscali give the same results: the loss in value is 95% and 74% respectively.

4.4.5 Net Worth Based Methods and Cost Based Methods

The final group of methodologies that can be used to determine the equity value of a firm is that of net worth based methods.

Net worth based methods are typical of Continental Europe countries and have been widely used in the past to estimate companies operating in industries requiring a high intensity of capital invested (mainly fixed tangible capital): manufacturing, real estate, banking and insurance were the most suitable candidates for such valuation methods.

Turning back to figure 4.1 and to the classification of business models presented in paragraph 4.3, practictioners have argued that if a company has its success deeply rooted in its capital invested, this variable should count in the final estimate. More precisely, capital invested now is the basis for gaining a future stream of earnings, for this reason its present amount should influence the final valuation of the firm.

Net worth based methods are used for company valuation in Continental Europe still today although not in their simplest form but in association with earning based methods. If used in such way, the difference between the value calculated with earnings based estimates and the basic net worth valuation is attributed to a residual value of goodwill (badwill if the result is negative).

It's easy to understand that net worth does not mean pure book value (the difference between the accounting value of assets and liabilities): it is calculated by referring to updated estimates of the current value of all assets and liabilities. These estimates are referred to, alternatively, current market values for similar assets or to reproduction cost of a given asset (Guatri 1998; Dallocchio 1996).

Referring to figure 4.1, we can certainly affirm that a pure net worth based method is unsuitable for the estimate of equity firm value as long as a new company firm is concerned. Looking at the figure, in fact, we can notice that only a very small part of company value is explained by the amount of accounted assets, even if they are re-evaluated at their current market value. Most part of the value lays in intangibles and internally generated intangibles aren't recorded in the asset side of the balance sheet.

However, net worth based methods can be useful for the analyst because they are based on methodologies that take into account also the value of internally generated intangible assets.

Equity value of the firm would then be calculated as follow:

$$\text{Value of Asset Place} + \text{Value of Intangibles} - \text{Outstanding Debt} \qquad (4.14)$$

Literature regarding the valuation of intangibles (brand equity, patents portfolios, customers portfolios, workforce value etc.) is large, but for our purposes we can summing up the basic methods of valuation in two large groups (Renoldi 1992):

1. income based or cash flow based methods;

2. reproduction cost methods.

Income based and cash flow based methods share the same principles of valuation discussed in the preceding paragraphs. They try to estimate the flows that can be enjoyed thanks to the possession of the intangible in order to discount them to present and calculate the value the intangible would have if bought or sold today. Their approach is clearly forward looking and rely heavily upon forecasts.

Cost based methods are instead focused on historical data. They consider the value of the intangible as the cost the firm would incur if it had to repurchase the asset in the conditions in which it is today. Examples are many: referring to new economy firms, take the case of a profitable portal.

If a portal is lucrative, he had passed successfully the start up phase, has consolidated its market position, has built a strong and stable customer base: these results come at a cost. Advertising, promotion, marketing expenses had been incurred by the firm in order to become a leader. The brand, at this point, can benefit from network effects and scalability of business. Another example is that of biotechnology laboratories: a strong pipeline of on-the-way projects is particularly appreciated by business analysts.[21] The highest the breadth of the portfolio, the highest the value attached to the lab.

Reproduction cost methods try exactly to reconstruct the amount of money spent in order to obtain the availability of the intangible.

The reconstruction is based on the following data:

1. costs incurred and recorded in the P&L of the firm: provided that the costs haven't been capitalized (otherwise the amortized value would be already accounted in the balance sheet), the analyst has to split the cost items in order to identify the not-recurring quote of costs sustained for the intangible development. For example, if the maintenance of a well know brand requires a 10% of sales for advertising budgets and the firm has spent a 15% percentage over the past 5 years, the exceeding 5% can be computed in the calculation of the reproduction cost. The same reasoning holds true for R&D projects.

2. Related investments recorded in the balance sheet: technology investments, hardware and software, equipment for research are all items to be considered in the final valuation of the intangible. Related working capital investments should be considered too.

According to German Association of Biotechnology Industries (1998), reproduction cost is often the only viable way to determine the value of a biotech lab especially if it has been set up recently. However, the method faces severe pitfalls. First of all, it is based only on accounting data re-

---

[21] In the pharmaceutical and biotechnology industry, drugs sold on the market usually pass through the following phases: 1. Discovery (basic research); 2. Preclinical (in vitro tests and trials on animals); 3. Clinical trials phase I (test on a restricted sample of human beings); 4. Clinical trials phase II (tests on larger samples of human beings); 5. Clinical trials phase III (test on large samples of at least 3.000 patients); 6. Approval of the competent Authority for the sale on the market. A biotech firm is valued not simply on the basis of the number of research projects in the pipeline but on the breakdown of the whole projects portfolio according to the development stage achieved. That's because the conditional probability of success is obviously dependent on the stage of the project (Myers and Howe 1997).

ferred to the past. Secondly and more importantly, it doesn't tell anything about the efficiency in spending. A 5% more than average marketing budget paid by a portal doesn't mean necessarily that the larger budget will contribute to the generation of a strong brand and loyalty effect. So, it's difficult in a historical reconstruction of values to discriminate between value generating and unefficient/not value generating costs and investments.

4.5 Conclusions

The valuation of new economy firms is essential to the success of a venture capitalist. The ability to pay the right price for a would-be successful company affect straightly the performance of the investment portfolio. Paying too much for an equity stake depresses the final IRR reaped by investors.

There isn't a general agreement on which is the best method to determine the correct acquisition/selling price of firms working in high tech or very young industries. In the previous pages we have presented all the alternatives available. Some of them are rooted in the anglosaxon tradition (cash flow and market comparables), others are more known and used in Continental Europe but any of them fit perfectly to the special features of a new economy firm.

We are firmly convinced that the value lays in the cash a firm is able to create in the future and that basic DCF models are the best way to evaluate high tech firms. If the forecast of cash is too hard, earning based methods and, as a last chance, net worth based methods can be applied.

What probably is still the missing link is the availability of methodologies that can evaluate properly the value of flexibility that several new industry are experimenting. We have cited in different parts of this chapter the term "option". Indeed, the school of real options has developed dramatically in the past few years but, oddly enough, after a period of intense use by investment banks and pratictioners, today we see the coming back of more traditional methods. This is the clearer sign that value is still based on the fundamentals of the firm: good management, cash generation, clever management of business and financial risk.

5 Specialties in Managing Closed-End Funds

Gino Gandolfi

5.1 Introduction

Financial publications highlight how Italian companies (small, medium and large) are characterized by a "scarce" financial culture, both in terms of knowledge of the possible financial instruments available, and their exploitation (Caselli 2001). The objective of the present discussion is not to examine the different causes that have brought about the present situation. However, it is not possible to ignore that this phenomenon has negatively affected the development of closed-end funds and all the other instruments used by investors in risk capital, as well as the growth of financial markets as a source for the development of the business.

Certainly, favourite channels can be used to obtain a financing depending on the size, age, reputation of the business, and on the availability of information (to the market and to the intermediaries). Figure 5.1 (Berger and Udell 1998) illustrates the overview of the process, in relation to the type of company and the following distinctive features.

The closed-end fund is a financial instrument that raises capital from institutional investors (e.g. companies, foundations, insurance companies, pension funds), and private investors, in order to invest in non-quoted companies with high growth potential (A.I.F.I. 2000). Clearly, it relates to a particular instrument comprehending two investment philosophies: if the fund invests in small sized companies that, after the first steps, require a strong hand to guide them, one can affirm with no doubt that it relates to a venture capitalist, while if the target is constituted by "young" companies, which are not mature enough to move in autonomy, the closed-end fund assumes the typical characteristics of private equity investors.

Being at the same time a product, an instrument and an investor, it is not possible to trace an objective economic environment or a business sector for the closed-end fund, thus making it impossible to provide a clear and all-embracing definition of the phenomenon. In addition, this high "normative" uncertainty makes a correct and meticulous analysis of the real evolution that the product is undergoing very difficult, as it can not be limited to specific characteristics. The "closed-end fund" phenomenon cannot be defined, without having first gathered further information on the object of the investment.

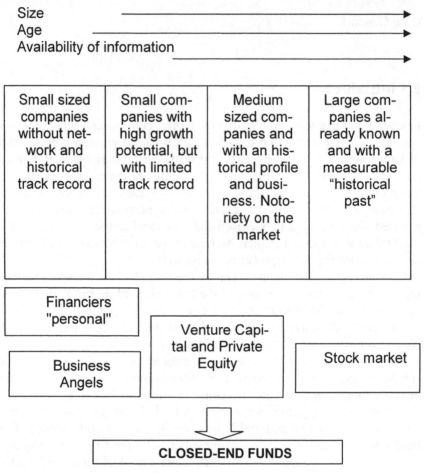

Fig. 5.1. Different channels to obtain financing

This environment of information "opacity" becomes even more obscure if the attention is shifted towards the resources that an investor is capable of offering to interested companies. Certainly, a part of the contribution is of pure financial nature: a capital able to support the predicted growth. Further necessary contributions although difficult to quantify economically, are represented by the management, the know-how, the image, the reputation, and by the network of relationships. In addition, the degree of importance among these is never constant and varies according to the age of the company, its size and its changing requirements (see Fig. 5.2).

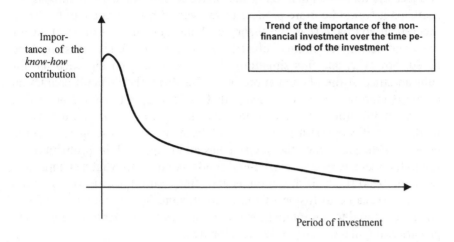

Fig. 5.2. The role of know-how contribution

However, it is more than legitimate to agree with the thesis according to which venture capital funds, although they bring a smaller flow of financial resources , in practice they "invest as much" as the private equity funds. It can also be claimed that the non-financial investment assumes less importance in the later years, when the projects have already started. In other terms, it is particularly at the beginning of a relationship between company and investor that the quality of the latter should emerge within the organizational structure, in order to reveal itself as indispensable "fuel" to the development of the projects judged suitable by the financial operator.

This short introduction leads us to the following conclusion: a closed-end fund is significantly different from an open-end fund. In particular, the first difference is related to the typology of "target" company. Open-end funds invest almost exclusively in quoted financial instruments relating prevalently to large-sized companies already known to the market. This is not the case for closed-end funds.

The type of contribution is also fundamentally different: open-end funds contribute only with financial instruments without performing any direct role in the management of the company[1], notwithstanding the voting right

---

[1] For a small-sized quoted company, the presence of an open-end fund as a shareholder assumes, in any case, greater significance: certainly, financial contribution, but also image to be exploited with customers/suppliers/market. For this reason, the strategies and all the other fundamental choices are taken after considering the opinions of these qualified investors. However, the situation outlined is fundamen-

connected with the acquisition of the shares. It follows that it is impossible (perhaps it would be more prudent to say very difficult) to adopt the same theoretical approach (e.g. the theory of "management portfolio") used for open-end funds for mutual closed-end investment funds. It should be noted, however, that this situation does not derive from the doctrinal insubstantiality of the phenomenon, but rather from the difficult adaptation to the operating reality; this means that the concepts theorised are valid, but the application and the consequential interpretation are more difficult in this specific operating context. Not least, it is interesting to observe how, over the past years, the doctrine has all but ignored the problem relating to the management of closed-end funds, contrary to what has happened with open-end funds. If this circumstance finds partial justification in Italy, due to the scarce diffusion of the phenomenon, it is less understandable when compared to the international experiences where venture capital and private equity have seen greater development.

These are, in substance, the principal reasons giving rise to this contribution that, without supposing to be exhaustive, wishes to provide a stimulus to the discussion with reference to the peculiarity and criticality typical in the management of closed-end funds. In particular, the objective of this section is to explore whether a common "modus operandi" exists between managers of closed-end funds. In other words, whether it is possible to trace a common line that helps us to identify which are the key steps in the allocation of resources among different companies seeking investment. Finally, the paper attempts to highlight, where possible, the most significant differences with the open-end funds. Comparisons will therefore be made with relation to: asset allocation; stock-picking; performance determination; evaluation of participations at the end of the investment; the management of liquidity and, finally, the function performed by the Investor Relations.

During the entire work there will deliberately be no reference to any type of closed-end funds classification or organisational model. This "radical" choice is a direct consequence of the considerations that generated the proposal and the successive carrying out of the research. The strongly operating slant of the project, together with the need to find a standardization or a unified vision of the phenomenon under examination, added to the need to create an institutional reference. This has resulted in the author deciding to analyse the existing situation by verifying the processes common to all funds, without dwelling on aspects that are clearly important but, in this context, scarcely significant. The concept of strategic asset al-

---

tally different "from the interference" typical of venture capital and private equity instruments.

location is in fact ascribable to specialised closed-end funds both by sector and by geographical area. The final implementation of the choices is obviously different, but the initial theoretical concept is the same. In fact, as better explained further on, strategic asset allocation "builds " the entire portfolio without entering into the specifics issues of the sector being analysed. If, for example, the analysis concerning the single categories of "closed-end fund" products should consider how asset allocation changes on a variation of the aspect that the instrument wishes to highlight. It could be said that this kind of analysis does not provide added value to the reader who, intuitively, would clearly understand that if the conditions are not the same, the final exit composition is certainly not the same even when equal samples of target companies are being used. The concepts that will be used in the following examples refer to a common step for all closed-end funds, independently of the individual structures of each instrument. In this research the existing relationships between the promoter and the manager of the fund[2] will not be taken into consideration, as it wishes to provide an opportunity for any kind of technical debate, free from direct confrontations between those who designed the investment and those who, effectively, proposed it. The objective is as a matter of fact to analyse the execution model of individual operations. The resulting relationships of force are already foreseen in the manager's level of decisional freedom with reference to the choice of investment.

Attempting a metaphor in the mechanical field, the objective of the work is to analyse how an engine works, regardless of the type of car and manufacturer . The variables excluded from the analysis are certainly very important, because the propeller of a utilitarian city car must be differentiated from that of a sports car, in terms of number of revolutions, speed, consumption levels, etc.; however, the so-called "testing bench" takes into consideration the same variables for all categories.

This is the reason why, with reference to the present dissertation, the perspective is that of a "testing bench". We wish to analyse the characteristics of a closed-end fund, independently of the underlying organisational structure that created it (e.g. bank or private company), the "net" of existing relationships between promoters and managers (same subjects, different subjects, etc.) and the variables they wish to maximise (e.g. project development speed , periodic cash flows, etc.).

Before entering into the specifics, one further aspect must be examined . The various points underlined as "significant" must be considered as an in-

---

[2] For an exhaustive definition of these subject-matters and the relationship problems found in the investments of risk capital, see Gervasoni 1989; Gervasoni and Sattin 2000.

tegral part of a single system, interdependent one from the other and, in the specific case of closed-end funds, strongly correlated. The objective of each single analysis is to determine the core problem or the specific peculiarities of the instrument being examined by comparing it with those normally found in open-end funds. In other words, we wish to demonstrate, in terms of operating model approach, that the "open" or "closed" adjective should not be considered as a simple indicator of different typology of funds, but as lexical expressions capable of distinguishing among different investment categories, corresponding to products constructed and managed in substantially different manners.

5.2 Asset Allocation Process

For all investments, strategic asset allocation implies a choice in the composition of the fund by "area of interest". Normally, managers of open funds attempt to solve this problem by deriving several more or less personalised models from those which, at a doctrinal level, are considered milestones and indisputable points of reference. That is, the modern portfolio theory introduced by Markowitz and successively developed by Sharpe and numerous other authors. The same methods can not be applied by the managers of closed-end funds for at least two reasons. T first relates to the fact that liquidity is not considered in the theoretical models an element which assumes considerable importance in the going concern and for the realisation of the fund strategies. The second relates to the impossibility of defining the elements that constitute the basis of calculation. Managers of closed-end funds do not have at their disposal all the necessary information to calculate the β and the σ of the shares.

It should be noted that this situation is very similar to the conditions of those operators working in emerging markets characterised, among others, by peculiarities such as high risk, limited availability of information and specific research, non-existence of a developed secondary market (with ensuing difficulty in attracting fresh capital), high sensitivity (but difficult to quantify compared to macroeconomic variables). However, a detailed analysis of the problem identifies other specific factors that make the business of investors of risk capital completely different from that of others such as the non-existence of historical data, the novelty of the business in question, the absence of experience (both of entrepreneurs and intermediaries).

As a result of the non-existence of all the data necessary to define a standardised course in evaluating the intervention a manager of closed-end funds finds himself in a condition of "information obscurity ". Moreover, another problem concerns the calculation of the contribution of all those non-financial values that are an integral part of the investment, constituting a more or less evident link between the financier and borrower.

It is evident that the strategic choices of a closed-end fund, are much closer to the logic of industrial investments than to the logic of financial investments typical of open-end funds. Hence why, before choosing by type or geographical area, it is necessary to define the level of involvement of the fund in the company's activity. Once the participation is taken, the instrument can activate a collaboration activity with the entrepreneur involving a simple intervention in case of necessity, up to the joint implementation of the development plans. Naturally, the greater the involvement, the greater the closed-end fund makes investments of an industrial nature.

A further direct consequence of the impossibility to perform a quantitative approach for the strategic choices of a closed-end fund is the importance of the manager and all the analyst staff who assist him in defining the composition of the portfolio by sector and/or geographic area. In fact, in this case, more than the simple correlation between the returns of the activities financed or the quantification of the risks, the specific experiences of the company's management and of the investment manager assume great importance. Thus, the process of strategic asset allocation between the different sectors and geographic areas is based on the growth potentials and on the "quantity" and "quality" of the experiences matured by the management of the fund and of the company.

With relation to closed-end funds, the concept of strategic asset allocation must, therefore, be widened, as the simple definition of "creating portfolio by area of interest " tends to limit its capacity. As previously mentioned, the strategic variable that strongly influences the effective possibility to carry out the objectives previously formulated directly depends on two factors:

– Capital

– Human resources

These aspects tend to travel on two parallel tracks: the more the manager or better, the team that is entrusted with the implementation of the objectives, has good reputation, credibility, experience, knowledge and expertise, the greater the amount of capital subscribers will make available for the investments. This relationship is completely different from what happens with (large) open-end funds and the institutions who head them: the greater the financial resources, the greater the decisions taken on the basis of quantitative analysis tailored on large volumes. On the other hand, this choice tends to limit the real contribution of the individuals who not only must be capable of "modelling" their own method and analysis capabilities, but must possess strong relationship capabilities being part of a large team. In closed-end funds, on the contrary, the main requirement is for people to be able to work on their own having a deep knowledge of specific companies and sectors. Ideally they should have a high professional background and being capable of relating, rather than to colleagues, directly to entrepreneurs whose projects are subject to a con-tinuous evaluation. We do not mean to express an opinion on the difficulty required in comparing one set of skills against the other, but we claim that, in relation to a manager of funds, the expertise required is strictly linked to the product that will be managed.

It follows that the process of strategic allocation of the resources of closed-end funds presents at least a double value: the classic, standard one referring to the fundraising policy, choosing the size of the fund, the investment policies (e.g. how to sub-divide the funds between geographic area and sector, small and medium companies), and establishing the priorities to be achieved; the second one, typical of the instrument, reflecting on the human resources side an aspect of strict criticality for reaching the objectives.

5.3 Investment Choices

In the classic portfolio theory, after having defined the composition of the investment in different asset classes, it is necessary to transform the strategic asset allocation of resources in real portfolios. In practice, the geographic area or sector must be turned into participations in companies. The principal objective of this phase, for the closed-end funds, consists of searching for and, successively, correctly evaluating the most important business plans to be transformed into effective development projects.

It consists of a very important and, at the same time, delicate phase because a private equity or venture capital operator does not have the same flexibility as that of a manager of an open-end fund: once the investment is made, a time period elapses within which the participation cannot be sold. As noted, in fact, the managers of open-end funds use the tactical asset allocation to exploit contingent situations that are created on the market in order to improve the results.

In practice, the tactical asset allocation of a closed-end fund includes several aspects: selection and analysis of the investment projects, acquisition, monitoring and management of the participations. Obviously, this step must also examine those aspects related to fundraising or exiting a quota, but both are one-off choices, while the former aspects must be continuously taken into consideration and timely re-evaluated. Especially in the initial phase, the flow of proposals (deal flow) is of fundamental importance as, in order to generate a "powerful effect" it is necessary not to run up against gross errors, which are not only related to the choice of individual company, but also to the real possibilities to contribute in a significant manner to the development of the business through direct participation in the management.

One can therefore sustain that the tactical asset allocation for a closed-end fund are still more important than the strategic asset allocation because it is the only vehicle that can guarantee a result. In other terms this function tends to have the same importance of stock-picking for the open-end funds, or rather with the choice of the individual shares in the portfolio (see Table 5.1).

Table 5.1. Investment choices in closed-end funds and open-end funds

	INSTRUMENT	ASPECTS EMPHASIZED
STRATEGIC ASSET ALLOCATION	OPEN-END FUNDS CLOSED-ENDS FUNDS	Risk and return profile Diversification (especially the qualitative aspect), degree of participation in the management, determination of "overall" policies
TACTICAL ASSET ALLOCATION/ STOCK-PICKING	OPEN-END FUNDS CLOSED-end FUNDS	Extra-return Maximisation of the value of the participation

During this phase decisions are translated into practice : e.g. insertion of clauses into shareholder agreements or in constitutional acts; decision on "interventionism" policy. Normally, besides the examples mentioned, this is realised through the direct participation on the board of directors, the possibility to nominate a number of directors, and the possibility to obtain information within a certain time frame. These choices not only better define what the manager of the fund intends by operating interference in the company, but also act as "limiting devices" of the risk. It has been said, in fact, that for closed-end funds it is impossible to define in a quantitative manner an appropriate measure of risk; that does not signify, at a tactical level, that certain choices are not made in order to limit the effective exposure to factors determining the volatility of the final portfolio. Guaranteeing himself the possibility of intervention, the manager limits the possibility of any incorrect behaviour by the entrepreneur, for example a "disregarding of banking covenants" from the agreements in place. An-other course that will give the same result provides, for example/instance, the participation of the fund in companies that, even if related to the same sector (or geographic area), are experiencing different stages in their life cycle (start-ups or companies with much greater dimensions). Hence, at the same time, the difficulty of classifying closed funds among venture capital or private equity instruments. The most important risk to be en-countered in this phase by the fund managers relates to the possible "ad-verse selection " of the target companies. Unfortunately, only later on one can understand the real value of each operation performed.

Even though the overview outlined up to now can suggest that the process is ascribable to a top-down approach, this affirmation can be extended to all managers as, in certain events, they could find it interesting to invest in a company belonging to an area not taken into consideration in the strategic plans. The same cannot be affirmed for the managers of open-end funds who, except in well defined cases[3], must adopt investment choices in line with pre-defined strategies, because the strategic asset allocation represents a precise "guide" for the portfolio management.

The tactical asset allocation for closed-end funds, in a context of open-end funds, does not exist and also assumes connotations relating to the choice of the individual companies that must be inserted in the portfolio. Given the objective difficulties connected to the definition of the overall problem, even the valuation of the individual participations to include is difficult to classify, and it can be affirmed that there does not exist any specific methodology for the managers to follow in valuing their own intervention in the management of a business. It is a subject fundamentally different from that which will be analysed in the following section dedicated to valuing the participation at the end of the investment. During the selection phase, the risks and scenarios are even more indefinite and require a much more detailed analysis and personalisation compared to the final phase when the projects undertaken give their hoped-for results, facilitating the definition of the value of a participation.

The fundamental points concerning the determination of the initial value of a quota are clearly related to the strengths of the business under examination. In fact, other than the development scenario, the business model

---

[3] As for the "flexible funds".

under consideration (naturally if existing and standardised, as in the start-ups), the management value, i.e. experience, capacity, reputation, etc.

This consists, obviously, of a process with little standardisation, in which the sensibility of the analyst assumes a fundamental importance. Therefore, the valuation of the individual activities take place through a series of considerations which, connected one to the another, are capable of defining the degree of desirability of an investment: first of all, the company must operate in sectors "of interest" for the fund (derogated condition); in addition, the potential market must not be a niche[4] as, in this way the obtainable results are less uncertain. Descending to the details of the individual single economic initiatives, the analysis relates to both the company's growth potential, through the application of theories related to realising cash flows and verifying the scenario, and the evaluation of the business model that must be strong and deep rooted in the organisation.

In order to avoid easy misunderstandings, the managers of the closed funds like to repeat the following sentence: "Finance only what you know". In fact, given that the contribution relates to aspects not only purely financial, the investment has sense only if the added value of the fund can be considered significant or fundamental for the development of the company.

5.4 The Liquidity Management

The liquidity management process is prominent and critical, because the acquisition of the shares in a non-quoted company is normally made in several stages. The only way for the financial operator (provided that there is no legislation to the contrary) to limit the importance of the phenomenon is to allow the subscribers to pay the amount they have pledged in different stages, reducing the "mass" of liquidity initially held. If this option is not taken into consideration, the problem becomes febrile only for the short period, necessary to make the investments.

As the objective of the closed-end fund is to invest in share capital of non-quoted companies, the liquidity is normally placed in non-risk instruments, which are easily convertible in legal tender. Once the investments are made, management problems decrease and become almost negligible, as the quota in the fund cannot be turned into cash at any given moment. Hence, an evident difference compared to the open funds, where

---

[4] This concept must be considered as "relative"; it is possible, in fact, that, in the operating world, markets that for some are considered niche and thus of little significance, are, for others, subject to detailed analysis.

the problem of liquidity management is more important and difficult to manage as it is subject to the unpredictability of the investors' requests. This particular situation establishes a different trend in the liquid reserves over the years in the two instruments. In the closed-end funds, for example, the trend is similar to a satellite turned upwards: in the first years of activity the research for opportunities and the technical timing relative to the investments mean that the "supply of liquidity" is high; in the short and medium term, liquidity assumes secondary importance as the participations must be left to mature slowly. The problem relating to managing the liquid reserves rises towards the end of the funds' activity, when the previous investments made by the fund are transferred into cash for the redistribution of the final result to the investors (see Fig. 5.3).

It is possible to draw a similar, but not homogenous, trend for open-end funds, due to several peculiar aspects: firstly, the search for activities to invest in is much easier when there exists a reference market. For this reason, the amount of initial liquidity (received from the subscribers) does not stay still for an excessive period. The same phenomenon happens at the end because the exit process of the quotas is simplified by the same variables. However, throughout the process the manager of an open-end fund must pay attention to the liquidity which is normally used to liquidate the quote to the subscribers, meet the commitments of the fund (e.g. the payment of taxes) and , respond to "tactical".

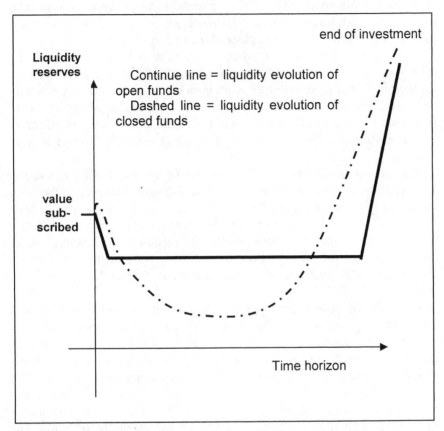

Fig. 5.3. Liquidity evolution

5.5 Evaluation of the Participation at the End of the Investment

Even if the problem of evaluating a participation only becomes fundamental at the end of the investment, it is useful to remember that, regardless of all other factors taken into consideration, the companies subject to analysis are not yet "mature" and may present some "critical areas" difficult to exploit, such as the sector, entrepreneurial profile, future projects.

Obviously, the estimate at the end of the investment will be far "simpler" compared to the same problem encountered by investors of risk capital facing the decision to invest in the company at the beginning of the relationship. At "that time", in fact, the risk associated with the entrepreneurial activity was certainly greater, as the dimension were inferior and the same standardization did not exist in terms of business model that, on the contrary, may be in place at the end of the investment.

Agreements undertaken between companies and investors normally make sure that at the end of the investment period there is a planned "exit" for the participation. In order to avoid possible tensions between entrepreneur and financier, it is in fact necessary that from the beginning of the relationship there is a provision for how and under what terms the exiting of the quotas that, in the majority of cases are a minority holding will be made.

The management company of closed-end funds or any other investor in risk capital establishes the objective of liquidating its presence in the company through a private sale ("Trade Sale") or through the company's quotation ("IPO")[5]. What is important is the presence of a third operator: as a matter of fact, a merchant bank, which determines the exit price, studies the valuation of the participation. Above all it is necessary to obtain a speedy exit is the presence of a series of exits all efficiently and easily usable.

The IPO, for example, represents a liquidation method that guarantees excellent image for the company and the private equity operator, but does not permit to obtain the maximum valuation possible (that is normally achieved in the trade sale operations) or guarantee the same strategic benefits or timeliness (e.g. the technical time necessary for a Stock Exchange quotation) (MSDW conference 1999). The same considerations can be made by analysing the basic logic for choosing a trade sale that is probably slightly different from the points considered in the IPO case. The trade sale strategy is normally used by venture capitalists when the company, although having reached excellent results, does not yet have or will never have the necessary dimensions to be quoted on the stock exchange; or when in order to reach this break even point it must consolidate its position through the intervention of another entrepreneur who may become the owner of the previous quota held by the closed-end fund. In addition, this opportunity is taken into consideration where company's growth process is slower than either what was predicted when the investment was initially made or compared to the expectations of the venture capitalist.

Regardless of the exit strategy taken, the presence of an investor of risk capital within the company simplifies the liquidation operation. One of these operators' particular abilities consists in reducing the information

---

[5] There exist other methods for the liquidation of the participations but the two shown here are those most frequently used by the operators. For further details see AIFI 2001.

asymmetry between the parties involved in the exit initiative. The presence of a venture capitalist in the company guarantees the quality of the operation (be it an IPO or trade sale) through their own financial investments and reputation and renders, ceteris paribus, less costly the whole liquidation process, compared to the same operation made by a company without the presence of investors of risk capital among its shareholders.

If, for example, the selected exit for the liquidation is an IPO, the problem of valuing those shares related to the fact that they are, for the first time, traded on financial markets. The expectations are less clear if the absence of information increases. An entrepreneur or a manager has, therefore, a strong incentive not to reveal or to reveal with delay all adverse information that could have a negative effect on the entry price on the Stock Exchange. This , however, cannot be the case for the investors in risk capital who, on the contrary, have a completely opposite incentive, as the market image is one of the competitive advantages of this particular industry. A solid reputation built on expertise, on the information provided to the market and the congruity of the prices adopted in the exit operation guarantees the private equity operators that continuity of relationships essential for its development. In addition, this reputation permits them to remain competitive both in the stock market (for the IPO operations), and the companies' market (for trade sale or successive investments) (Megginson and Weiss 1991).

Therefore the exit strategies allow the investor in risk capital to return in possession of the liquidity held before the investments were made. The intermediary, whether it is a closed-end fund or any other financier, is part of the whole initiative, and cannot be responsible for the valuation of its quota, as this situation would result in a clear "conflict of interest". For this obvious reason, the management of the whole activity is performed by primary merchant banks, which aim at providing a value for the participation that must be liquidated through the market or by the intervention of another entrepreneur. The objective of this part of dissertation is not to show the various valuation techniques[6], but to underline how managing these situations is not a direct responsibility of the closed-end fund, which has, in practice, the duty of "monitoring" the banker's operation who, from time to time, is required to intervene. Naturally, the choice of the merchant bank by the entrepreneur (or much better, on the advice of the private equity operator) is never random, as empirical evidence in the more developed markets, in particular in the USA, demonstrates that the close ties existing between these two intermediaries cannot be denied: the venture capital companies always tend to choose the same merchant banks, especially

---

[6] In particular, reference should be made to chapter 4 by Stefano Gatti.

for the IPO operations, while the merchant banks easily tend to commit themselves to companies with investors of risk capital (Megginson and Weiss 1991; Beattv and Ritter 1986; Rock 1986).

5.6 Performance Determination

Performance determinations follows the guidelines of by industry associations and specific regulations set by each single country where the "directives" are acknowledged by the regulatory organisations in the financial sector.

Nevertheless, for our purpose, it is not necessary to enter into the technical and legislative details of the problem and all the considerations will be based on the analysis of the most recent instrument: the guidelines proposed by EVCA[7] in March 2001.

The guidelines proposed by the association are a reference point for all investors and not only for the category that we are analysing at the present moment. In fact, the strong development in the investments in risk capital has made inadequate and impractical a specific type of approach that, even if capable of providing further indications related to the single operations and instruments, at the same time does not keep up with the level of diffusion of the phenomenon, or with the typical innovation in the financial world. The guidelines proposed by EVCA may not be acknowledged by the operators but, in a business where the operators' reputation plays a fundamental role for the success of the initiative, the danger that different managers ignore the association's "advice" can deteriorate its image, thus making the work of research and development of the financier/company relationship much more exhausting.

The problem of determining the performance arises when there does not exist an efficient market or where trading is carried out in a context of little transparency. In practice, when the concept of value can be freely interpreted, there is a problem of finding a route that, even if not objective, is capable of standardising the key passages in order to arrive at the final value and ensuring that this final value is measured in accordance to the

---

[7] The EVCA (European Venture Capital Association) is the European association of investors in risk capital and has the objective of standardising the processes of acquisition, management, sale, valuation of the quotas between the different operators. This situation, in fact, is capable of allowing a simplification of the relationships between all the participants and permits, in addition, a homogenous comparison at international level.

same principles by all the operators, in order to make the comparison easier and, consequently, allowing a valuation of the operator.

The concept of return can be calculated with reference to different nuances: EVCA has chosen to use the IRR (Internal Rate of Return) or rate of internal return (TIR) as a measure of performance; i.e. it relates to the net present value on the outgoings (in particular, the purchases of quotas) and receipts (dividends, exits) for one or more operations. This choice, even if it is from a mathematical point of view the most difficult to perform, can be considered as the most correct, because this indicator is the only one capable of taking simultaneously into account the time variable, and can be calculated with reference to a single investment or a number of operations.

The objective of calculating the performance is twofold: on one hand, it provides an indicator for measuring the ability of the manager to choose the target companies and manage the development projects; on the other, it obtains an indicator that, taking into account also the effective costs sustained by the subscribers, allows an evaluation of the effective "result" obtained by the investors. As the starting data are not the same, three different levels of IRR that are normally taken into consideration:

• GROSS RETURN ON REALISED INVESTMENTS: it gives the net present value of the entries and exits relating to the investments made (for example, the write-downs or bankruptcy not yet certain, are not included)[8];

• GROSS RETURN ON ALL INVESTMENTS: different from the previous, it takes into account the possible value that the operation will also give to all the interventions still to be made (for example, a quota still to be invested or a write-down still to be made) with exclusion of the liquidity reserve;

• NET RETURN TO THE FUNDER: this is the measure that is most interesting for the subscriber as it clearly shows the final result, net of expenses and commissions that the manager applies to the product. Intuitively, in this case, the liquidity reserve is also taken into consideration for calculating the return.

Each manager of closed-end funds can decide whether or not to widen the data series available to the subscribers: for example, if a closed-end fund has participations in foreign companies, the return calculation could take into account the exchange rates effect. The information of the net return, even if it comprises of all these aspects, which effectively have an

---

[8] For more detailed information reference is made to EVCA 2001.

influence on the management of a closed-end fund (or any other financial instrument), includes within itself a series of information that does not permit the comparison among several managers. If the weight of the liquidity is not the same or the commissions and estimates are not equal, the comparison loses its significance. For this reason, normally the indicator most frequently used is the first, as it offers, even if partial, a complete vision of the operation analysed[9].

5.7 The Function Performed by the Investor Relations

In the first paragraphs, one particular aspect in the operating environment of closed-end funds was highlighted, i.e. the need for a large network of contacts. On one hand, this peculiar aspect is vital to the business development and, on the other, it is necessary to giving life to that sense of "trust" that normally links companies, closed-end funds and investors. Obviously, the non-existence of a regulatory market means that the only vehicle capable of furnishing correct information on the investments depends directly on the parties who take part in the operations. In practice, in private equity a generator of information which is different from the business actors does not exist. For this reason, the role covered in closed-end funds by the investor relations is of fundamental importance in streamlining the state of information asymmetry and the risks of "adverse selection" that characterise the investment process in start-ups or in those companies that do not yet have the structure to proceed and grow autonomously.

The IR (investor relations) function, in this particular economic context, has two fundamental objectives: receive the financing and guarantee the continuity of the relationships. The achievement of this goal also depends on the ability of the companies subject to the investment to internally create a series of information flows capable of providing in a timely and clear manner the state of health and, as a consequence, the possibility for the manager of the closed-end fund to enter in possession of all the necessary data in order to "refer" to the subscribers. As the investments under examination have the final objective of obtaining a suitable capital gain at the end of the investment, even on a medium-long term time horizon, the role performed by the IR is fundamental for the maintenance of "good relations". In fact, they relate to operations that, especially at the beginning, do not give satisfactory results and, in order to limit the possible and natural

---

[9] Also the AIFI, the Italian association for investors in risk capital, in the periodic publications on the theme for the valuation of the results of the investments, utilise the Gross Return as the indicator on return.

state of apprehension of the subscribers, it is indispensable that the closed-end fund or intermediary following the investment offers a series of information showing to outsiders their commitment through constant and effective monitoring of the work performed.

The quality and quantity of information that must be made available directly depends on the particular operating context of the closed-end funds (see Fig. 5.4): in fact a substantial flow relating to economic and finan-cial results, strategies, efficiencies is not enough, but a commitment to cre-ate a range of indicators relative to an analysis of the market, managers, fu-ture projects, state of the investments, etc is also necessary. In other terms, the managers of the fund and, as a consequence, the entrepreneurs fi-nanced, must offer a large number of data including events that are not di-rectly ascribable to the business operations. As a matter of fact, this flow must convince, in the first instance, the managers of the fund and then the subscribers on the health of the investment and, independent of all other considerations, must demonstrate the growth targeted by the company. Since the projects in which to invest are numerous, the discriminating factor that influences the choice is the quantity, quality and speed with which information is available. Given that the IR function represents, in whatever context the analysis is made, the connection between the internal organisation and the external world, an actively supported IR development programme can be an instrument that, realistically, any company interested in the investors of risk capital and all investors themselves, must put in place in order to overcome the typical anonymity of the market and, therefore, obtain the investors' interest. Up to this point there has been no conscious wish to separate the IR between the company subject to the investment and the intermediary who decides to make the financing operation as the "semi-industrial" logic underlying all the private equity operations does not permit to define, in an objective manner, how much and which data are absolutely necessary in order to prepare the reports. The only matter that can be confirmed with certainty is that the amount of information that the closed-end fund must furnish to its subscribers is certainly superior to that which the company must furnish to the intermediary. This is easily understandable. In addition to the information on the individual investments, the subscribers have the right/need to receive information on the health of the closed-end fund itself.

Thus, for the closed-end funds the IR function must be analysed from two perspectives : the first sees the intermediary as a simple reporter of another person's data, and the second, on the contrary, sees the closed-end fund as an autonomous source of information relating to the state of its own business.

In particular, in relation to the first aspect, the correctness, quality and timeliness of data depends directly on the ties that each fund manages to contrive with the entrepreneurs subject to the investment and the ability of the fund managers to place some of their "own men" in the key roles or to insert specific obligations within the financing contract. All this information is used by the managers to better understand the state of progress of the operations undertaken and for the subscribers to learn the principal characteristics of their own investment target.

COMPANY
Data on: economic and financial results, strategies, market positions, managers, future projects, state of investment in course, etc.

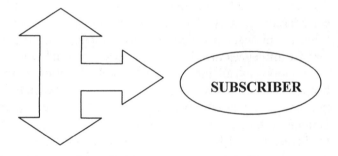

CLOSED-END FUND
Data on: number of investments made, number to be made, cash requirements, IRR, research synergies, exit strategies, etc.

Fig. 5.4. The information level requested by the subscribers

The IR represents one of the "channels" a company possesses in overcoming the anonymity typical in the financial markets and in attracting the investors' attention. In other terms, any entrepreneur wishing to get the investors' attention, to increase the liquidity of its shares (especially for the quoted companies), and create the basis for attracting capital in the future must provide to the market all the information requested for achieving these medium-long term objectives (McBride A S and McBride R G 2001). On one hand, in the closed-end fund managers' perspective, it is better to invest in companies that believe in this function, as, at the end of the investment, it will be much easier to exit liquidating the participation.

On the other hand, in relation to the closed-end fund as an autonomous generator of information, the attention must be turned to the distinct characteristics of the fund such as the number of investments to be made or the quota invested by sector, and on more specifically strategic aspects such as the control of management or intangibles.

For a closed-end fund, the IR function assumes different meanings:

- It increases the subscribers' knowledge of the companies in which investments are held

- It "guarantees" the solidness of the project even in uncertain moments (e.g. at the beginning of the operation when the results get slowly)

- It strengthens the image and credibility with other financiers (e.g. with other venture capitalists or banks)

- It improves the image and reputation in the market (as not all the companies interested by investors of risk capital are effectively subject to successive operations)

It is likely that it relates to a series of activities that can be included within the context of marketing operations: for a closed-end fund, IR is essentially a function related to image. In a company's perspective, it is clear that, in the latter case, the fundamental marketing objective is to make its products better known to the acquirers. In the closed-end fund, one can adopt the same viewpoint: the managers of the fund want the details of the instrument better known to the subscribers.

At the same time, however, one can affirm that the IR of private equity intermediaries do not have such a limited meaning. First of all , assuming a purely marketing vision one misses what in reality distinguishes such an investment compared from a purely financial contribution, i.e. know-how and reputation. In addition, it is not possible to obtain all the qualitative aspects connected to an investment's efficiency. It follows that, in this specific case, the IR can also be a strategic guide for the companies and, if necessary, for the other investors. The ability of the fund managers to correctly interpret and "carry" the information to the subscribers, with respect to the aspects not easily measurable or connected to non-physical aspects, represents a further effort to strengthen the relationships between the parties involved in the operation. Factors such as the loyalty of the management or the workers' "contentment" and participation or the effects of social and environmental policies involving the company can be explained and brought to attention only in a qualitative way as they are not easily measurable. This is why the more these aspects assume importance in the investment of a fund, the greater must be the relationship effort made by the managers who must report these aspects, which normally lie outside the remit of standard reporting, clearly and correctly.

The IR function is evident of strategic guide is evident in relation to those companies that are targets for the closed-end funds and are characterised by elements of innovation, quality, new products, technological evolution, social and environmental policies, improvement in internal relations, etc., or rather phenomena where it is difficult to find a classification in the financial or cash flow statements. And as a matter of fact , these variables play the most important role in determining the suitability or otherwise the investment decision, by both the subscribers and managers of closed-end funds. However, a valuation that does not take into consideration these drivers would consider, without any doubt, these companies not interesting as they are devoid of results, even if in reality characterised by projects, strategies and objectives worthy of consideration.

In order to avoid limiting conclusions, it is useful to highlight one last aspect: the "operating" management of the relations is not easy at all as it shows in a series of facts that, if taken singularly, are not significant, but play a vital role if considered in the context discussed above. For the managers of closed-end funds, the concept of relationship is much wider and goes from a simple telephone call from a subscriber who is anxious for the investment made, to the direct participation in the board of directors of the companies in which investments are held; to the preparation of the periodic financial reports; to the road-shows in order to find new financiers, relationships with the specialised press or research centres; to the updating of the internet site, etc. It therefore consists of a series of operations that must be carefully planned where the "time management" takes on unavoidable importance: looking at the investments where the payment of dividends or cash flow are not the only decision drivers, every activity must be seen, organised and endured as a strategic key because it has a fundamental impact on the development of the next one. One thinks, for example, of the "future effects" of an absence at a social assembly or at a meeting of the board of directors, or simply, to an impolite response to an investor's telephone call.

5.8 Conclusions

The analysis carried out allows to draw several conclusions on the methodology that should be followed by the management of closed-end funds. A special focus has been placed on those factors that have influenced these choices. We will first propose a brief list of the facts that, according to the author's point of view, are fundamental to a correct evaluation of the closed-end fund phenomenon. These should also be taken into account when conducting further studies on risk capital investments:

- closed funds bring partnership capital and not only financial capital;

- venture capital and private equity are "relationships businesses based on the managers' reputation";

- the importance of the human element is one of the critical success factors;

- the absence of a market strongly influences the methodology adopted when choosing the participations.

It is evident that the success of any investment directly depends on the strategic decisions and the management activity, i.e. the coherence of the strategic formula and of the ability/expertise of the manager's human resources in translating in operating behaviour what was previously fixed as an objective. The workforce is as a result the natural basis of the fund management activity. It is in fact e responsible for all those steps that, if correctly carried out, bring the results to the subscribers and guarantee the achievement of the expectations.

The centrality of the human resources also estimates the real contribution that the various phases bring to the final performance. In fact, the attempt to allocate the result to each different phases of the process (strategic asset allocation, tactical asset allocation, etc.) is clearly impossible.

More that the financing ability what counts for the closed-end funds are diverse activities such as the ability: to attract and maintain the management for the fund; to create a team of people who work well together and characterised by integrated acquaintances, consisting of long term interconnected relationships with the target market. The centrality of human contact can be found in all the construction and monitoring phases of the instrument. This is in line with the concept of partnership capital, according to which the financing is not conceived as a mere operation in itself, with the sole objective of remunerating the investment made, but as an entry in a company with an integrated vehicle of instruments (not only of a financial nature), in order to support its development. Being the nature of the contributions a "hybrid" one, also the relationships between company, fund and subscribers cannot be explained by using just one variable. What connects the closed-end fund and the company is in fact not only the financing in the same way that what connects the subscribers and the fund is not only return. The results achieved are certainly important, but those who acquire the quota of an instrument want, at the same time, to be informed in a direct and complete manner on what the managers are proposing. Transparent information certainly makes the relationships less tense in case, for example, of a write-down in the participations. At the same time, the entrepreneurs financed must find in the managers of closed-end funds a source of continuous dialogue, as the possibility and effectiveness of any interventions are critical factors for the success of the entire operation. The situation just outlined changes significantly if compared to that of open-end funds, where there is a well defined "distinction" between subscriber, manager of the fund and companies. This factor could however lead to think that, in this business, the relationship capacities are not of fundamental importance for its development.

The activity of investment/financing in projects with high development potential are not only characterised by a high risk/return profile, but also by the long time horizon requested from the investors in order to deliver the results. For this reason, the closed-end fund is comparable to wine that, in order to become truly superior, necessitates a slow "ageing" process; In business too there are entrepreneurial enterprises that, before giving concrete satisfaction to their shareholders, require particular attention and time in order to be transformed into concrete operations.

Essentially, all the decisions relative to the portfolio depend on the market opportunities, the ability to develop the companies, the fund mission. All these aspects require a very strong relational channel between financier and financed, in order to be taken in serious consideration or just to be visible. Small and medium sized companies subject to risk capital operations do not enjoy the notoriety that is typical, for example, of the quoted companies. to The absence of institutional channels of communication outlines how the relationships between companies and investors are still based on reciprocal acquaintances (personal and business) rather than on standardised information. As personal relationships can be developed only if reciprocal trust and credibility exist between the parties. It is evident that a business with no trading market characterised by the high operating risk of the investments must be based on strong interpersonal relationships among the operators.

The absence of a place in which to trade and the lack of standardised processes play an important role in determining the value of the participation. Even the less complex valuations, that is those related to the sale of the quota invested, are not characterised by a common course by the operators, but rather by the importance assumed of the "personal variable" or, simpler, by the analyst's sensitivity in interpreting the activities and the company's projects. However, for the instruments under examination, the problem of the underlying methodology of the valuation criteria adopted in order to determine the value of a participation is particularly felt only in two stages of the fund's life: at the beginning of the activity, when acquiring the quota, and at the exiting, when the capital gain of the fund's intervention in the company is determined.

6 How Does a Venture Capital Work: Case 1 – Pino Ventures

Sonia Deho' and Elserino Piol

6.1 Pino Venture Partners and Pino Partecipazioni S.p.A.

Pino Venture Partners is an Italian company for operations in information technology, communication and media. Pino Venture Partners S.r.l. and its partners specialise in venture capital, strategic advisory services and private equity.

In venture capital Pino Venture is Advisor to Kiwi Management Company Ltd, which manages Kiwi I Ventura Serviços SA, a Euro 110 million investment company. Pino Partecipazioni S.p.A., an affiliate of Pino Venture, is Advisor of Kiwi II Management Company, Luxembourg, which manages Kiwi II Ventura Serviços de Consultoria SA, a second European venture capital initiative in the fields of telecommunications, media and information technology.

Pino Venture Partners, founded in 1997, capitalized on the industry knowledge and the international venture capital experience of its two founders, Elserino Piol and Oliver Novick. Pino Venture is considered as the first mover in the Italian market, at a moment when the liberalization of the telecom market, the growing supply of high bandwidth infrastructure, the Internet and new technologies created opportunities to develop new and innovative companies, the European market was underserved by venture capital providers and the growth of European stock markets (EASDAQ, AIM, Nouveau Marché, Neuer Markt and Nuovo Mercato) catering to smaller companies increased exit opportunities.

Pino Venture is not an investment bank, or a fund operating in the private equity field. Nevertheless, in its area of specialisation, Pino Venture may operate to find and/or support private equity investors. The core business is to provide equity capital to unlisted companies, both on a majority or a minority basis, with the final objectives to maximise their value for the medium-long term. Through various co-investment agreements with leading qualified institutional investors, Pino Venture has privileged access to a large amount of capital resources, which allows Pino Venture to suggest transactions of almost any size, being active also in structuring and finalising large initiatives in the IT/Telecom business.

Pino Venture brings its industry knowledge, international business experience and venture capital know-how. This helps Kiwi's portfolio companies to develop and maximise the potential of their ideas. Pino Venture helps companies to formulate strategic and marketing plans, break into new markets, arrange business and professional contacts and recruit key personnel. Pino Venture can provide not only partners with financial resources, but also qualified and reliable strategic and professional support for the realisation of an expansion strategy or for the access to stock exchange markets, both in Italy and abroad, with the final purpose of maximising the shareholders value. Pino Venture provides, on a selected basis, a range of advisory services in the corporate finance area, including advice on company valuations, mergers, acquisitions, disinvestments and capital restructuring.

Pino Venture offers consulting services to the following target client segments:

- Seed stage (Founders/Entrepreneurs), assisting the development of business plans (strategy and business model), helping to recruit management and/or supplying interim managers, helping to raise finance;

- Start up and early stage (small companies), refining strategy, helping to recruit management and/or supplying interim managers, helping raise finance;

- Non-European (particularly US) companies seeking to establish Italian or European subsidiaries, JVs, or licensees, assisting the development of Italian/European entry strategy, helping to recruit management and/or supplying interim managers, helping to find investors if Italian/European business is to be a separately funded venture, locating potential licensees;

- Major telecom or information initiatives in Italy, assisting to assess the opportunity, assisting the development of business plans, helping to recruit management and/or supplying interim managers, helping to secure strategic investments and to deal with Government Constituencies;

- Financial institutions seeking to invest in information and communication companies, providing commercial due diligence or investment/financial appraisals and assisting the development of the investment strategy.

6.2 Kiwi I and Kiwi II

6.2.1 Mission and Structure

Kiwi I Ventura Serviços S.A. is a Euro 110 million venture capital initiative based in Madeira (Portugal), focused on telecommunications, the Internet and new media. Launched in October 1998, Kiwi I is managed by Kiwi I Management Company Ltd, Jersey, and Pino Venture is the advisor to the Management Company.

Kiwi II Ventura Serviços de Consultoria S.A. is a Euro 500 million venture capital initiative based in Madeira (Portugal), focused on telecommunication, hardware and software infrastructure, the Internet and new emerging technologies. Launched in April 2000, Kiwi II is managed by Kiwi II Management Company, Luxembourg, and Pino Partecipazioni, a wholly owned subsidiary of Pino Venture, is advisor to the Management Company. Pino Partecipazioni is also investor in venture capital initiatives in Italy.

Both Kiwi I and Kiwi II have Advisory Boards, which assist the Management Companies with definition of the strategy, deal flow and industry contacts. Leading investors are represented on the Advisory Boards. The Advisory Board is also a means by which investors can evaluate the activity of the management company and the formulation of the global portfolio strategies. There is also an Executive Committee of the Advisory Board, representing investors to deal with conflicts and strategy matters, and a Valuation Committee to deal with valuation issues. The general meetings of the Advisory Board are usually held twice a year, generally in parallel with the Investors' Meeting, in order to inform investors about the strategies, the investments made and the related performance.

The financial objective s of Kiwi I and Kiwi II are the same as the one of the entrepreneurs which they support: substantial long-term capital gains. Kiwi I and Kiwi II encourage substantial ownership by the company's founders and key personnel and back management teams which combine entrepreneurial talent with business skills, in order to exploit opportunities for fast-growing innovative products and services.

While investment criteria may vary depending upon the type of transaction, Kiwi I and Kiwi II believe that the following elements are critical to the success of their investments: thorough due diligence prior to investing, close monitoring of Investee companies, identifiable exit strategies, and effective communications among Kiwi, the Management Company, the advisors and investors.

In Europe, US and Israel, Kiwi initiatives have access to the partners' network of venture capital relationships, including 4CV, Alta Berkeley,

Amadeus, Invesco, Part'com, Quest of Growth, UBS. This translates into better deal flow and co-investment opportunities for Kiwi.

6.2.2 Key Terms and Investors Relations

Kiwi I and Kiwi II key investors are financial institutions, major companies, pension funds and insurance companies.

Investors invest in shares of Kiwi; each investor agrees to a "Total Commitment", that is the maximum amount for which it will be responsible to Kiwi. At subscription, investors paid into Kiwi 6% of the Total Commitment, and for the rest the Management Company makes periodic draw downs as investment opportunities are identified; draw downs are made on call of the Management Company in proportion to the shares owned by each investor.

The duration of Kiwi I and Kiwi II is fixed as ten years (agreed extension is possible), while the draw down period is fixed in 4 years with extension of up to 2 years, depending on market trends, deal flows and investment opportunities. The general rule is that as soon as Kiwi sells one of its participation, the proceeds are paid back to investors (i.e. proceeds are not reinvested).

The Management Company produces a quarterly report to investors, to permit them to evaluate the performance of their investment and to provide for their own reporting requirements. The quarterly report is sent within sixty days from the end of each quarter.

The quarterly report gives information about the fund performance and the portfolio companies. As regarding fund performance, the principal information are represented by the statement of the overall position of the fund – i.e. total commitments, total drawn down and invested to date, total distributed, current investments, new investments, total committed or reserved for follow-on investments, total remaining available for draw down -, the statement of the management fees and carried interest, the IRR fund calculations, the portfolio valuation and the overview of the economic and venture capital environment. As regarding the portfolio companies, the main information given in the quarterly report concern: general information on the portfolio company - i.e. name, location, mission - and the investment - stage, amount invested, percentage ownership, board representation, valuation at time of investment -, significant events and issues during the reporting period, budget, profit and loss of the period.

The investment focus: stage and size of investments (see Fig. 6.1).

The mission of Kiwi I and Kiwi II is equity financing in initiatives in seed money phase, start-up stage as well as emerging growth stage, where the Management Company and its advisors can provide concrete contributions based on skills, knowledge and contacts; the investment focus is on small and medium companies able to grow (investing in "baby" companies, not in "dwarf" companies!). Pino Venture supports with advisory services in financing needs even in later stages.

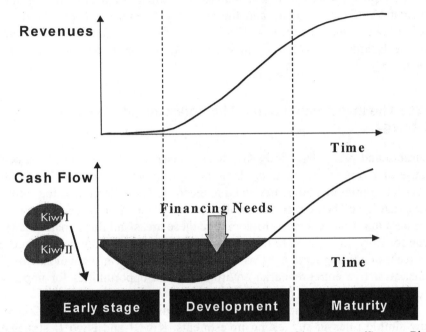

Fig. 6.1. Kiwi I and Kiwi II investment focus: stage of investment (Source: Pino Venture Partners)

Kiwi I and Kiwi II investment strategy is based on active involvement with emerging and growth-oriented private companies, having a well focused business, where there is a well defined market need, strong differentiation and significant barriers to entry. Kiwi's investment program is proactive, so that investments are not only based on "traditional" deal flow – developed trough their Advisors' extensive relationships in the target industries and in venture capital community - but also on the identification of opportunities and creation of green field initiatives. "Waiting for business plan is not enough! Venture capital of the digital age moves from being a mainly financial play to being a "full service" operation" (Piol 2001).

The size of investment in any individual Investee company is usually between Euro 1,000,000 and Euro 5,000,000; no more than 15% of the Total Commitment of Kiwi's investors is invested in any single Investee company. Kiwi I and Kiwi II also make seed investment of less than Euro

1,000,000 to support promising projects in the embryonic stage. Investments in seed money and start up projects are usually split into two or three successive rounds, being the second and the third rounds submitted to some milestones by the Investee company. Depending on the circumstances, Kiwi I and Kiwi II seek co-investors after the first round.

The equity ownership in each Investee company depends on both the evaluation of the company and the investment required. Typically Kiwi hold minority share between 15% and 35%. If Kiwi is lead investor, a seat on the Board of Directors of Investee companies is normally a requirement.

6.2.3 The Investment Focus: The Reference Market and Strategy

Europe, and particularly Italy, has been chosen as reference markets, because of the presence of many areas of high growth in services (particularly telecommunications, new media, electronic commerce and the Internet), that have been either monopolies or non-existent until recently; there are no natural players ready to dominate these areas and this is particularly true for Italy. In addition to this, Italy is characterised by lower penetration of technology applications, less developed (but growing) financial markets, attractive entrepreneurial environment and opportunities for importing business ideas from other markets.

Pino Venture and Pino Partecipazioni, as Advisors to Kiwi funds, follow few simple rules in suggesting investments. Kiwi I and Kiwi II strategy can be summarised in the following:

– Invest only in initiatives operating in areas deeply known by the partners of Pino Venture and Pino Partecipazioni (i.e. information technology, telecom, Internet, media). Elserino Piol always says: "We may make mistakes in investing in our areas of knowledge. Imagine what can happen if we invest where we do not know anything. There are very important and innovative sectors, like biotechnology, that are not considered by Pino Venture because of the lack of knowledge and, as a consequence, Pino can not add value".

– In accordance with the previous point, giving the knowledge of Pino Venture of the Italian market, Kiwi I and Kiwi II are lead investors, "promoters and entrepreneurs" in Italy, taking the major responsibilities toward the company, the management and the other shareholders. Kiwi I and Kiwi II are also active co-investors in Europe, US and Israel, focusing on deals where the lead investors are other venture capitalists known and trusted by Pino and where Kiwi's Italian presence is seen as added value. No more than 15% of the total commitments may be invested outside Europe.

– Every company in which Pino Venture recommends to invest must be very focused, having a limited, well defined mission, and a break-through model (frictionless scaling and exponential growth). Pino is reluctant to suggest investments in companies that try many different businesses at once. When a company grows it may diversify, but during the start-up stage it must have only one mission.

– The market to be addressed shall be large; if the company addresses a large market, it may growth even with a limited market share, while if the target market is limited, the growth requires a relevant market share, and this is too risky. The market to address shall exist. Many Net Economy companies failed because they thought their idea could develop a new market; even if this could be possible, it is too risky. Kiwi I and Kiwi II generally focus on initiatives applying concepts and technology already successful in more advanced markets, like USA, in order to offer new services, rather than developing new technologies or products requiring the development of new markets.

– The market of products and/or technology (hardware and software) is global. Therefore, as an example, an initiative in the product area must be able to compete also in USA requiring investments in marketing, sales organisation in US, otherwise it does not have any chances. The situation is different in the Italian market of telecom services, infrastructure and the Internet, where operators must be local, even if subject to the rules of globalisation. Pino Venture advises investments believing that major opportunities are represented by service companies where, to have a defendable competitive advantage, local knowledge is of paramount importance – differences in culture, language and business practices - and new services may apply advanced technology already proven in other markets.

– The strategy of Kiwi in the next months is basically to concentrate on the present portfolio, investing only in portfolio companies where there is potential value and helping these companies to get additional financing and building lasting organisations. The first decade of the new millennium[1] might be the decade of growth capital managed with the VC style, where the thoughtful and well implemented adoption of technol-

---

[1] Eighties were the decade of buyouts and Nineties were the decade of venture capital.

ogy will lead to industries transformation, competitive positions improvement and sustainable value creation and where good businesses will absorb the technology advances of the past two decades to continue the innovation push and to create growth and lasting value.

6.3 From Deal Flow to Investment

Deal flow to Kiwi comes from entrepreneurs as well as from other venture capitalists. Deal flow analysis is made by Pino Venture in its role of Advisor; some deals are originated inside Pino Venture, when an opportunity is individuated in the market. In this case, Pino Venture operates establishing the company, doing the business analysis and the business plan, individuating management team and eventually other investors.

All the deals addressed to Pino Venture are registered in a database and classified by source, geographical distribution, phase and industry. Deal flow, especially coming from USA, is an important source of information on market trends.

During year 2000, there was a sharp increase in deal flow, reaching 1,500 deals. Since the beginning 2001, deal flow showed a decreasing trend in number, obviously due to the negative mood characterising the market (Fig. 6.2). During 2000, as a consequence of the high number of deals received by Pino Venture, the definition of deal flow has been narrowed to exclude contacts and initiatives which Pino Venture decided not to examine in any detail because the opportunity was outside of Kiwi's strategies focus or because the information provided was unsubstantial. Even in these cases the policy of Pino Venture is to reply to all contacts (and often this results in best formulated proposals in a latter date).

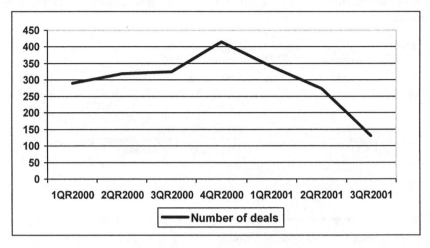

Fig. 6.2. Deal flow Pino Venture and Pino Partecipazioni 2000-2001 (Source: Pino Venture Partners and Pino Partecipazioni)

As regarding the source, consolidated data show that entrepreneurs represent the major source of deal flow (ranging from 65% to 80% of the total deals); deals submitted by other venture capital companies are generally in the order of 15-20%, having had an increase during year 2001, as a consequence of an increased international visibility of Kiwi. Other sources of deal flow are represented by portfolio companies contacts, lawyers, consultant companies.

Because of the focus of Pino Venture, Italy has always represented the major source of deals (usually more than 50%), followed by Europe (from 12% beginning 2000 to 24% in 2001) and USA (ranging from 8% to 18%). More than 60% of the deals are related to seed and start-up initiatives, reflecting the focus of Kiwi; the average size of investments requested has been steady, ranging from 2 to 3 million Euro. Breakdown by industry has shown the dominance of Internet initiatives (with a consistent decrease from 2000 to 2001) and telecommunication, while infotainment, media and mobile Internet project are a minority. During the third quarter 2001 the breakdown by industry showed a particularity: a great number of proposals had an industry focus external to the areas in which Kiwi II invest, such as real estate, semiconductors, medical devices, biotechnology, energy.

Deal flow is analysed by Pino Venture as far as the proposals are received. After analyses and meetings with promoters, Pino Venture submits the investment proposal to the Management Company, whose Board of Directors decides if approving the investment or not (see Fig. 6.3). The average timetable of operations from deal flow to investments are as the following:

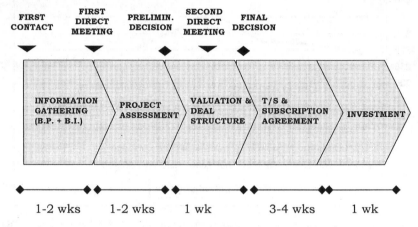

Fig. 6.3. Timetable from deal flow to investment (Source: Pino Venture Partners)

6.3.1 Information Gathering and Project Assessment

This phase lasts from two to four weeks and covers the whole analysis of the project through business plan analysis and meetings with founders. The decision drivers of investments can be summarised in as following:

– Business plan. The ideal business plan is formed by: executive summary, management team presentation, company presentation, market and competitive analysis, products and services, sales and marketing strategy, financial and economic projections. The executive summary is basically a resume of the key elements of the company, some of which will be better analysed in the rest of the business plan: mission and business model, products/services offered by the company, target market, competitors, management, financial needs and financial plan, sales. The management team presentation should underline the experience and skills of key managers, stating also the who are the key external contributors and the members of the Board of Directors. The company presentation includes mission, legal information, global strategy, technology, value proposition, human resources organisation, partnerships and agreements, risks. Products and services offered by the company are to be presented together with the target market and competitive analysis , marketing and communication plan, sales strategy, distribution channels, pricing, research and development plan, product roadmap. Economic and financial projections include financial summary, revenues, financing needs, profit and loss, balance sheet and cash flow analysis.

– As business plans are often incomplete and/or subject to revision, Pino Venture gives usually more importance to the business idea and business model and operates with founders and entrepreneurs to review the original business plan.

– Competition arena. The valuation of a new opportunity is made by Pino Venture together with the analysis of the target market (and competitors) to which the company addresses. Strategy of the company and barriers to entry are of key importance, as well as customers acquisition previsions.

– Management team. If you ask Elserino Piol which are the decision drivers of investments, he will surely answer: "firstly management, secondly management, thirdly management". Management is the key element for the good performance of a company, being leadership, knowledge of the market, experience and ability to build a motivated and valid team essential. The Net economy enterprise is usually without middle management and requires small "hands on" entrepreneurs, speed in decisions, flexibility and healthy lack of respect for established constituencies. Venture capitalists and investee companies must have confidence in young people, who are more willing to take risks and work harder to achieve their vision.

– Business idea and business model, which translate in value of innovation of products/services, business model scalability, business margins (in order to better valuated and recognised by the market), speed of execution of the business model (the speed of growth of the business is critical for a new initiative: in particular, speed and timing of revenues generation and operating and financial break even in short time), sustainable growth ("value addition" growth sustainable for a long time – customer fidelity, technological advantage, offering differentiation, good margins per client, etc.) and capital intensity (what is the company's burn rate? More cash is needed to reach a competitive dimension more the investment is risky!).

– Exit: when considering an investment, venture capitalists carefully screen the possible exit from the investment, which derives from the analysis of the previous points.

6.3.2 Valuation, Deal Structure and Investment

The valuation of the investment is mainly the result of the negotiation with the entrepreneur/founders, as the business plan is often incomplete or unsubstantial. Negotiation is based on two principal activities, interface with entrepreneurs (presentation of the initiative and the team) and background work (analysis based on the decision drivers indicated in the former paragraph and due diligence activity). The valuation is not typically based on "traditional methods" (i.e. discounted cash flow) due to the high level of uncertainty of the values used for the calculation (the only "certainty" is the initial financing needs!).

The negotiation leads to the "pre-money" valuation, which is the value attributed to the company/project before the investment, and it is determined also considering "intangibles", such as level of innovation and potential of the idea, value of the management, and so on. The ratio between pre-money and investment determine the percentage of participation of Kiwi in the capital of the company. In many cases the investment is split in two or three successive rounds, being the second and the third subject to milestones that the company must reach. When the investment is spread in more tranches, the level of pre-money valuation of the next rounds is sometimes determined in the investment agreement, at the reaching of the agreed milestones.

When the valuation is agreed, the following step is the term sheet and the investment agreement. The term sheet summarise the key terms on which the investment agreement will be based. In some cases, when the parties (Kiwi and founders) agree on all the key terms without a long negotiation, there is only the investment agreement.

The investment agreement states the terms of the investment and contains the shareholders' agreement, that regulate the private relationship among shareholders. The investment agreement contains some clauses that are typical in venture capital investments: the investment is subject to due diligence, the presence of members appointed by Kiwi in the Board of Directors of the company, Kiwi's veto rights on Board of Directors and shareholders meetings decisions, relationship with the founders (non concurrency agreement, stock option plans, etc.), exit rights (i.e. "tag along" clauses, "drag along" clauses, co-sale rights, "put and call" clauses) and regulation of IPO process (i.e. mandate to Kiwi to assess IPO likelihood and to choose the Advisor bank).

The due diligence is made mainly by external consultants with Pino Venture's support. The due diligence is aimed at analysing: corporate information (corporate books, deed at incorporation, articles of association, chief executives powers), staff analysis (staff costs, employment agreements, fringe benefits), agreements and commitments (with suppliers, customers, banks, leasing companies, consultants), guarantees and insurance, intellectual property rights, trademark, litigation, fiscal information, economic and financial data (fixed assets, balance sheets, P&L, day-by-day administration).

When the investment agreement is closed and the due diligence has given positive result, the investment takes place. Kiwi, through Pino Venture's advisory, actively work with the investee companies' management by contributing its experience and business savvy gained from helping other companies with similar growth challenges. Pino Venture support the companies even in later stages, helping the company in strategy definition, business planning, industry networking, fund raising, M&A activities.

6.4 The Economic and Market Scenario From a Venture Capital Prospective

It probably does not exist a moment as inadequate as this to talk about Information Technology, telecommunications and venture capital. After seven-years period of excessive optimism, we are now witnessing a period of deep pessimism with differing messages coming from the "crystal balls" of visionaries.

The ones who believe, as we do, that the Information Technology and telecommunications industries will be the drivers of the world economy, face the risk to be considered excessively optimistic or unable to understand the current trend of the financial markets.

We propose to evaluate the evolution of the industry, identifying three different stages:

– the enthusiasm and hyper-valuation phase;

– the present situation;

– the future.

6.4.1 The Enthusiasm and Hyper-Valuation Phase

This phase started at the beginning of Nineties, with a strong acceleration in 1994 (when Netscape was listed on the Stock Exchange) and finished in the second half of year 2000[2].

During this phase the the high tech sector contributed one third of the US GDP growth. The economic explosion due to the high tech sector partly explains how the US was able to maintain low unemployment and inflation rates together with high economic growth. The structure of the financial market also helped the US economy. During this period, more than 50% of the new companies listed on the stock market operated in the high tech sector and a majority were backed by venture capital.

Year	Number of companies	Estimated Venture Financing	Annual change (%)	Venture Financing per Company	Annual change (%)
2000	5,557	$103,323	76	$19	27
1999	4	58,758	158	15	114
1998	3	22,755	33	7	16
1997	3	17,173	46	6	20
1996	2	11,734	101	5	25
1995	1,340	5,832	77	4	40
:	:	:	:	:	:
1985	845	2,473	-3	2	4
1984	902	2,555	3	2	-10
1983	801	2,486	100	3	41
1982	574	1,241	58	2	10
1981	403	786	137	2	33
1980	221	331	62	1	15

Fig. 6.4. Explosion of venture capital funding in USA ($ millions) (Source: Venture Economics/NVCA/Thomson Securities Data)

---

[2] In fourth quarter 2000 venture investment decreased for the third consecutive quarter, falling from $16.76 billion in 937 deals during the third quarter to $13.71 billion in 853 deals in the fourth. Despite a significant decline in venture capital spending during the fourth quarter, 2000 as a whole was a banner year for venture-backed companies. Venture Economics, "VentureEdge", Spring 2001.

Before the impact of Information Technology, the experts view was that the US economy would not be able to grow at a rate higher than 2-2.5% per year without generating inflation. In fact, the economy grew without inflation at a rate higher than 3%, thanks to the role of Information Technology[3], which substituted the traditional manufacturing industry as the reference strategic industry in the economy.

In recent years, if the US economy has grown at a 2.5% rate, instead of growing at a 4% rate, the US would have lost USD 600 billion with 2 million less people employed. But understanding the Net economy was not easy. Not everybody was willing to recognise that there was a revolution behind the convergence of computers and telecommunications: Internet increased productivity at a rate of 2%, generating a higher growth rate and creating further conditions for productivity growth while companies restructured.

Venture capital funds played a central role in this development process, focusing on high tech sectors and early stage companies (Fig. 6.4). Before year 2000, more than 65% of venture capital investments in the US were addressed to high tech companies, and of these more than 50% were early stage.

Following the US, European venture capital and buyouts boomed, even if the size remained ¼ compared to the US. During the enthusiasm phase, European investments in high tech increased five-fold in five years, with particular emphasis on mobile–wireless technologies, and Europe has seen a strong upward trend in early stage investment in high tech companies[4]; venture capital strongly contributed to the rapid growth of companies[5].

It is proven in many countries (the US, the UK and Israel for example) that venture capital has been the most effective means to manage innovation and creation of new companies; this is a condition to increase em-

---

[3] Economic growth over the past decade has been propelled by corporate spending for information technology. Such IT spending boosted productivity even while it provided jobs for millions of tech workers.

[4] In 1999 the amount invested in the technology sectors increased by 94%, EVCA Internet site (www.evca.com).

[5] EVCA-Coopers & Lybrand Corporate Finance, "The economic impact of venture capital in Europe", September 1996. Over the four year period 1991-1995, European venture backed companies' sales rose on average by 35% annually (twice as fast as the top 500 European companies), employment increased by an average of 15% per year (versus only 2% for the top European companies) and R&D expenditure represented on average 8.6% of total sales compared to 1.3% for the top European companies. 81% managers of investee companies surveyed in this study believed that their company would not have existed or would have grown less rapidly without venture capital.

ployment. The competitiveness of a country in the Net Economy is largely dependent on the quantitative and qualitative level of the existing venture capital; the Internet revolution would have proceeded far more slowly if it had not been led by venture capital funded companies.

The following figures give a clear indication of the value created by new companies backed by venture capital (DRI-WEFA and NVCA 2001):

- the billions of dollars of venture capital pumped into U.S. companies over the past three decades has created 1.6 million jobs and USD 1.3 trillion in revenue;

- as of the end of 2000, 5.9% of the jobs in the U.S. and 13.1% of the U.S. GDP were created by USD 273.3 billion of venture capital created companies;

- for every USD 36,000 of venture capital invested from 1970 to 2000, one job existed in 2000.

The "genetic secret" of Silicon Valley, the cradle of U.S. venture capital, can be explained at two levels:

- "ecosystem": capital market, market of ideas, market of talents;

- single company (e-company): innovation and/or product value, business model (revenue model), brand, management, customer base (quality and dimension).

The "ecosystem" is the base on which new companies are born and grow; the market of ideas, talents and capital represent the component parts of it, and are strictly related. The market of ideas[6] is the result of Silicon Valley's belief that innovation is the only means for value creation, and entrepreneurship and value creation are part of the Valley's culture. At the same time, good ideas attract talent[7] and entrepreneurial success stories serve as inspiration for others.

All of this, combined with the network of venture capitalists[8] and business angels, with their risk-taking attitude and tolerance to change and

---

[6] In USA, venture capitalists receive on average more than 5,000 business plans a year.

[7] As an example, 35% of Stanford MBAs accepted to be employed in start-ups in 1999. Near Silicon Valley there are two of the famous US universities, Berkeley and Stanford, with a strong link between universities (students and teachers) and companies. Also important are infrastructure of support, (lawyers, fiscal consultants, head hunters) specialised in venture capital operations.

[8] 40 venture capital funds are based in Sand Hill Road (the Silicon Valley "Wall Street") and control 30% of the venture capital market.

failure, create the ideal environment for the birth of new companies. New companies, that must be able to transform "ideas" (innovation and/or product value) in structured businesses and exploit opportunities through resources management and "value added partners" relationships; speed and quality of execution of the business model are critical success factors. The "good" company has also to be open to change the business model in terms of positioning, margins and competitive advantages.

6.4.2 The Present Situation: the Explosion of the Bubble

The Net Economy enthusiasm[9] created the illusion that "everybody could easily and rapidly become rich and famous". People have talked about a "New Economy" as opposite to the traditional economy, as though the fundamentals of the economy (revenues, profits, return on investments, etc.) are less important than "intangibles" (vision, new business model, new market scenarios, etc.). As a consequence, we have witnessed the enormous growth of the stock exchange value of Net Economy companies, which was often linked to unattainable development rates and unrealistic future scenarios, instead of being linked to economic and financial realities.

All at once, in year 2000, the bubble exploded, at first affecting Internet companies, and successively influencing also telecommunications[10], Information Technology and semiconductors companies. The crises did not make a distinction between good and bad companies and contaminated the whole high tech sector, without regard to the documented successes of the Nineties. During the enthusiasm phase, venture capitalists were caught up in the momentum and were afraid not to invest, scared that they would miss out on something. Now they are concerned about making the opposite mistake[11]. The venture capital situation is basically the same in all coun-

---

[9] The Internet boom of the past three years was indeed one of those gold rushes that Silicon Valley experiences every so often – like the chip boom of the 1970s and the PC splurge in the 1980s. The time round, though, things were more extreme.

[10] "World telecom sector has shed 290,000 jobs in three months", Financial Times, "Telecoms job cuts watch", 27th July 2001.

[11] In the second quarter 2001, equity investments into US venture backed companies fell to $8.2 billion compared to $10.4 billion in the first quarter. This represents a decline of 21%, significantly less than the 41% decline experienced in the prior quarter, indicating than investment levels may be settling around historical norms. In the second quarter, 669 companies received funding, down from 752 in the first quarter, a decrease of 11%, VentureOne Corporation, Pricewaterhouse-

tries: as a consequence of limited exit opportunities (no IPOs and limited trade sales), it is hard to find co-investors – only fully funded companies attract potential investors -, so that venture capitalists are focusing on supporting portfolio companies[12]. As venture capital firms are unable to take companies public as quickly as they did in the late Nineties, and fellow VCs are reluctant to participate in late stage funding syndicates, the result is that early stage VCs need to invest more of their money over a longer period of time to sustain their promising start-ups and to maintain a healthy equity position (see Fig. 6.5).

Nobody believes in the Net Economy any more; this is a surprise, because many of us, opinion leaders, financial analysts and investors, believed in the Net Economy, wanting to be part of the play. Suddenly the scenario changed, and it seems that all the expert professionals went wrong.

Coopers, "The PricewaterhouseCoopers MoneyTree Survey in Partnership with VentureOne", San Francisco, August 2001. As regarding US venture capital returns, they slumped to an average of only 6.4% in the third quarter of 2000, down from 1.6% for the second quarter, and 23.1% for the first three months of the year, while in 1999 they ranged from 15.6% to 59.4%.

European venture capital investments in the first half of 2001 fell 27% by amount invested (from Euro 1.3 billion to Euro 5.3 billion) if compared to the same period of 2000, but only 9% in terms of number of companies backed, EVCA, "EVCA Mid Year Survey of Pan-European private equity and venture capital", press release, Helsinki, 17th October 2001.

[12] Many venture capital funds are implementing "annexe funds". As an example, Accel, Battery and NEA are going back to their limited partners to get more capital to invest in follow-on rounds on their portfolio companies. "Simply put, we went through our follow-on capital faster than we thought", said Nancy L. Dorman, General Partner of NEA, which is trying to raise $150 million, that will be used to invest in about 20 portfolio companies of NEA VIII, a $556 million fund closed in 1998. Palo Alto's Accel is trying to raise $50 million to make follow-on investments, to support companies in Accel VI, closed at $275 million in 1998 and Accel VII, which closed at $600 million in 1999. Battery Ventures, Wellsley, Mass., plans to ask permission from its LPs to raise the cap on the amount of profits from Fund V that it can re-invest in that fund's portfolio companies.

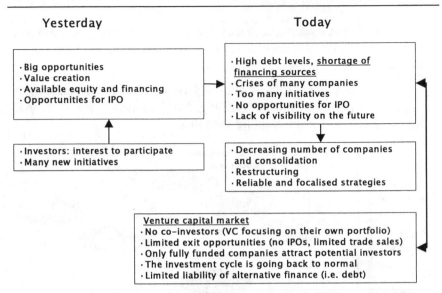

Fig. 6.5. Enthusiasm phase versus present situation (Source: Pino Venture Partners)

This is partly explained by as series of events, that we call "Net Economy killers":

1. the tumbling of the stock markets: the market had to return to realistic values, and this happened when a number of companies highlighted their inability to realise their mission;

2. it was discovered that the Net Economy is subject to the business cycle just like the traditional economy. The Net Economy is capable of higher sustained growth rate over the long term but it has not abolished the business cycle (i.e. in US computers have reached near-saturation as a consumer durable). It is necessary to notice that, as opposed to the opinion of many, the high tech revolution increased volatility, instead of reducing it. Excessive investments in telecommunications[13] and high tech contributed to the sudden and severe profit collapse of these sectors. Chief Executives, who formerly declared their ability to control the business cycle, admitted their inability to forecast to have the next quarter;

---

[13] Among the "killers" of the telecom markets: excessive indebtedness (UMTS, vendors financing, etc.), shortage of focused strategies, in particular wrong imitation of winner companies and role models, shortage of aggressive commercial-marketing strategies.

3. the get-rich-quick-game (i.e. many dot.com initiatives) gave birth to a huge number of new companies, but, as we have seen, the future is not big enough for all of them;

4. the acceleration of the IPOs, that often have concerned companies without a proven business model[14] (see Fig. 6.6). Practically, the stock market was asked to become venture capitalists;

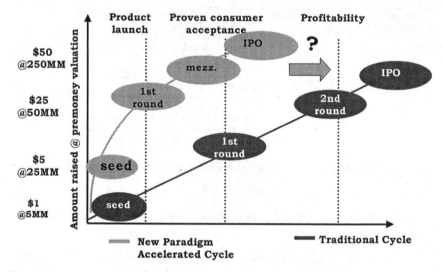

Fig. 6.6. Internet accelerated financing cycle (Source: Maveron)

5. the attitude of investment banks and financial investors, that have often supported non solid initiatives[15];

6. the overcapacity of many industries, caused by the presence of too many initiatives (i.e. long distance fibre optic networks in USA) with the consequent price pressure, which is therefore not sufficient for utilising the whole capacity of the realised infrastructures;

---

[14] Large capital availability and market liquidity caused accelerating investment cycles and increasing valuation (Fig. 6.4).

[15] "A lot of money came into the industry that filtered into start-up companies in 2000...and I think it's fair to say that a lot of companies that got funded where "me too" companies. I think everyone has wised up to that, and there has been a fundamental correction of valuation, and so going forward you will see a lot more discipline and a lot more selectivity in terms of kinds of companies that get funded", Haque P., General Partner, Norwest Venture Partners, March 2001.

7. the overcapacity of hardware companies (information technology and telecommunication), due also to the fact that their main customers, the services companies, reduced investments or faced crises[16];

8. the attitude of some European governments (the UK and Germany in particular), which have tried to speculate on the Net Economy, through the auctions for UMTS licence assignment. The principal consequences of this process have been: winners have been chosen for their financial commitment, and not for the quality of the infrastructure and services (the economics of the future UMTS operators may be jeopardized), increased debt of telecommunication companies, and, finally, the beginning of the financial crises of the telecom sector (see Fig. 6.5). In addition, as the financial market is global, the financing of the auction winners is reducing the financial resources available for other telecom initiatives[17].

DJ EURO STOXX TELECOM-PRICE INDEX DJESTEL

Fig. 6.7. "Oranges and lemons: the UMTS story"

This negative scenario hides also positive and encouraging signals: technology is "innocent", and it is not the cause of the crises it could repre-

---

[16] In fourth quarter of year 2000, US capital investment in equipment and software actually dropped at a 4.7% rate, after years of double-digit increases.

[17] "When six bidders paid crazy prices for third-generation mobile licenses in Germany, it set off a chain reaction which is still causing explosions: a scramble to borrow money, debt downgrades, fire-sales of assets, frantic attempts to stuff shareholders with stock in mobile companies - culminating in this month's Orange flop", Hugo Dixon, Chairman of Breakingviews Ltd, January 2001.

sent the means for an upturn. The analysis of the situation underlines that the crises is mainly financial and concerns the need of IT and telecom companies to act in the short term for the reduction of excess inventories and the adjustment of the productive capacity, and in the medium-long term to restructure (process, costs, etc.) to join adequate profit levels. In the meantime, the competitive scenario increases and non solid initiatives disappear.

The scenario changed, but the promises of the Net Economy, born at the end of Nineties, are still alive. There is no technology crises, so technology preserves its power to manage innovation and better productivity. History is full of innovations that are often misunderstood and underestimated at the time[18], but in spite of the reservations of the opinion leaders, most innovations are generally absorbed over time and will forever change the way the world works and businesses function[19]. As an example, Internet development continues at high growth rates, allowing the belief that the Internet is a revolution that creates radical changes and improvements in all the sectors (from the private-social to the world of companies and institutions).

The principal Net Economy technology users (i.e. Internet related technology) are not still new companies, born as a consequence of those technologies, but "old economy" companies, which were afraid of losing business to nimbler competitors. The new Internet-based companies, born

---

[18] Mankind has been poor at assessing the impact of technological innovations. In 1878 a special British Parliamentary Committee set up to evaluate the electric lamp, concluded: "(Edison's lamp)...is good enough for our transatlantic friends...but unworthy of the attention of practical or scientific men"; all the internal Western Union memo which was a leader in communication in the US at the time – dated 1876: "this telephone has too many shortcomings to be seriously considered as a mean of communication"; engineering editor of The Times said in 1906: "All attempts at artificial aviation are not only dangerous to human life, but are foredoomed to failure from the engineering standpoint"; even a group of the brightest investors meeting with Alexander Bell in Philadelphia asked while ridiculing his business plan: "what are you planning to do Mr Bell...wire up every house in the country?".

[19] Christiansen 1997, demonstrates why outstanding companies that did everything right, lost their market leadership when confronted with disruptive chance in technology and market structure. New companies will emerge to lead and market the innovation. As an example, at the beginning of nineties leaders in Information Technology (IBM, Digital, Olivetti and many others) faced their business model crises and did not react with a rapid change of it. New companies (Microsoft, Intel, Compaq, etc.) born and grew, imposing new winner business models and making the crises of the existing companies.

during the "euphoria" period, proved to be a destabilising element for the market scenarios and for existing companies. In front of this real or supposed threat, the old economy reacted by utilising the same innovative technologies as the Net Economy in their process management, marketing activities and so on. Even if traditional companies have been slow in understanding innovation, they reacted as soon as they realised the impact of the Internet: good management teams and adequate economic resources have been the key factors during the shift from "old to new" or the co-existence of the "old and new".

In ther words, the old economy has been the real killer of the Net Economy, when the old economy companies decided to utilise the Net Economy technological bases.

6.4.3 The Future

We have no doubt that information technology, telecommunication[20], Internet and multimedia will be the key industries for the next ten years all over the world, and not only for more industrialised countries.

Every company, in order to be competitive, must utilise the Internet. Through the Internet Public Administration will reach higher levels of customer (citizens) satisfaction, individuals will use better their time and will be more informed. We will turn again to the Nineties technological enthusiasm, but only for the spread and application of technologies. We will probably not witness again a period of crazy stock market valuations. The process of net usage has just begun, being education and infrastructure the main intervention areas. Only with high bandwidth will it be possible to completely face all the application needs and it is necessary that bandwidth enter homes. We are in the "telecosm era", the era of bandwidth: fibres are acting as the connecting nerve tissue, the neurones and "real life" is cre-

---

[20] The future of telecommunications in the next 12-18 months could be represented by (optimistic scenario - 70% probability): a general upturn of the markets, debt reduction, re-equilibrium of the market structure, strong increase of high bandwidth, many start ups leaving the market, consolidation of some new companies, GPRS development and concentration (new M&A in the wireline, wireless and infrastructure producers markets); if we consider a pessimistic scenario (30% probability) future for telecom could be: weakness of the market, high indebtedness level, shortage of success of high bandwidth (contents scarcity and high prices), investments decline, infrastructure producers crises due to start up failures and investments reduction, slow GPRS take-off, and concentration (new M&A in the wireline, wireless and infrastructure producers markets).

ated, and the Internet is the architecture or the "real life" waiting for "infinitive bandwidth" to be fully implemented (Gilder 2000).

Technologies that are now available are sufficient to face the present application needs, but in the future it will be necessary make them available to everybody, without having a "digital divide". New technologies, that could give further contributions to the evolution, are therefore on the horizon[21]; there are many analysis on the impact of new technologies. One of them stated[22]:

- 2005: network TV all but ceases to exist;

- 2007: last music CDs roll the assembly line-all music now downloaded off the web;

- 2008: people regularly implant chips in their bodies as part of their personal area;

- networks;

- 2009: most household appliances plugged into the Net. A broken toaster in Maine can be;

- diagnosed with a faulty spring-by a company in Arizona (any appliance breaks can be distance-repaired);

- 2010: a half –dozen satellites, probes, and landers, each equipped with special net-like;

- protocols, form the backbone of a new interplanetary Internet; mobile handset penetration reaches 99% in the United States, 65% world-wide. The handset which comes in hundreds of forms is now the No 1 portal to the Net, and includes voice, data;

- video, GPS, PDA, ATM, and many other functions;

- 2019: most business meetings take place virtually over the web.

To understand the future, and evaluate how things quickly change, we do not have to forget that ten years ago www, html, browser, java, dsl, did not exist, and people looked at the mobile phone as a niche product.

---

[21] Technologies will be available, confirming that services companies should not invest in infrastructures.
[22] Business 2.0, January 2000 and McCue J., UBS Warburg, "Telecommunications and the New Economy", Lucent Client Success and Partner Conference, Doral, Florida, 30th October - 2nd November 2000.

To sustain the development of what will be the key industries of the future, venture capital must be proactive, as it has a profound impact on the technology sector and the global economy[23]. The impact of venture capital on innovation is four to five time greater that corporate research development (Lerner 1999). Innovation in market and technology come as waves and the success of the venture capitalist depends on the ability to surf the waves of innovation – go on then when they are far away from the shore, go down when the wave is near the shore and loose energy – and catch week signal to identify the formation of new waves.

It is important that Europe continues the build-up of the venture capital environment, because two dangers lie ahead. The first risk relates to Europe's entrepreneurial culture, because a whole generation of motivated, excited managers have poured their energies into starting Internet companies. There is a danger that the continent's businesses may assume, that, after all, nothing needs to change. That would be a shame: an overly protected, over-regulated Europe has long needed more, and fewer, risk-takers[24]. A second risk is that Europe's imported American-style venture capitalism, might be thrown side almost as soon it arrived; it is easy to forget that America's investing infrastructure took nearly three decades to develop. Over the past two years, Europe has seen a remarkable burst of activity, but venture capital firms have not had sufficient time to foster the dense networks of contacts and knowledge that they need. Europe could not hope to replicate in two years a process of tech-company formation that has taken more than three decades to develop in the U.S. (Europe needs to develop entrepreneurial management!). The market crash was probably a good thing: it came before much money had been throw at European start-ups, and it imparted loads of valuable, if painful, experience to entrepreneurs, investors, and regulators.

As the technological development is largely driven by the US, it is now probable that, after the 11 September 2001 events, defence will have a role in the economy more important than before; as a consequence, many R&D activities and their related investments will be concentrated in high tech sectors that are of interest for the Defence Department. This will probably generate more innovations in some high tech areas, connected to security and defence, penalising other entrepreneurial activities. The system "venture capital/IPO/Nasdaq", which has represented the base of the technological evolution of nineties, face the risk of having less financial re-

---

[23] "Venture capital contributed to the largest legal creation of wealth in the history of the planet", Doerr.

[24] How many failed entrepreneurs get a second chance will be an important test of whether European business has really "got" the new economy.

sources to be addressed to new innovative companies, leading to a significant loss. Investments in R&D could be more oriented towards security systems technologies than towards technologies aimed to improved companies and organisations productivity.

6.4.4 The Future is the Internet

There is a huge confusion about the Internet[25]. People do not really know its magnitude, where to apply it, who will be the winners and, mainly, how to make profit with the Internet. That does not mean that we can ignore it, as it represents a real and major innovation and will be a presence in the future. We do not have to forget that even if the Internet came on the scene five years ago, it is the result of two decades of technology advances.

Internet is just starting to have an impact on businesses and the economic environment and should not be underestimated[26]; at the same time, that does not mean we should get caught up in the irrational excitement of this innovation. Internet is not an end in itself, but, like all great innovations, its impact will be how extensively and effectively it is integrated into business and ultimately into the entire economy. The Internet is just a "channel", for communications, information, distribution and entertainment. The future will show whether if it is a channel for other functional uses.

---

[25] "Imagine a sphere; at the centre of the sphere is a pinpoint of enormous density: a black hole. Imagine bodies orbiting within the sphere – these bodies are companies in our industry. The sphere is constantly expanding as more and more and more bodies are created. Those companies that orbit closer to the centre are low margin businesses. Those that orbit near the surface of the sphere are the highest margin businesses. When a company touches the black hole it creases to exist. So there is a constant struggle to orbit near the surface of the sphere. As the bodies orbit, some combine with each other. Pieces also break off and orbit on their own. The orbits appear random. Like flashes of electricity between neurones in the human brain. And yet, there is some higher order. But what drives this process of creation? Who is the choreographer of this dance? This dance is mankind's increasingly intimate embrace with itself using its greatest technological creation, it's Internet". McCue J., UBS Warburg.

[26] "We are witnessing a social, cultural and economic revolution in which the "flow of power" is being replaced by "the power of flows"...because information is now the supreme power, the optimum structure for information "flows" is the network. So, our society will increasingly take on the characteristics of the networks. This is why technology, i.e., the "tool" which will advance informationalism, is a networking technology", Gilder "Telecosm".

We have gone through the Internet "innovation phase", also the Internet "bubble" is non over. Now we have just entered the "adoption phase", that, for businesses – generally established enterprises ("old economy companies") – means becoming better competitors, making the Internet an integral part of the economy. The drivers of the adoption period are simple: the acceptance of standards which invites mass usage, the coming of broadband which will unlock the full potential, and acceptance culturally by the general population - not only the X and Y generation but also our parents and grandparents!

The Internet is agent of change that presents wonderful opportunities for businesses to capture or defend market shares, to strengthen the link between companies, customers and suppliers, to provide greater access to information and to strengthen communication and, most importantly, to create value. We do not have to forget that in this scenario the Internet can create industrial chaos where innovative companies can change the competitive dynamics and the threat for new innovative companies will be ever present.

Enterprises and investors should be prepared for the adoption period and learn from it. Depending on the company, it can be a defensive or offence tool to attack competitors or avoid being attacked. Internet is not something to run away from, but it is quickly become a permanent feature of the economic and industrial landscape and should be incorporated by companies. Having not the Internet is like not having the phone: we wish to ask investors if they would consider investing in a company that did not have a phone system.

Adoption of the Internet is obviously in function of the kind of company involved:

- for some industries it is critical, where controlling the "channel" is a critical part of their strategies (i.e. entertainment industry);

- for other industries it is important to move quickly, where the customer dynamics can be affected, and with it, market share and profitability (i.e. financial service businesses);

- finally for a group of industries it is less important but interesting; those companies that innovate can leave slow competitors behind (i.e. restaurants).

Companies that do not have one important strategy might be at a competitive disadvantage[27]; any company hoping to survive the innovation and

---

[27] When, during an interview, the interviewer asked Andy Grove, Chairman of Intel, how many companies in five years will have an Internet strategy, he answered

change that web threatens to unleash will need more than a "digital business plan". It must mirror the Internet itself to be open, democratic, tightly networked, non-hierarchical, experimental, endlessly adaptable, and utterly restless.

With regard to opportunities for investors in the Internet sector, companies adopting the Internet to improve competitive advantages and market position as well as profitability, should lead to market premiums. The common theme will be that the application of technology to fundamental businesses will result in innovative companies enjoying increase market share, strengthen market position, improve competitive dynamics and, finally, better margins and reach more value creation for shareholders. The initial Internet wave (the innovation wave) is over, now we must identify new Internet waves (the adoption waves).

6.4.5 The 21st Century Corporation

For nearly all of its life, the modern corporation has made money by "making things": it has done so by amassing fixed assets, organising large workforces, and managing hierarchically. The 21st century corporation will do little of that; it will make money by producing knowledge created by talented people working with partners all over the globe.

In an idea-driven economy, the Internet does not change just some things. It changes everything, the rules, the players, the organisations, the public policy. Only the corporate goal remains the same: profits[28]. And profits are strictly related to innovation, because innovation builds profits. In an information economy, companies can gain an edge through new ideas and products that increase in value as more people use them.

Information-based products can reward early leaders with temporary monopolies and winner-take-all profits, but the emphasis is on "temporary". Knowledge-based products and networks can quickly disappear in a burst of creative destruction[29], so that corporations must innovate rapidly

---

in few years all companies will have an Internet strategy, because otherwise will be dead.

[28] Cutting costs is the answer. In an economic universe of downward pressures on margins, one path to profitability will be to reduce expenses.

[29] Foster and Kaplan 2001:"Creative destruction: why companies are built to last underperform the market and how to successfully transform them", "To create new businesses at a faster rate, corporations need to ponder the details of divergent thinking. Divergent thinking is a prelude to creativity managing for divergent thinking – that is, managing to ensure that the proper questions are addressed early enough to allow them to be handled in an astute way – requires establishing a

and continuously. Redesigning the corporation to evolve quickly rather than to operate well requires more than simple adjustments; the fundamental concepts of operational excellence are inappropriate for a corporation seeking to evolve at the pace and scale of the markets, in order to long-term improve performance.

Human capital becomes the key asset; globalisation and the Net will allow corporations to seek out the best educated and trained around the world. In the 21st century, corporations know that creativity is the sole source of growth and wealth. The value of education rises exponentially in an economy based on ideas and analytic thinking.

Successful businesses must be built on exploiting scarcity not abundance (Gilder 2000). Today the two greatest scarcities seem to be: the limitation imposed by the speed of light and the scarcity of human attention. Therefore in the bandwidth, the winners will be the companies that can minimise the limitation of photonic speed; and about the scarcity of human attention, people are not interested in technology, or in more information. They want solutions, in the form of services, to fulfil their needs. The killer business model is one that builds internal processes that optimise the delivery of complete solutions and services to the customers (all business are now servant to the web!).

We are just at the beginning of the beginning. The 21st century is going to be hard on corporations, governments, and all the rest of us. The changes will be astonishing and unthinkable[30].

"rich context" of information as a stimulus to posing the right questions." See also Csikszentmihalyi 1991 "Creativity: flow and the psychology of discovery and invention".

[30] McNamee 2001: "...as scary as the stock market is now, the biggest mistake would be to ignore it. The boom did not last forever, and the bust will run its course as well. If you are selective and careful, you will find profits in the New Economy II...".

7 How Does a Venture Capital Work: Case 2 - Intervaluenet

Claudio Scardovi

7.1 Consulting Intelligence in Venture Capital Transactions

"Consulting intelligence", runs an old saying, "is a contradiction in terms". "The lack of it", the joke goes on, "has about the same impact on consultants projects pipeline as striking oil have on drilling dentists". A joke, that carries a lot of truism, given the recent e-start up business history, which, amongst other things, is littered with disastrous consulting intelligence mistakes.

Surprise, innovation and speed of execution were some of the cardinal principles of the so-called "New Economy". From the very "world's reigning strategy guru", professor Gary Hamel, to every Principal consultants and even Junior staff in the world, anyone of them have preached the need to achieve surprise in the market – and to guard against it – as rule number one to reap increasing returns and network effects in the electronic marketplace. Despite this, would be e-entrepreneurs and their consultants appears recently to have been caught out with almost predictable regularity by pre-existing incumbents. As a result, not only it has been a lot harder for start-ups to take established market leaders by surprise, but also, many of these promised call options of exponential wealth creation have expired worthless.

By and large, strategic consultants are not stupid. Even the most intellectually challenged consultants, just fresh in the profession from his MBA, can fully appreciate that a venture financed start up is a risky matter and that go to market strategies have at least two sides victory bringing him honours, cash, career opportunities, academia and media attention and defeat which typically results in a major and at times uncompromised blow to his reputation. Why then, with such incentives, do the vast majority of them get it consistently wrong? In the great majority of cases, and with particular reference to venture backed start ups, business plan's defeat can be traced back to a lack of knowledge of the basic rules and practices of the market it sets out to revolutionize (products, channels, competitors, potential clients etc.). In fact, whether from overconfidence, ignorance, gullibility, impatience or just a failure to comprehend the facts, venture backed start ups defeats appear to be almost invariably associated with a "market intelligence" defeat and more specifically, given the roles and responsibilities in the start up process, with the consultants' "market intelligence" business plans often titanic both in size and destiny failings and failures. Another frequent cause for many strategic blunders is the way venture capitalists, consultants, start-uppers and their stakeholders at large have interpreted the "strategy is (just) execution" motto/proxy that so often has been stated by business literature Inklings. An interpretation that has been taken to extremes especially for e-start-ups on the commonly held belief that, for this type of ventures, the paths of value creation and wealth multiplication invariably lies in the field of technological operations and web design excellence. So true (things have to work) yet so false and lethal especially in absence of true "market intelligence" (you have to know what things have to work for ... otherwise they don't work).

To "force" would be e-entrepreneurs to acknowledge the market truth that often stares them in the face and to manage the greedy urgency to deal making of overly-financed venture capitalists some very simple analytical exercises and strategic reasoning have to be made firstly to transform information into "market intelligence" thus turn sound value propositions in smartly executed go to market strategies/projects.

In was in order to supply this unique value proposition to both entrepreneurs and venture capitalists, that Intervaluenet S.p.A. was incorporated on March 1st 2000 from an earlier spin off from the Andersen Consulting (now Accenture) and other leading consulting practices. Some 18 months into Intervaluenet's life, it is interesting to illustrate and understand the basic principles and processes which determined the first eight months of Intervaluenet's positioning in the marketplace and why and how its value proposition has evolved from an "incubator/market intelligence provider" model, to an "accelerator of entrepreneurship & innovation" one and is thus considering a further evolutionary phase projected towards a "fully-integrated incubator-accelerator-venture capital" business model.

7.2 Phase 1. Incubation - Intelligence Cycle for Internet Start Ups

Intelligence is nothing more than information that has been systematically and professionally accumulated, processed and analysed. There are many definitions of intelligence. The one that I like most is "the capability of defining cause – effect relationships in uncertain environment subject to limited information available". For the professional consultants, focused on offering new venture advice to both entrepreneurs and venture capitalists, intelligence could be simply defined as "processed, accurate information, presented in sufficient time to enable a decision maker (the entrepreneur, the venture capitalist) to take whatever action is required".

With this in mind, in the first months of Intervaluenet's existence, all my energies were devoted to adopting (and encouraging my colleagues to do so too) an entrepreneurial approach to traditional issue based analysis of the new venture business plans we audited and/or developed ourselves both for single un-funded professionals turned entrepreneur and fully-funded major industrial and/or financial groups that at the time were planning some "click and mortar" business model spin offs.

It was clear from day one that this "entrepreneurial approach to consulting" (that's how we defined it) was really coming to the "no time, no space, no chance to fail" bottom line both for ourselves (having left the re-assuring Big International Consultancy brand name, my colleagues and I had to manage our personal reputation as our most valuable, though intangible, asset) and our client base which, incidentally, were often in very similar conditions if not worst given the revolutionary (thus risky) nature of most e-initiatives they were embarking on.

The intelligence cycle concept, usually represented graphically as a continuous, never-ending circular process, was by no means an Intervaluenet, and in general, a management consulting industry's creation. As it happens for most of the strategic, organisational, operational and logistic concepts often (mis)used in management consulting, the concept itself was borrowed, with minor adjustments, from the military doctrine and warfare prevailing theory. Wasn't in fact professor Lester Thurow talking about a third industry revolution taking shape? Not to mention Gary Hamel's "Leaders of the revolution" manifesto. Taking advantage on the shoulders of these two giant gurus, we just thought convenient to calibrate the concept of a grand (French) revolution to a (Cuban) guerriglia-based one. In other words Intervaluenet would have fought in the New Economy arena through surprise fights and even more speedy retreats in friendly and know environment. In order to pursue our historic mission ("offer tangible value to the entrepreneurs and VCs of the intangible") we would have focused on the management of an intelligence cycle mostly made of five critical conceptual phases:

– The direction phase – the manager in charge of the audit/development of the new venture business plan would have stated his basic and ancillary intelligence requirements, usually in the form of one or more questions.

The main questions was usually drilled down through an issue tree analysis made up of at least 3 to 5 levels.

- The collection phase – our consulting analysts would then translate the manager's intelligence requirement into a series of essential elements of information (deterministic inputs of single analysis) usually involving market scenarios, the status of relevant technological innovation and deployment, the potential pent up demand.

- The collation phase – our consulting analysts would have then collated the information from the various sources into a readily accessible database. The database would have been cross checked for consistency both through entrepreneurs, venture capitalists and experts' interviews and (when applicable) focus group.

- The interpretation phase – the manager in charge would then have proceeded in analyzing the information collated, possibly turning them into intelligence. This was usually done by asking basing questions such has: who should be willing to pay for that? How much? How frequently in time? At what margins? …

- The dissemination phase – the manager and myself would have communicated to the entrepreneur and/ or the venture capitalist, through a written brief, an informal meeting or a formal presentation the synopsis of the analysis that were developed and the executive summary of the intelligence that was deducted.

Intervaluenet pursued this "incubator/market intelligence provider" business proposition for at least 8 months, sometimes on a traditional fix time–fix price fee based revenue model; sometimes through innovative value pricing ("sweat for equity" or "carried interest"). Part of these value added offering is still in place, though promoted almost exclusively towards (guess what? It's recession time) fully-funded industrial and financial groups striving to get fit, thin and agile and to "turn revolutionary", following the advice of London Business School professor Gary Hamel.

Looking back at the definitely mixed results of our "incubator/ market intelligence provider" positioning I could hardly state everything but truly self-reinforcing bits of truism:

- Inaccurate information or analysis speaks for itself; no one get credit for wrong facts. This is particularly true in case of value pricing (a 30% stake of worthless paper, whatever the market turns to be, is still worthless). Equally, it is hardly timely intelligence to shout "look out!" after the piano has crashed to the ground as it is to state "watta a wonderful business innovation that ama-something website that sells books! Could we not make an Italian based clone of it? How about tuscanforest dot it?".

– Of greater value to both entrepreneurs and VCs turned to be the intelligent understanding of the difference between the pent up demand capabilities and its intentions (everybody has a TV, but not everybody wants a set top box as a portal to web surfing and e-commerce). At the same time, new competitors or old incumbents capabilities are relatively easy to measure, but determining their true intentions is quite difficult to quantify.

– The final part of the intelligence cycle (dissemination) is perhaps the most fraught with risks. Strategic consultants, principals and business analysts alike are (though you won't believe it) as human as the rest of us. A nervous, unsure or ambitious consulting manager, even Intervaluenet ones, will always have the temptation to fake the business plan to suit an entrepreneur's whishes, to please a grand financial group whose 20 million dollar equity venture still represents penny money, or avoid the displeasure of the venture capitalist desperately willing to close the deal and mature the management fee applied on the total capital he employed.

Having got were we got it was clear that Intervalunet's business model needed that extra humph to meet its wealth creation and long term sustainability objectives. We moved on.

7.3 Phase 2. Creative Thinking and Innovation Agenda to Wealth Creation

Armed with a better understanding of the mechanics of the "incubator/market intelligence provider" business model, and able to perhaps discern the true role of "entrepreneurial intelligence" – getting things done, with a focus on liquid, tangible, transferable on check account "wealth" as opposed to academic and far from pragmatic definitions of economic value added or risk adjusted value added – we were thus able to start looking more closely at the truth behind some intelligence and business model blunders that were applicable at that time (late 2000) to both "New Economy" venture funded start ups and "Old Economy" industrially funded spin-offs.

As a result of this careful observation of ongoing business practices and of the murky world of the business innovation offerings promoted by all the major international consulting companies, some 4 months into its existence Intervaluenet eventually repositioned itself with a new mission in mind this being "to drive entrepreneurship and innovation within and in partnership with the best industrial and financial groups of our country (Italy)".

This new "accelerator of entrepreneurship & innovation" model was distinctively focused on managing the so called "intangible capitals" that can be defined, traced and analysed in any industry and for any company. These intangible assets could, for example, refer to the know-how already compenetrated in the "structural" organisation of a growing company, to the talents' portfolio of a recently-started service firm, to the client portfolio of a need-to-be-revamped old companies or to the potential target of new clients that could have been reached through new market offerings and targeted value propositions. Leveraging on these intangibles assets to get tangible economic results, we thus focused our competencies and energies to drive creative thinking and innovation agenda for all the major financial and industrial groups that chose to be our partners. This was done for several clients in the financial services and in the industry/manufacturing sector both for click and mortar and (more so) for brick & mortar, yet mostly innovative, ventures.

One of our clients once stated "Innovation projects are too serious a matter to be left to consultants. They prepare for the last innovation project in order to fight for the next one". Though there is obviously some truth in it, I still believe that our consultants' creativity, their ability to identify, understand and solve problems both logically and through "non-logical/out of the box thinking", and their consistency in coming up with new sustainable business ideas are (intangible) wealth creating drivers as strong as (tangible) equity capital and venture financing. In fact:

– the simplest meaning of business creativity is the ability to bring to market something completely new and thus able to outperform incumbents either through the restructuring and bisociation of existing ideas and paradigms into new, viable business models or through completely new insights on clients, markets and products;

– the innovation agenda simplifies the concept that, for wealth creation to be completed, there must be intelligent thought in the R&D Lab, but also perfect execution in the market (the key performance indicator of innovation is demand, non utilised/demanded innovation is non-innovation).

In our view, in the age of intangible assets and of customer-focused organisations and of talent based competition, new wealth must/will be created by, and old wealth can/will be migrated to, the more creative and committed competitors in the marketplace. Moreover we believe that creative thinking and an innovation agenda will be some of the most important defining competitive advantages, as capabilities to conceive again existing or non-existing business models in ways that create new value for customers and new wealth for investors. We have pursued these business priorities ourselves from day one gaining significant insights on the workings of mostly inefficient markets and all-to-human organisations. We felt it was right and logical enough to pursue the same business priorities for our existing and prospect clients. We did so.

To date the "accelerator of entrepreneurship & innovation" business proposition has proved quite successful and has driven Intervaluenet's growth, both in terms of professionals employed (approximately 30 people), revenues (yearly turnover of approx Euro 6 million) and franchise (Italian client portfolio comprising a most top bank groups and a number of top industrial players). More importantly, our financial and industrial partners have confirmed to us that in their mission of driving wealth creation through venture capital structures and innovative/revolutionary entrepreneurial ventures the "accelerator of entrepreneurship & innovation" mode (i.e. think and execute with a hands-on/entrepreneurial approach) has helped them more than the "incubation & market intelligence" one.

7.4 Phase 3 (May Be). Extracting the Value (on the Trail of Wealth Creation)

Notwithstanding the satisfactory profit & loss results obtained so far, both due perceived market needs and to our obsession with the identification of sustainable yet revolutionary (i.e. radically differentiating) go to market strategies, it is more than likely that the "accelerator of entrepreneurship & innovation" model is soon going to become history. Which direction/strategic mainstream are we going to pursue ?

We started by analysing ways to obviate to the blunt and in many concerns, so true "innovation projects are too serious a matter to be left to consultants" perception by our clients/prospect partners which de facto appeared to hinder the sustainability of the existing model (not to speak of traditional consulting models). One of the strategic options we have come up with is the evolution towards what so far has at Intervaluenet have baptised as the "fully-integrated incubator-accelerator-venture capital" business model.

This model entails Intervaluenet acting not only as a consultancy firm (though really dynamic and fully market intelligent) but also as a committed investor player able to bring new opportunities, sharing the same long run and industrial vision with our industrial partners and our former consultancy clients (i.e. the entrepreneurs or the industrial groups looking for great potential value added spin offs/ventures).

Clearly, we recognize (and most importantly understand) that new business model will be a very different and risky one from the one pursued so far and that we might have to fill significant competence gaps to fully succeed in this new entrepreneurial feat. Yet, whist recognizing that venture capital and consulting are very different and often conflicting business we are driven to pursue the project further by the fact that putting them under one roof could lead way to some interesting/value adding synergies in screening, financing and overseeing the targeted companies. Indeed, a synergic management of all the different components of our previous business models (market intelligence, creative thinking, etc.), plus a direct knowledge of the venture capital business, could provide extraordinary opportunities to Intervaluenet's quest for value and wealth. But how these new "fully-integrated incubator-accelerator-venture capital" business model is going to be like?

As one venture-backed e-entrepreneur once stated to me, "venture capitalists don't create business, they steal it". But, if your typical client is an established and consolidated industrial group, you'll make a bad job if you get only a VC short run approach. The lack of an industrial approach could be a deadly pathology for your client's new business plan. Of course, it's ok for everybody to steal wealth from a tired business to a new one, but for an industrial group it's really crucial to arrive to sustain this wealth in the long run. The alternative, as consultant, would be to suggest to your client a very ontological change in his life this being to shift its focus from industrial entity to a financial entity: that's always possible, but not always realistic.

There is a similar joke about incubators and accelerators that bites even harder on their promises of wealth creation (distinguished by their wealth "migration" from venture capital funds and entrepreneurs). In fact, the negotiating process that usually gets going before deal making is sometimes alternatively epitomised, in the best Hollywood tradition, as "The Quick and the Dead" or "the Good, the Bad and the Ugly":

- The first epitome refers to the "greed is good" philosophy sometimes associated with the venture capital (and in general, investment banking) industry. This philosophy was superbly satirized in the film "Wall Street". Its anti-hero Gordon Gekko (played by Michael Douglas) was the "quick as death" archetypal corporate raider (or possibly, venture capitalist) with his unforgettable motto "greed is good".

- The second epitome refers to the role game that sometimes takes place among the VCs (the Goods), the incubator consultants (the Bad) and the accelerators consultants (the Ugly). Humane nature being what it is, those outside the business secretive scenes, usually get to hear only about the deals that go well. Few thoughts are usually spared about the deals that went wrong because of these inefficient Good-Bad-Ugly role game.

Many of these inefficiencies, in fact, are driven by conflicts of interests ("agency problems") that can seriously impair the wealth creating opportunities of a business plan/new concept idea. In our personal experience, we found that an inadequate appreciation and management of these different expectations is usually the surest recipe for failure. In many cases, where information asymmetries between incubators/ accelerators and VCs were particularly important (in favour of the former), we found out how these "agency problems" were managed by them to drive to earlier conclusion the deal making process, thus negatively affecting the willingness of venture capitalist to provide equity financing. In other cases, particularly when the incubator/accelerator consultants' equity stakes were essentially stock options, the latter's often wanted to swap them to receive cash benefits from managing successive, never-ending innovation projects, while the formers tempted to pursue highly, indeed excessively, risky strategies. The game has, in fact, three different and sometimes conflicting players with different interests and goals:

- The venture capitalist typically focuses on the trade off between liquidity and uniqueness of the business plan. The more unique is the business plan, the greater the chances of a thirty fold increase in its market value through an early Initial Market Offering in the stock market; the more unique though, the greater the chances of its illiquidity onto the private placement market, if it doesn't get to IPO).

- The incubator usually focuses on the maximum diversification of its portfolio of start-ups' minority equity interests, achieved though at the price of wild defocalisation of its operating management. Moreover, the diversification effect of its equity stakes is more than matched by the excessive leverage and risky profiles of the different entrepreneurial initiatives that were incubated.

- Finally, the accelerator focuses mostly on the equity financing of the business plan operated by the VCs. Only then the accelerator can push its only-for-fees wrap up market offering, thus managing a new trade off between the interest of the entrepreneurs (seen as a partner during the negotiation with the VC) and its short term interests (once the entrepreneurs gets the money thus becoming a fat-cash-cow client).

Indeed, the entrepreneur's willingness to develop a business plan that makes industrial sense isn't quite fulfilled from anyone of these three approaches, except, possibly, in the case when there's a unique player in the place of three. In the same way, the value embedded in the market opportunity/ business plan isn't fully realized if there isn't an integrated player, able to address sinergycally an effective and a creative approach. In our approach, where the accelerator/ incubator focus ends, a VC focus begins, as well as where the fee-based job ends, a capital gain job begins, and vice-versa. Only in these way, we can support our clients with the complete range of the three tangible and intangible capitals a new business venture needs in order to grow, prosper and outperform the market: a three capitals approach that blends together knowledge capital (the accelerator/ consultants' asset), the financial capital (the VCs' asset) and the franchise capital (the networking asset of incubators).

Intervaluenet believes that overcoming these "agency problems", by changing ourselves from agents to a sort of co-principal, will be a good start towards wealth creation. We believe that there is still some room venture for venture capitalists, incubator & accelerator consultants to provide value added to the entrepreneurs. We think that, time will tell, one single entity that integrates both incubating/ market intelligence, accelerating/creative thinking consulting capabilities and venture capital funds, will help setting up ourselves better on the trail of new wealth creation opportunities.

Part 2

Venture Capital in the Financial System, Market Trends in Europe and the Relations with Banks and Stock Exchanges

1 The Special Role of the Venture Capital Industry

Stefano Caselli and Daniela Ventrone

1.1 The Structure of the Financial System

The analysis of the characteristics of the financial system is based on the observation of the organisational elements, entrepreneurial processes and contractual forms existing, in order to transfer financial resources within the market over time. In particular, in line with the main approaches, we will refer to the definition according to which a financial system is the combination of institutions, financial markets and financial instruments that, through reciprocal and complete interaction, assure the connection between the units in surplus and the units in deficit within the community (Onado 1992; Goldsmith 1969).

The institutions are the organisational units that participate in the financial exchanges over time, equipped with autonomy in the economic choices of their investments and savings in order to distinguish between the operators. With reference to accounting the aforesaid exchanges, traditionally these units are divided into macro categories, identified as families, companies, Public Administration and financial institutions (Onado 1992; Cotula and Filosa 1989). The distinctive criteria are related to a functional logic, i.e. the distinctive activity performed by each operator. However, other organisational units must be added: the supervision bodies, that independent of their positioning in the financial exchange circuit, cover an intervention role, with varying graduation and configuration, on the structure of the financial system.

The financial markets are the place where the demand and offer exchange their surplus and deficit positions. The definition of the market must however be strictly in line with the characteristics and the organisational configuration of the exchange: as a matter of fact, the trade of financial resources is characterized by different graduations that extend from the spontaneous transfer, to the direct (assisted or not) and to the intermediaries one. The exchange may occur within a rigid and disciplined market structure, or according to the free negotiating choice of the parties.

The financial instruments are the technical-legal models through which the transfer of financial resources is realised. The commitment of a definite contractual form, apart from offering certainty to the exchange and protection to the parties, allows an allocation of the contracting parties' rights and obligations coherent with their specific needs. This results in defining contractual solutions that, combined with the risk, return, cost and liquidity factors, respond effectively to the objectives of the borrower and lender of funds.

The presence of a financial system permits the operators to have an infrastructure at their own disposal, which is functional in the exercise of the fundamental investment and savings choices. In other terms, the financial system is complementary and, at the same time, indispensable to the real system for the proper organisation of the economic activities. Hence a relationship of mutual dependence between the real and the financial economy, so that each financial system takes its own distinctive characteristics from the history, choices and behaviour of the economic operators in the community and, vice versa, the real system is influenced in its capacities and investment choices by the financial system's structure. This latter relationship and, in general, the ability of the financial system to support the operators' real activity, can be verified with reference to three diagnosis parameters that emphasise the intensity and quality of the service offered by the financial system to the real one. They refer to the concepts of allocation, technical-operative and functional efficiency (Cesarini et al. 1982).

The allocation efficiency signals the ability of the financial system to assign the resources to projects with the highest expected returns in real terms. This occurs when in the distribution of the resources an order of priorities is used that awards the projects capable of offering the highest performance, so that the marginal productivity of capital is equal for all the types of commitment selected. In this situation a further allocation of the resources would not be convenient as it would not produce any incremental benefit for the system. However, the concept of allocation efficiency is strictly related to that of information efficiency. This relationship is justified by the fact that the criteria of investment selection are based on a system of prices able of reflecting at every moment, with timeliness and completeness, all the available information for the receiver of the resources. Therefore the prices assume an allocation, distributive and informative function.

The technical-operative efficiency underlines the ability of the financial system to minimise the unit transfer cost of the financial resources. The condition of technical-operative efficiency is realised where the cost of provisioning the resources by the units in deficit and that of investment by the units in surplus are near to zero. This occurs if all the costs related to the exchange process, i.e. the transaction, research and valuation costs of the project, disappear. The concept of technical-operative efficiency appears therefore strictly related to the allocation and information efficiency, as it presumes that the reduction in the organisation costs for transferring the resources arises from a high degree of transparency and visibility in the characteristics of the projects to be financed.

The functional efficiency signals the ability of the financial system to allow the matching between the demand and offer for financial resources. This represents, in fact, a pre-requisite for the other forms of efficiency as it highlights the reciprocal visibility of the borrowers and lenders of funds, the concrete possibility of accessing the market, as well as the awareness of this possibility. The aforesaid aspects are more relevant the more the operators are acknowledged as "weak" or small parties. In this case, even with high levels of efficiency in the allocation and low costs in the transfer of resources, the "weak" operator may not be capable of signalling his own surplus or deficit position to the system or may not possess adequate knowledge to make the signalling itself. Therefore, a non optimal condition of functional efficiency results in either the exclusion of operators from the financial system, or in a costly access that negatively effects the technical-operative efficiency of the system.

The overall structure of relationships and the characteristics of the financial system require a series of specific valuations in order to state, in applicable categories, the aspects of the different forms of efficiency. In relation to this, the definition of the financial system as the infrastructure of the real system focused on the transfer of financial resources, appears too broad if the final objective is to understand the competitive dynamics in place and the characteristics of the operating choices made by the different organisational units. A broad definition of the financial system results, in fact, in considering the system itself as a place in which acquirers and sellers exchange an homogenous product constituted by financial balances (Forestieri 1980). In reality, different factors permit the exclusion of this homogeneity and result in the emergence of strong diversity and distinction within the financial system. In particular, the necessary aspects for a correct reading of its articulation are: the typology of operators as borrower and lender of funds, the kind of circuits for the transfer of resources and the market forms, the type of financial intermediaries operating and the characteristics of the instruments negotiated.

The aforesaid elements represent the basis of internal differentiation and segmentation of the financial system. The identification of several subsystems, although not helping the representation of the financial system structure in its entirety, offers, on the contrary, a key to a closer reading of the competitive characteristics of the market. As a matter of fact, each subsystem is characterised by a different structure in relation to the demand and offer of the financial resources. This has a strong impact on the competitive behaviour models, the market policies and the individual operators' structure of the production process. In this perspective, the organisation of the financial system in several sub-systems must in any case be an operation whose target is represented by the investigation and understanding of the dynamics in place within a specific environment. Hence, it is possible to compare the degree of efficiency and the characteristics of the different components of the financial system.

1.2 The Determinants of the Specialisation of the Financial System: SME and New Companies

In order to focus on a competitive diagnosis and on the operators' choices, the distinction among several significant sub-sectors within the financial system requires, first of all, the identification of a dominant criterion of segmentation. As previously noted, this depends on the objectives of the valuations made and on the relevance, significance and importance elements for the operators. In other terms, the selected sub-sector must represent an area of the financial system perceived as totally or potentially autonomous by the main subjects in the market, and where it is possible to distinguish between a specific - actual or potential - dynamic competitiveness and singular competitive strategies in place, or to be activated, by the operators. If these conditions are not holding, the segmentation assumes the function of mere classification of the financial system in sub-categories.

A particularly efficient and easily applied criterion for structuring the financial system is based on the demand driven approach. Here, the critical element in order to distinguish and organize the operators, financial markets and instruments in different segments, is the typology of the demand of financial services. The selected perspective is that of the economic subject who, against specific needs, compares its demand with the structure of the offer in the financial system. The presence of macro institutional sectors in this specific case indicates the segmentation models of the financial system: the families, non-financial companies and Public Administration show behavioural characteristics and strongly distinctive requirements that stimulate a differentiation, with varying intensity and rapidity, of the financial markets, instruments and intermediaries.

In relation to this, in the market economy the companies' sector represents the principal component of the system for realising the real investment process and, as a consequence, assumes the most important role in comparing and verifying the degree of allocate efficiency of the resources. Therefore, the financial system characterises its intervention as efficient the more it is capable of supporting the companies' sector in the realisation of industrial projects, taking into account the expected profitability and the objective of reducing the costs of the resources. The importance of the companies' role as the connection element between the financial and the real circuit requires a detailed evaluation of the different categories of operators. This is important as the presence of significant differences within the sector can justify an analysis of the financial system which is differentiated in relation to the characteristics of the categories identified. Hence, the perceivable distinctions in the companies' sector are numerous depending on the commodity sector they belong to, the type of economic parties present, the number of employees, value of turnover and degree of internationalisation. Although each of the criteria may have a high degree of explanatory efficiency in the companies' universe, it appears necessary to use those factors that produce the best categories in order to justify and explain the differentiation process within the financial system. As a consequence, the size of the number of employees, that tells the small and medium sized companies (SME) apart from the large ones, appears particularly efficient under the importance profile of the category of reduced size operators. In particular, the aforesaid importance can be examined under two different aspects. First, the high diffusion of small and medium sized companies appears as the structural phenomenon in all the market economies in the western world. In the specific Italian case, this characteristic also assumes a country specific aspect so that, in terms of unit productivity, the degree of diffusion of the SME in Italy is largely superior to the European average. Second, the significant number of small sized companies causes this category to have a strong role in the occupational process of the resources and triggers the development processes of new companies. The analysis of the data from the latest industrial census, confirming the most recent significant findings, signals that the category of the companies with less than 50 employees absorbs in Italy 40.4% of all employees (see Table 1.1). This percentage shows a specific concentration in the class of companies with a number of employees between 10 and 19 units in that the number of employees represents 17.7% of the system.

Table 1.1. The dimensional structure of the principal European economies (share in percentage of employee terms. Data referring to the latest census carried out in the different countries)

Number of employees	France	Germany	Italy	United Kingdom	Spain
1-9	22.03	21.38	45.81	26.61	29.80
10-19	7.03	10.02	11.18	6.41	12.52
20-99	21.01	17.77	15.48	16.,05	23.07
100-499	16.25	17.46	9.88	17.18	14.56
500 and more	33.68	33.37	17.65	33.75	20.05
Total	**100.00**	**100.00**	**100.00**	**100.00**	**100.00**

Source: European Commision 2002

The quantitative significance of the phenomenon requires detailed consideration and to verify the fact that the SME system can build a strong differentiation in the financial system. Otherwise, it is possible to identify, in the presence of this type of companies, a system of operators, markets and instruments that, actually or potentially, underline this grouping with an aim at making market strategies and designing dedicated market circuits or specific contractual forms.

In order to evaluate the degree of internal differentiation in the financial system, with respect to the presence of SME, numerous factors must be taken into account. These, with different intensity, have a direct influence on the characteristics, operating rules and market policies of the players and financial instruments servicing all those companies (not only SME) that exercise their activity in the real system. The most important ones are as following:

1. intervention strategies by the institutional policy makers;

2. degree of exploitation of the potential economic space;

3. overall degree of development of the market;

4. contestability of the market.

1.2.1 Intervention Strategies by the Institutional Policy Makers

The first factor refers to defining the financial system's characteristics for the SME: the choice of systems made by the institutional policy makers, represented by the State (exercised at central and territorial level, in its legislative and governing function), by the dedicated public bodies, international bodies and institutions of control and supervision (Corbetta 1997; Caselli 2001). The intervention of the institutional players can be developed through their direct presence in the financial system or through an indirect role in the exercise of a planning-regulatory function.

In the first case, the policy makers characterise their intervention in the financial system as agents who directly assume the vest of "operators of the system", rather than operators of the market circuits and direct distributors of financial instruments. Leaving aside a judgement on the merits of the opportunity and efficiency of the direct intervention models, the tendencies in place in the advanced economic systems, and in particular in the European Union, in general provide a significant limitation. As a matter of fact, the trend can be observed with reference:

1. to the operators: the community regulatory structure on the corporate governance of the financial intermediaries excludes the presence of significant participation stakes by state authorities in their control and management functions;

2. to the markets: the most recent regulatory European provisions on the matter provide the institution of regulatory markets in the form of private market-companies that exclude the presence of public bodies;

3. to the financial instruments: the form of assistance provided by the State must be coherent with the level of maximum intervention defined under community regulations.

Table 1.2. The intervention models of the policy makers in the financial system in support of SME and new ventures

Methods of intervention by the policy makers	Components of the financial system		
	Operators	Market	Instruments
Direct intervention	Creation of intermediaries and specialised operators for SME. Participation in share capital of intermediaries and specialised operators for SME.	Control and management of the markets dedicated to SME	Direct payment of forms of support and financial instruments for SME
Indirect intervention	Definition of the legislation and regulation structure on the typology of operators and on the areas of activity for the SME. Supervision on the activity of the operators. Sponsorship and promotion.	Definition of the legislation and regulation structure of the SME market. Supervision on the activity of the SME market. Sponsorship and promotion.	Definition of the legislation and regulation structure of the financial instruments for SME. Sponsorship and promotion.

In the second case, the intervention of the policy makers assumes an indirect characteristic as it does not operate within the financial system but, according to a classification defined by levels of decreasing intensity, it depends on the design of the regulatory environment and on the logic of control and sponsorship of the initiatives. The design of the regulatory environment as an indirect activity with strong legislative content because it defines the structure of the components of the financial system. This type of intervention, if properly developed, is as incisive as the direct intervention because it allows the public authorities to govern the structure of the market in its entirety. However, in this case the space for intervention is limited by the value of competition and the mobility of resources, and is restricted by community regulations, which leave reduced intervention possibilities to the state authorities. This causes a less significant form of intervention, as it aims at exercising a control activity on the financial system, according to the logic of levelling the playing field and prudential supervision. Here, the action of the policy makers, typical of the supervision bodies, has a supervision content and a fine tuning function of the competitive system with the objective of assuring, with respect to a pre-established level of stability in the financial system, increasing levels of efficiency in the exchanges. The indirect intervention actions can assume, finally, a less meaningful configuration, that is based on the sponsorship and, as far as possible, on the moral persuasion exercised on the market forces. The aforesaid action characterises the intervention of the policy makers as promoters of initiatives, not only in the areas of activity of the financial intermediaries but also in the structure of the markets and in the characteristics of the financial instruments. The efficiency of the promotion is therefore related to the parties' reputation and its capacity to create consensus with reference to a specific business idea.

However, the relationship between the type of intervention put in place by the policy makers and the degree of internal differentiation of the financial system for the SME is also related to the number of players with the powers of effective intervention in the market and to the characteristics of their respective objectives. An increasing number of players can, in fact, exercise a divergent effect on the objectives and thus cause a loss of specialisation in the intervention of the financial system to the advantage of the SME.

1.2.2 Degree of Exploitation of the Potential Economic Space

The second factor in evaluating the attitude of the system to meet the particular requirements of the SME is related to the degree of exploitation of the economic space available. In general, this analysis for a specific financial instrument, market or intermediary, must start from the existing and potential economic space available[1]. The first relates to all the transactions already existing and those that can be activated without a significant modification in the productive and technical characteristics of the operators, instruments negotiated and markets used. This means that the growth in the exploitation of the existing economic space increases, ceteris paribus, the degree of differentiation and characteristics of the financial system in favour of the SME. The second relates to all the exchanges that can be activated only with a significant modification of the product-market-operator combination referring to the competitive forces operating within the financial system. The exploitation of the existing economic space is therefore related to two distinct factors: the pre-existing degree of specialisation of the financial system for SME and the constraints to the possible exploitation by the SME and the operators in the market. In the first case, the economic space is larger the more the financial system is characterised by attributes of completeness and significant specialisation, and the more the policy makers aim at creating financial instruments focused on the needs of the SME. On the other side, it is particularly limited when the two components do not combine significantly in the aforesaid direction. In the second case, the possibility of exploitating the economic space relates to the knowledge gap and the asymmetric dimensions that may characterise the SME. Under this profile, the SME may not be able to make recourse to specialised financial intermediaries, dedicated market circuits or specific financial instruments, or may not have the necessary technical or basic knowledge, or its reduced size does not consent a direct access. Therefore, the characteristics of reduced size companies may represent a structural constraint to the differentiation of the financial system as the SME is not able to signal its diversity to the operators in the market. From this point of view, the potential economic space is very important, or rather that specific market area which can be used only by significantly modifying the existing product-market combinations. The "diversity" of the SME can be recognised and exploited by those financial intermediaries that, adjusting their production processes and the financial products offered, take an additional competitive advantage compared to the starting position.

1.2.3 Overall Degree of the Development of the Market

The third differentiation factor for the SME relates to the overall development level achieved by the financial system. In general, it is possible to af-

---

[1] For a complete and exhautive evaluation of the "economic space" concept see De Laurentis 1995.

firm that the greater the financial system presents requisites of intensity, articulation and completeness, the more the operators can find in it an effective support in engaging their economic choices. Therefore, a greater development of the system leaves space to a more intense internal differentiation by type of operators, financial markets and instruments in relation to the different categories of units in surplus and in deficit.

In particular, the financial intensity of an economic system, measured as the relationship between the variation of the overall financial activities and the return produced, indicates the ability of the financial system to sustain the production processes. This means that when the ratio is high, a greater quantity of financial resources is "mobilised" to produce a unit of return. Vice versa, the articulation of the financial system refers to the number of financial intermediaries whose performance belongs to different intermediary activities referring to credit, stock market and insurance area. In this environment, an increasing articulation increases the possibilities for the operators to find in the market the ideal party in order to satisfy their specific financial requirements. However, the valuation criterion can be focus on the institutional aspect, based on regulation characteristics and not only on the financial intermediaries. This means that although in the presence of a high level of articulation, determined by the entire range of financial intermediaries, the financial system is not necessarily able to satisfy the users' needs if the activities that can be exercised are not actually carried out. For this reason, it is necessary to analyse the development level of the financial system also through the completeness parameter, whose valuation criterion is the functional prospective. Here the completeness, independent of the institutional model, highlights the ability of the financial intermediaries, and broadly of the financial system, to assist and perform the matching between the borrowers' and the lenders of funds' goals. These can be characterized by high levels of complexity as the lender's risk, return and liquidity parameters and the borrower's cost and sustainability can assume different graduations and articulations stretching from the simple financial exchange to the establishment of complex relations, that aim at satisfying situations and financial problems over time. More specifically, the characteristics of the SME as borrower of funds stimulate the presence of differentiated institutional operators, in order to obtain a more intense satisfaction of the distinctive requirements of reduced size companies and of emerging complex ones. In this case, the differentiation of the financial system depends on: the ex ante design of intermediation forms tailored on reduced size companies; and the ex post market legitimation of those operators focusing their activities on one or more businesses specifically organised for the SME.

1.2.4 Contestability of the Market

Finally, the forth factor relates to the contestability level in the market. This approach, although suffering from numerous limitations especially when applied to broad financial markets, allows to shed light on the functional character, the inherent level of development and the competitive structure of the market, therefore becoming an effective criterion for analysis and overall judgement. The presence of a high contestability level signals the presence of a market circuit potentially flexible and consistent with the company's requirements, as it is able to stimulate the entry of new operators, their selection (through competitive comparison) and the exit of unsuitable ones. Vice versa, a low level highlights the inability of the market to activate the necessary circuit in attracting external parties, and a scarce competitive comparison and operators' turnover, therefore indicating the ability to satisfy the final users.

The contestability judgement of the financial market for SME must be made with reference to analysing the intensity of the entry and exit barriers for the actual and potential players of the "system". Therefore, a comparison must be performed between the elements of the specific SME market and of the financial market in general. In the first case, the entry barriers blocking the contestability of the SME market mainly relate to the downsize effect or rather the dimensional asymmetry between the lender and borrower of the funds (high information asymmetry). This effect can compromise the market entry as the total cost of activating and managing the production process for reducing the information asymmetry could be not compatible with the expected remuneration flows of the financial intermediary's investment activity. On the other side, more complex appears the valuation of the exit barriers, or rather the sunk cost connected to abandoning the chosen productive combinations. This peculiar type of cost is not differentiable compared to the structure of exit barriers related to the product-market combination in the financial system[2]. In the second case, the tendencies in place in the financial services market signal an overall fall in the entry and exit barriers and the development of a dynamic equilibrium. In the latter case, the intermediaries' efforts for protecting the competitive

---

[2] This signifies that the structure of the *sunk costs* relating to the financial system assumes a strong immaterial character not being present in a significant manner components of fixed capital. The *sunk costs* are related to: the capitalised costs for acquisition, maintenance and production of the *know how* necessary to compete in the context of the specific product-market combination; the expenses of research and development; the expenses relating to the *brand* factors.

advantage within a specific sector is rapidly eroded, driving the research of new protective factors, thus effecting:

1. the intense development of the information technology as the vehicle of transmitting and spreading information;

2. the growth of contract banking, i.e. the greater possibility of integrating and de-integrating the financial production process through in- and out-sourcing (Llewellyn 1999);

3. the differentiation of the "competitive models" (from financial to industrial and commercial parties) performing important functions within the banking and financial industry[3];

4. the fall in the barriers intra and extra country, natural consequence of the progressive and continuing deregulation action of the financial system.

1.2.5 Conclusions

In general, the overlapping of the elements referring to the SME and to the financial system signals, on one side, the decrease of the factors restraining the rising markets contestability and, on the other, the stronger importance of firm specific factors determining the intensity of entry barriers. Hence, the elements and relational components related to the commitment and partnership assume a strong significance. These factors directly affect the information asymmetry, therefore generating areas or domains of reserved and captured information, which allow to make investment decisions with sufficient certainty. However, the intangibility of the relational elements must find an adequate contractual place and be developed according to specific implementation models. On the offer side, this can result in developing banking oriented solutions that, with reference to the economies of scope and relations, identify the credit intermediaries as the central point in the offer of financial services to the SME[4]. This approach can reach strong intensity levels, typical of the "housebanking relationship". Vice versa, it is also possible to develop specialised suppliers oriented solutions, where specific financial intermediaries organise production processes focused on provisioning one or more services for the SME. In this case, the production process forsters economies of scale, in terms of reaching the desired eco-

---

[3] On the theme of the factors that strengthens the contestability of the financial markets see Llewellyn 1999; Santomero and Babbel 1997.

[4] See Berger and Hannan 1997;Hannan 1991; Peek J. Rosengren 1996; Keeton 1996; Strahan and Weston 1998; Walraven 1997.

nomic equilibrium and compatible with the satisfaction of the client company, in a professional relationship.

Although opposing, the two aforesaid approaches present space for complementarity and strong contamination, so that the structure of the final market appears clearly influenced by historical and specific market factors. In particular, this means that the presence of successful banks offering a range of articulated services for the SME does not preclude the presence of specialised financial intermediaries.

The combination and the intensity of the four elements examined contribute in determining the degree of differentiation, development and specialisation of the financial system towards the SME. The analysis, however, must be verified and applied in relation to the characteristics in the different systems in each country, in order to evaluate and understand the country specific factors. This means verifying their contents on the basis of historic and prospective elements in the medium term, with the objective of examining the effects on the provider of the financial services, and on the financial behaviour of the SME.

1.3 The Internal Differentiation of the Financial System for the Development of the SME and the New Companies

The structure of the financial system for developing new companies and the existing SME, determined by the intensity and interaction of the differentiation factors previously described, assumes an overall structure, that is therefore observable with reference to the financial instruments, operators and markets. The three aforesaid components are not exhaustive and can show broad complementary space with the other segments in the financial system, where the single operators, rather than the markets or instruments, assume a generalised character of utility and exploitation by the end users. Hence, it is important to verify two distinct aspects:

1. the degree of specialisation related to the needs expressed by the SME, the operators, instruments and financial markets;

2. the overall structure of the financial system for the SME.

Therefore, this verification must be made: ex ante, with reference to the structural characteristics of allocating, effectively and potentially, the components of the financial system for satisfying the SME's needs; ex post, i.e. the effective allocation of the components. Therefore, the ex post valuation can be made only when applied to a specific environment. This means proceeding to understanding and adapting the country specific factors in order to perform a correct diagnosis of the overall structure of the financial system for the SME. It follows that this diagnosis is necessarily sequential to those of the differentiation factors of the financial system and represent both the complete and the concrete application.

Under the degree of specialisation of the operators, instruments and financial markets, it in necessary to make an exhaustive mapping of the components dedicated and functional to the SME's financial requirements. In other terms, it is necessary to identify, according to an ex ante valuation, all those elements concretely involved in the process of providing financial services in the environment of a fully developed economic system.

1.3.1 The Characteristics of the Operators

With reference to the operators, they mainly refer to three categories:

1. financial intermediaries;

2. professional agencies;

3. mutuals.

The financial intermediaries perform, first of all, the function of satisfying the deficit position of the SME, that is the external financial needs determined by activating and developing the purchase, production and sales activity (see Table 1.3). The production and provision of complementary and autonomous services and products can overlap and interact with this function, in addition to the main financing one. These operators can be classified under a functional or institutional profile, thus producing the categories of banks, credit intermediaries and other financial intermediaries. With reference to the first two ones, neither the institutional elements nor the functional development of the activity allows to highlight the characteristics and profiles of the intermediaries' structural specialisation towards the SME. However, the banks may have an internal organisational specialisation that identifies the SME as an important segment or market, thus performing an internal differentiation of roles, organisational units and production processes (Hunt 1996). Only specialised Italian banks for medium/long term credit, within the category of the non banking credit intermediaries, have a structural and constitutive mission of their activity with reference to the SME.

The category of other financial intermediaries appears more significant, as they are able to interact with the system of SME that determines their operating boundaries, the content of the activity developed and, in the final analysis, their reason to exist. This is evident with reference to the closed-end funds and the venture capital companies, whose mission and core business is offering financial resources as risk capital to small sized companies and in the constitution phase too. Here, the presence of several models for transferring financial resources and different types of companies' needs determines a further internal operating segmentation of the financial intermediary that, in some cases, becomes a true and proper functional specialisation[5]. It should be recalled that, according to a logic of activity "contamination" towards different categories of customers, the closed-end funds and venture capital companies tend to broaden their activity also towards non small sized companies. This occurs in particular in case of debt and turnaround restructuring operations and/or for leveraged and management buy outs, i.e. events where the size of the company is no longer a differentiation element in the intermediation process of the financial resources.

The merchant banks category is more complex, as the range of activities and the relative boundaries are rather wide and not necessarily clearly identified[6]. Therefore, the relative business areas extend from capital markets to corporate finance, according to different models and graduations. However, differing from the other financial intermediaries examined, the presence of SME does not represent a factor of autonomous differentiation as the merchant banking activity presents a clear orientation towards corporates and large corporates. This does not exclude, however, the possibility of focusing on the SME sector, as a market segmentation with an internal organisational and production differentiation of the operators.

The professional agencies are a heterogeneous grouping of operators that do not directly perform an intermediation activity of the financial resources, but participate facilitating, assisting and supporting it. Common elements are the specialisation in dedicated business areas and, in general, their non focalisation ex ante towards the SME; therefore, the choice of specialisation itself may represent a market policy examined later on. The extension and depth of the operators' intervention cannot be precisely defined beforehand, given its qualitative characteristic; however, it can be related to observing several categories typically present in the market. The first refers to the companies' consultants or rather those parties which are external to the company and connected through specific contractual forms, whose target is carrying out an intervention in one or more areas of the op-

---

[5] See: Block and Mac Millan 1993; Evca-Coopres and Lybrand 1997; Venture Capital Report 1998.
[6] See: Gatti 1996; Kuhn 1990; Kay 1992; De Cecco and Ferri 1994; Autorità Garante della Concorrenza e del Mercato – Banca d'Italia 1997; Caselli 2000.

erating activity[7]; this intervention may heavily affect not only the company's investment choices, but also the selection of the technical forms of financing and/or the counterparts involved in the fund-raising stage. The second category refers to the brokers, or more generally those parties that carry out a brokerage and assistance function in the financing process of the company and during the intermediation of the resources. Here too, it does not seem possible to define the content of the brokerage activity beforehand, as it can have articulated meanings that adapt to the specific choices of the broker's market, as much as to the nature of the companies' requirements. The third category finally refers to data providers, parties performing an activity of production and provision of information to the companies, functional to undertaking decisions in the financial management context. This activity can be mainly developed in relation to two directions. The first is the production of a wide-spectrum of information on the commercial and, partly, financial counterparts under the soundness and solvency profile in the medium term. This is especially functional to a more effective management of the complete working capital cycle, on the investment as much as the financing side. The second is the production of a company's rating, that is the emission of a concise and authoritative opinion on the credit risk. This activity is indispensable and functional to the construction of a general signalling process to the players in the financial market with reference to the company's credit standing. Here, the data providers' activity assumes a configuration which is very close to the financial intermediaries' one, as it is centred on the organisation of a production process finalised to the acquisition and production of aggregated information and added value, due to the superior ability in managing the market's asymmetric information.

Finally, the mutuals category includes those operators that carry out a non profit based function in the financial system - meant as co-operative rather than focused on intervention - without any remuneration in the transfer of financial resources to the companies[8]. The distinctive factor of the aforesaid operators is firstly represented by the differentiation of origin compared to the for profit ones, which can however find different application paths with heterogeneous graduations depending on the SME's characteristics.

A first type of operator is formed by mutual guarantee consortiums ("confidi" in Italy or "socama" in France) or rather by consortiums and co-

---

[7] On the role of the consultants with reference to the activity of the company and especially the interaction with the intermediary circuits of the resources see De Cecco and Ferri 1994; Caselli 1999; Rybczynsky 1996.
[8] See the valuations of Arnott and Stigliz 1991.

operatives that collectively guarantee the loans. They arise from the spontaneous initiatives of small entrepreneurs and local bodies with the objective of overcoming the difficulties that the smaller companies meet in the access to financial resources, which are offered by the banking system through the principle of solidarity. Therefore, the economic reason and the mission of the mutual guarantee consortiums to the SME is clear: the co-operative element and the provision of a guarantee. The first highlights the solidarity component that, aggregating small and medium sized companies, allows a reduction in the asymmetric dimensions of the lending banks. As a matter of fact, this often causes a less favourable relationship for the companies due to the quantity and relative price of the credit provided. The second represents an instrument for reducing the information asymmetry between the bank and the small company, as the consortiums' interventions constitute a greater and more qualified signal of reliability that the one the company can provide to the lender. Moreover, they also present a peer monitoring characteristics: a company's participation generates an "implicit social control" by the members of the same group, as they all have an interest in promoting the solidarity action effectively, and therefore preventing the emergence of a crisis. However, if the disciplining effect arising from belonging to a group represents a fundamental element in improving the quality of the relationship between bank and company, it is also necessary to highlight the emergence of a lobby effect that can strongly distort and compromise the co-operative's mission.

The second category of mutuals are the business angels. These operators exercise in a professional and solidarity manner the start up financing activity of the entrepreneurial business ideas, with the mission of contributing to the growth of the industrial system and thus to the overall social welfare of the community. The area of intervention is therefore represented by the supply of risk capital, and the relationship with the SME is not only an element of differentiation but it is in fact constituted by the existence of the business angels.

Table 1.3. The operators of the financial system and the degree of differentiation with reference to the structure of the requirements connected to the companies' development

Operators of the financial system	Degree of structural differentiation with reference to the development of the SME	Possibilities of differentiation and adaptation to the requirements connected to the development of the SME
A. Financial intermediaries		
Banks	Low	General supplier or specialised supplier not focused ex ante on the SME. Possibly an internal organisational specialisation towards the SME through a process of divisionalisation and identification of target segments.
Leasing and Factoring	Low	Specialised supplier not focused ex ante on the SME.
Specialised banks for medium/long term credit	Medium-high	Specialised supplier tendentially focused ex ante on the SME.
Closed-end funds	Medium-high	Specialised supplier tendentially focused ex ante on the SME.
Venture capital	High	Specialised supplier focused and structurally dedicated ex ante to the SME.
Merchant Banks	Medium-low	Specialised supplier tendentially not focused ex ante on the SME. However possible either an effective specialisation or an internal specialised organisation towards the SME, through a process divisionalisation and identification of target segments.
B. Professional agencies		
Consultants	Medium	Specialised supplier tendentially focused ex ante on specific customer segments, among which SME.
Brokers	Low	Specialised supplier not focused ex ante on the SME.
Data provider	Low	Specialised supplier not focused ex ante on the SME.

Table 1.3. (cont)

Operators of the financial system	Degree of structural differentiation with reference to the development of the SME	Possibilities of differentiation and adaptation to the requirements connected to the development of the SME
C. Mutuals		
Mutual guarantee consortiums	High	Specialised supplier focused and structurally dedicated ex ante to the SME.
Business Angels	High	Specialised supplier focused and structurally dedicated ex ante to the SME.
Bank foundations	Medium-low	General supplier not focused ex ante on the SME. Possible identification of line intervention in favour of SME.

The third category of mutuals refers to the bank foundations, i.e. non profit parties that can operate with solidarity purposes and develop towards a very wide range of economic and financial counterparts. In this case, the SME represents an important and potential target due to the effect of their relative weak financial profile and their requirement of aid and support from the economic system. The intervention of the banking foundations can happen through a direct or indirect approach. In the first one, the foundation, appealing to its "contribution capacity", supports the projects, activities and companies that are worthy and requiring assistance. The financial resources are thus transferred directly to the SME, according to the models and allocation determined by the overall contribution strategy. In the second one, the foundation, even if contributing the necessary capital, is promoting the instruments that professionally and institutionally operate in the market in support of the SME. These prevalently relate to closed-end funds and venture capital companies that develop their activity of equity financier exploiting the sponsorship, and eventually also the critical mass of financial resources provided by the foundations, that undertake the path realised by those projects requiring the maintenance of specific expertise.

1.3.2 The Characteristics of the Financial Markets

When examining the financial markets, the traditional distinction between regulatory and non regulatory markets is mainly used. The first have the characteristic of exchange systems between demand and offer within the regulatory environment, which is defined and recognised by the economic system. The second is an informal exchange, not subject to fixed rules at a regulatory level. However, this distinction only consents a general overview of the market structure, and does not allow to understand the possible differentiation processes constructed on the structure of the SME's needs. For this reason, it is necessary to proceed through an internal articulation of the two categories mentioned, in order to define a sufficiently analytical and detailed picture (see Table 1.4).

The regulatory markets must be distinguished, first, into the the primary and second markets components, and then into the other ones. The primary markets represent the principal list of the individual "country systems". It follows that their function and constitution characteristics are not originated from the presence of the specific SME's needs, but from the more general requirements of the (large) companies. Very frequently, their true nature of "primary list" generates significant entry barriers for the smaller operators mainly due the direct and indirect costs of access and permanence in the market. However, this fact does not exclude ex post the internal differentiation of the list in segments of specialised quoted stocks with reference to reduced sized companies. Although this evaluation is much more a theoretical view rather than an empirical one, given that within the main lists of the more developed financial systems this solution is not encountered, it should be noted that the integration process of the financial markets in place in the Euro area shows a possible evolution option for realising a single principal listing, which includes the companies from each individual country. Within this structure, the segmentation into categories – by industry or size – constitutes a physiological process for reducing the complexity and adapting the structure of the risk capital demand to the offer, in terms of more appropriate signalling of the companies's commodity-industrial profile[9].

The secondary markets and the other ones represent those regulatory circuits performing the specific function of widening the base in order to match the demand and offer of risk capital, with reference to companies that would not find in the primary market the appropriate space, due to asymmetric size, information and knowledge gap, or structural constraints related to these costs. In analytical terms, the distinction between "secondary" and "other" markets is specifically determined by the practices of the developed financial systems rather than by a theoretical concept, as it is related to the company target. The second markets are small and medium

---

[9] On the evolution prospects of the European financial markets, on the agreements of collaboration and integration in place, on the processes of successive internal segmentation of the Euro-listing see the scenarios traced out by in Bank of England 1999.

sized oriented ones so that the SME's financial needs become factors of institutional differentiation and justify their existence. The other markets, instead, are specialised customer oriented so that the financial needs of specific segments of customers become the determining factors for the markets' existence. Here, the important segments are identified on the basis of the industry the company belongs to and of the evaluation of the profitability prospective: that is why they are specialised market for companies with high growth potential and/or belonging to a specific commodity sector. Differing from the secondary market, where the fundamental factor is the requirement of the system to find outlets for the demand of risk capital for small and medium sized operators, in the other ones the triggering factors can be multiple. By observing the financial systems it is possible to note that the propelling element in developing the market has a competitive, or rather confrontational symmetry to the primary one, though in a restricted area of companies. In other terms, the other market tends to perform a riding activity against the primary one: its competitive element is the greater adequacy and conformity to the companies' profile, therefore contrasting its tailor made logic to the primary market's de-specialisation one. Hence, the distinction between secondary and other may become less clear and, in the operating reality, processes of reciprocal contamination and overlapping of functions can be verified. Significant examples are the experiences developed in the Anglo-Saxon contexts with reference to the Nasdaq circuit in the United States, and the Aim and Techmark ones in Great Britain[10], that perform at the same time the secondary market function, as their target is the SME, and the other market one, as they are oriented towards technology based companies and sectors showing strong growth potentials. A similar situation could be recently found in the Euro area with reference to the Euro.Nm circuit, that stands at an intermediate level between the typical secondary market logic and the others' one, as it is oriented towards both small and medium sized companies, and those competing in the high technological intensive sectors and those with high growth potentials.

---

[10] See in relation to this London Stock Exchange, "Techmark, The technology market", London, 1999. In particular, the AIM appears a possible *benchmark* due to the success it has achieved. At December 31, 2000, in fact, it counted 524 quoted companies – of which 30 foreign – for a capitalisation equal to approximately 15 billion sterling and a large representation of commodity sectors. In these terms, this markets places itself in competition to markets once considered global and more specialised, such as Nasdaq, Easdaq and the European Network of New Markets.

The non regulatory markets are positioned within the financial system in a different way with respect to the regulatory ones. This is due, first, to the formal and substantial requisite of the absence of regulatory controls defining the framework and content of the exchange activity. This characteristic allows the participating financial intermediaries and companies to create a de-structured space for transactions, with reference to the securities exchanged, the operative models and form of safeguard and risk protection. A second distinctive element relates to the market's functional use: apart from minimising the costs and regulatory restraints for the demand and offer, it widens the company's spectrum of funding choices and the financial intermediaries' space for services and possible investment. The presence of third markets appears therefore as a physiological and dynamic character of the financial systems, and combined with the regulatory and normative activity, it helps define the structure of the primary and secondary ones. Moreover, the operators react according to an operating "deregulation" line. Overall, the system's development can be considered as an alternative to regulation and deregulation cycles that modifies the old equilibrium reached and imposes a "new definition" for regulation. Therefore, in the third markets the differentiation factor related to the SME's satisfaction is weak ex ante, if not totally absent. However, this situation does not preclude the presence ex post of smaller sized companies within the informal listing.

Within the non regulatory markets it is necessary to identify a second type of circuit for transfer the resources: captive or inner markets. They are established within a group structure or network of several financial intermediaries who require the development of space for autonomous exchange, relating to equity investments mainly realised in companies that are not quoted on the main listings. The exchange activity, given its strong private element, can be subject to more or less stringent regulations or specific moral obligations.

Table 1.4. The financial markets and the degree of differentiation with reference to the structure of the requirements connected to the companies' development

Markets	Degree of structural differentiation with reference to the development of the SME	Possibilities of differentiation and adaptation to the requirements connected to the development of the SME
A. Regulatory		
Primary markets	Low	Generally not specialised towards specific types of companies. Possibility of developing international listing specialised by company segments.
Secondary and other markets	Medium-high	Tendentially focused and structurally dedicated ex ante to the SME.
B. Non regulatory		
Third markets	Medium-low	Generally not specialised towards specific types of companies.
Captive markets (inner markets)	Medium-high	Generally de-structured. Probable however a focus ex ante towards the SME.

There are two fundamental constitution elements of the captive markets. The first relates to the presence of highly illiquid securities, requiring a market mechanism that allows the investors to unfreeze their positions and obtain the capital gain on the investment made. The second is the strong commitment and alliance of interest that is formed in the investment activity of the companies' risk capital. The requirement to have and realise a strong convergence of interest acts as a justification to the presence of several intermediaries belonging to the same banking group, association or consortium (for example, business angels networks) operating in the market. This is why the development of the captive markets is strictly related to the presence of SME, which justifies their formation and differentiation. As a matter of fact, the captive market is the natural way of transferring an effective liquidity to the start up market, that shows restricted and limited manoeuvre in finding an entry within the regulatory markets. However, the connection between captive markets and SME is not exhaustive as the realisation of internal risk capital circuits can also be finalised to exchanging securities negotiable on the regulatory ones, or it can relate to investments towards large companies not quoted on the Stock Exchange.

In more detail, a further element differentiating the third markets from the inner ones relates to the specific operator who forsters its creation. In the first case, the financial intermediaries (thus the offer), having the necessity to move quotas of participation held in their portfolio with greater flexibility than allowd by the existing regulatory markets, search other operators with the same requirements. For this reason it is not certain whether there exists a specific focus on the SME.

In the inner markets, on the contrary, the incentive towards the conclusion of the operations derives directly from the demand side, which requires securities with particular characteristics. It therefore relates to an even more "private" market as its aim is satisfying the particular requirements of specific operators.

1.3.3 The Characteristics of the Financial Instruments

The degree of ex ante differentiation in relation to the SME's financial requirements is tendentially limited, due to the function of transfer of financial resources performed by the financial instruments. This, according to different contractual solutions, segment the categories of instruments in relation to the needs of transfer, or rather the contracting parties' rights and obligations, and not to the user's economic profile (see Table 1.5). The evaluation is not significantly modified if one also adopts a wider, and appearently less rigorous, meaning, including financial consulting and the production of guarantees (Llewellyn 1992).

The first is the physical and logic support, and the client's content in relation to undertaking the financial choices. The incisiveness of this action can generate a few possible effects: the general integration between the instrument and the consulting support, resulting in the production of a new kind of financial instruments; the maturation of consulting as an autonomous product, clearly discernible and necessary to the production of the financial instrument. The first situation is verified with special reference to the application of information technology, that from a simple transmission vehicle of the financial transactions becomes a constitutional and distinctive part. The second situation is verified in the advising field, where the consulting support acquires the characteristics of a distinct product, however necessary for developing the equity operations.

The production of guarantees is part of the financial instrument category as it signals the distinctive characteristics of the company requiring credit and it is a necessary component of valuing the credit risk by the financial intermediary (Bester 1987; De Bonis 1996). Therefore, the guarantee becomes an integral and constitutional part of the concession and acquisition of the - debt, bank and, partly market - financial instruments. This characterises the functional importance, autonomy and thus, value of its use and exchange within the market. These aspects have a rising importance because the financial intermediaries acquire an increasing ability in distinguishing and measuring the collateral components in the environment of credit concession. Moreover, the concept of guarantees evolves from a purely material characteristic to a more articulated and even immaterial one, especially in the direction of auditing, peer monitoring and certification.

Table 1.5. The financial instruments and the degree of differentiation with reference to the structure of the requirements connected to the companies' development

Financial instruments	Degree of structural differentiation with reference to the development of the SME	Possibilities of differentiation and adaptation to the requirements connected to the development of the SME
Equity	Low	Not specialised towards the SME if not through an assisted financial deal or the action of a specialised operator
Mezzanine and subordinated	Low	Not specialised towards the SME if not through an assisted financial deal or the action of a specialised operator
Market debt	Medium-low	Can be specialised ex ante in function of the SME. In any case specialised through an assisted financial deal or the action of a specialised
Bank debt	Low	Not specialised towards the SME if not through an assisted financial deal or through the action of a specialised operator
Financial consulting	Medium-low	Possibly specialised towards SME
Guarantees	Low	Not specialised towards the SME if not through an assisted financial deal or through the action of a specialised operator

Overall, the space of ex ante differentiation of the financial instruments can only be referred to specific regulatory interventions of the policy makers. These interventions, identifying categories of users on which to focus and to whom dedicate tailor made contractual forms, modify the structural characteristics of the financial instruments adapting them to the particular needs of the smaller sized companies. This possibility is verified in the assisted financial actions that, although not originating autonomous financial instruments, constitute a model of ex post differentiation of the entire financial instruments category in relation to the SME's needs. In particular, the different assisted interventions affect the economic conditions rather than the provision of the individual financial instruments, leaving unchanged the company's acquisition function, but modifying the impact of several "key elements ", i.e. reducing its costs and lowering the entry barrier.

1.4 Managing the Market Relations and Production Specialisation in the View of the Financial Intermediary

The overall evaluation of the characteristics and degree of differentiation of the operators, markets and financial instruments allows to broaden the view from an analytical and detailed description to a general aggregated logic, that fosters a co-ordination between the different components in the financial system for the SME's development. A judgement on the potential structure of the financial system for the SME can only be made by connecting the type of operators to the offer of financial instruments and their effective entry in the different market categories.

1.4.1 The Relationship Between Operators, Markets and Instruments

With reference to the connection between operators and financial instruments, it is possible to identify significant relationships and important spaces for operative specialisation. Each operator-instrument combination must be stated within a country specific context in order to be more effective. However, there clearly emerge several basis characteristics on the structure of the relationships in the market:

1. the distribution of productive specialisation relating to certain categories of financial instruments appears wide and complex ;

2. no operator has full specialisation on the entire range of financial instruments;

3. the space for complementarity between different operators and categories of operators is significant;

4. the operators that have a strong and intense ex ante differentiation in relation to the SME's needs tend to be specialised on a restricted number of financial instruments;

5. the operators that have a low ex ante differentiation in relation to the SME's needs, especially if belonging to the financial intermediaries

category, tend to develop the abilities to offer a rather wide range of instruments.

Analytically, the operative specialisation and the differentiation on the SME's needs tend to be centred on several guiding principles (see Table 1.6). First, the financial intermediaries have a superior ability of production and offer of financial services compared to the professional and mutual agencies. This is not however consistent, apart from a large part of the non credit financial intermediaries (closed-end funds and venture capitalist), with a specialisation structure focused on the SME's demand. Second, the consulting and support activities as well as the production of the guarantees appear to be "scarce components" within the market. Moreover, consulting appears to be an asset held by only few operators - mainly concentrated in professional agencies and non credit financial intermediaries - as a production instrument suitable for risk capital. Other evidence relates to the guarantees cluster, that belong to only one category of operators: the mutual agencies, and precisely the mutual guarantee consortiums that, in fact, present a productive specialisation in this area. An intervention by the professional agencies (consultants and data providers) is also possible if the provision of the guarantees has a greater immaterial significance and thus is considered as an autonomous "certification" of the company's credit standing.

Overall, the structure of the relationships existing between the financial system's operators and instruments signals the presence of significant trade-offs, whose characteristics and intensity generate important effects on managing the approach to the SME. In particular, the first trade-off shows a contrast between the broadness of the product range and the level of ex ante specialisation of the operator. This means that the greater the operator disposes of a broad ability on the offer side, the greater the focus on the structure of the smaller sized companies' needs is lost. The second trade-off is connected to the first one, and it compares the degree of differentiation relating to the SME, to the ability of controlling and governing the relationship with the company. In this case, a high ability to manage the relationship, or rather to satisfy the entire range of the company's needs, is not compatible with the focus on smaller sized companies. This is especially so when the companies in the system show an autonomously search the financial resources, and use, as a consequence, bank debt as the principal source of financing. As matter of fact, the greater the ability of the offer – on the products relationship management side - ex ante, typically for the banks, is opposed to a limited vision of the smaller companies' needs, the more the bank itself proceeds to a segmentation of the market and a differentiation of the production processes.

In conclusion, the first and the second trade-offs show the need of co-ordination for the SME's offer system in order to reduce and reconcile, where possible, the spaces of incoherence in the overall process. This generates, on one side, important spaces and opportunities of cooperation and the connection between expertise and different operators. On the other, significant risks of a differentiation between the entire SME's needs and the ability of an organised offer by the financial system. The co-operative element requires a contractual and productive intervention that defines the organisational model, the boundaries and contents of the co-ordination of the expertise in order to guarantee a strong ability of control of the relationships, together with a strong capacity of offering specialised services to the smaller sized companies.

The system of existing relationships between the financial system's operators, markets and instruments can also be observed with reference to the specific connection between market and operators, according to a complementarity logic and a complete vision of the operators/instruments relationship (see Table 1.7). This approach allows to highlight the presence of productive specialisation, i.e. the operators' ability to maintain the offer of services in oder for the companies to gain access to different market circuits, and exchanging instruments, either debt or risk capital. In this case as well, although it is necessary to proceed to an appropriate and successive comparison with country specific elements of the economic system examined, there emerge several basic structural characteristics:

1. the complete and exhaustive control, with reference to the model of access, the ability of offering adequate services and the market's entire spectrum, is an asset mainly of the non banking financial intermediaries (closed-end funds, venture capitalists, merchant banks) due to their productive specialisation in equity interventions;

2. the professional agencies perform, although differentiated case by case, a role of primary importance in the process of bringing the companies to the market: assistance, connections and networking, production of information and consulting;

Table 1.6. The relationship between operators of the system for the development of the SME and the financial instruments: the specialisation space

Operators of the system for the development of the SME	Instruments				
	Equity	Mezzanine	Debt	Consulting	Guarantees
Financial intermediaries					
Banks	▒	▒	█		
Credit intermediaries		▒	█		
Other financial intermediaries	█		▒	█	
Professional agencies					
Consultants				█	▒
Brokers				▒	▒
Data Providers				▒	▒
Mutual agencies					
Mutual Guarantee Consortiums					█
Business Angels	▒			█	▒
Foundations			█	▒	█

██ : Intense Specialisation – Core ; ▒▒ :Limited Specialisation – Activity Performed; ☐ : Activity Not Performed

3. the mutual agencies assume a marginal role due, largely, to their productive specialisation orientated to consulting and guarantees. Notwithstanding, a significant role can be covered by the business angels, due to the closeness of their activity and mission to the process of bringing the companies to the market, with reference to the entire range of markets but, in particular, to the non regulatory ones;

4. the banks cover a position of potential and general connection with the market system, due to the effect of their relational ability with the companies and the access to the different market circuits. However, this potential position must be included in specific categories of the intervention that are produced by the individual banks' strategic choices and by a more general orientation of the banking system, aimed at interacting with the stock markets, in terms of assistance to the company (large and SME).

The "market" analysis of the operators' activity in the financial system for SME completes and confirms the "instrument" diagnosis, highlighting and emphasising the managerial and process trade-offs previously underlined. Therefore, the line of production, distribution and acquisition of the debt with non-market characteristics show a strong internal coherence, specifically against a productive specialisation of the banks and of the non banking credit intermediaries. In this case, the strong relational ability and low requirement of interaction with the market circuits, plus the diversification of the system in the direction of instruments other than bank debt, generate strong loosening between productive specialisation, relations' management and market interaction. This causes an "institutional" efficiency and an effectiveness problem for the offer process at aggregated - and for the entire financial system - as much as at individual and single operator level.

The first point of view, in fact, explains the impact of the overall qualitative structure of the financial system on the allocation process of the financial resources with reference to SME. Vice versa, the second relates to the impact of the individual competitive strategy of the offer of the financial services to the SME. In both cases, the critical element and the questioning at the system's policy level is represented by the identification of the productive structure, by single operator and as a whole, that allows the increase of the technical-operative as much as the allocate and functional efficiency in the system for the SME.

Table 1.7. The relationship between operators of the system for the development of the SME and the financial markets: the specialisation space

	Primary Markets	Secondary and other markets	Third markets	Captive markets
Operators of the system for the development of the SME				
Financial intermediaries				
Banks	▨	▨	▨	▨
Credit intermediaries		▨	▨	
Other financial intermediaries	■	■	■	■
Professional agencies				
Consultants			▨	▨
Brokers	▨	▨	▨	
Data Providers	■	▨		
Mutual agencies				
Mutual Guarantee Consortiums				
Business Angels	▨	▨	▨	■
Foundations				

■ : Intense Specialisation – Core ; ▨ : Limited Specialisation – Activity Performed; ☐ : Activity Not Performed

1.4.2 The Value Drivers to Manage Specialization Towards SME and New Companies

The focus on the relationship between the system's structure and unitary intermediation costs and the ability of conforming to the SME's needs states a more general problem of governing the relationship between financial system and smaller sized companies, as it is an essential element in the configuration of the overall efficiency in the transfer of resources. In particular, the elements that emerge from the relationship's complete diagnosis among operators, markets and instruments shows the presence of three essential elements for governing the customer relationship:

1. governing the production specialisation;

2. managing the relations with the clients;

3. managing the relationship with the markets.

The heterogeneous distribution of the aforesaid elements between the different operators and the relative control produced, generate specific costs under the organisation and productive profile, at individual as much as at aggregated level. At the individual level, for each operator it is possible to observe that:

$$K_i^S = K_P + K_R + K_M \qquad (1.1)$$

where the complete cost for the offer of the financial services to the SME for operator "i" arises from the sum of: productive specialisation costs (Kp), that are a function of the area of products managed; the costs of managing the relations with the clientele (Kr); and the operating costs of accessing the market (Km)[11]. Vice versa, at the aggregated level, the complete cost of the process for the offer of the financial services arises from the sum of the costs of the offer process for the individual operators. This means that, within an interval of a pre-defined time period, the average unit cost of intermediation of the financial resources for the SME arises from the relationship:

---

[11] The model proposed has some simplifications: the possibility of being able to determine in an unequivocal manner, for each operator, the quantity of costs for specialised production, management of the relations and interaction with the market; the possibility of isolating and of correct attribution of all of the components of cost to each managerial area.

$$K_{unitary} = \frac{R}{\sum_{i=1}^{n} K_i^S} \tag{1.2}$$

where the numerator R represents the value of the financial resources acquired from the SME as debt and risk capital in the different possible contractual forms, and the denominator shows the sum of the costs for the operators within the financial system[12].

In analytical terms, the stylisation of the relationship between costs - per operator, aggregated and average unit - of the intermediated resources highlights a series of important elements of opinion:

1. there exists a strong and convincing similarity between structure of the financial system, characteristics of the operators present and overall cost of intermediation;

2. the competitive strategies of the individual operators, as they determine a specific combination of the components governing the productive specialisation and the management of the relations and interaction with the market, generate a cost that modifies at aggregated level the unitary cost of intermediation of the financial resources;

3. each operator must find a structure which is coherent with the three components governing the customer relationship;

4. the research of the coherence between the three components governing the customer relationship leads to internalisation choices of the productive activity rather than externalisation choices and thus contract banking. The first, although directly influences the function of the overall cost with reference to the productive specialisation K_P, can generate cost savings effects due to economies of scale. The second, while generating a certain cost determined by the contractual solutions adopted, can produce such "effects" as the reduction of the relational capacity with the customer;

5. the heterogeneous distribution of maintaining the product-market combination increases the importance for each operator of the choices to be

---

[12] The numerator R of the relation shown is a flow indicator in that it makes reference to the value of the financial resources acquired by the companies within an interval of pre-established time. For a summary of the criteria in determining the level of technical-operative efficiency of the system and of average unitary cost of the financial resources see: Fama 1991; Santomero and Babbel 1997; Tobin 1984.

made between the processes of agreement for external lines and internalizing the production activity;

6. the possibility of choice between external lines and internalization are not equal for all operators due to regulatory, supervision and operative constraints. This means that the selection process of the activities performed and of the contracts with other operators is restricted ex ante depending on the operator's nature.

In any case, the process of reading the cost structure governing the relationship with the clientele and the development of the dynamic interaction among different operators of the financial system cannot be performed without a further evaluation on the efficiency of the offer process. Moreover, the relation between the companies' needs and the models used to satisfy them must also be considered. In other terms, the evaluation of the overall structure of the system under the policy profile, must be widened from the technical-operative efficiency perspective to the allocate and functional one. Optimal positions in terms of average unitary cost of the financial resources may not be compatible with a system that is not as strong in the resources allocation and adequate instruments positioning in relation to the clients' needs. It follows that reading the individual operators' dynamic strategy and organisation under the internalisation and contract banking profile must be combined with the perception of the overall utility to the company. This passage must again be proposed at individual and aggregated level. In the first case, the cost K of operator "i" must be compared with the benefits produced to the clients; this means that the expected utility of operator "i's" offer for client "k" is equal to the relationship between the benefits received by the operator in the process of acquiring the financial instruments and the cost pro quota transferred to the client[13]. Considering equal to t the number of clients of operator "i", it is possible to identify the following elementary relationship:

$$U_K^e = \frac{B_K^e}{\dfrac{1}{t}K_i^S} \tag{1.3}$$

---

[13] The model shows several simplified hypotheses: an even distribution of the costs on the customers; the full transfer of the costs to the customer. However, it should be noted that the first simplified hypothesis results partly reduced considering the context under examination or rather the inquiry of the financial system for SME. This permits the reduction of the segmentation-differentiation process and the cross-subsidation to the system examined.

where the expected benefits represent a complete function whose independent variables are the value indicators capable of generating the perception of the products quality and the relationship developed by the operator[14].

Vice versa, in the second case, the client's expected utility in relation to the offer developed by the financial system as a whole is represented by the relationship between the benefits received from the financial system and the cost pro quota transferred to the client. In this specific case, if the denominator is made up by the sum of the costs transferred by each operator to the client. The expected benefits cannot be determined with the same calculation model, due to the effect of a possible external matching between financial services and products offered by different operators, so that the overall benefit perceived by the client is superior to the sum of the benefits perceived by the individual operators.

---

[14] See: Eiglier and Langeard 1991; Simon 1989. For a more general valuation of the relationship between indicators of value and perception of the benefit received see Norman 1984.

2 Competitive Models of Corporate Banking and Venture Capital

Stefano Caselli

2.1 The Evolution Process of the Strategies in the Corporate Banking Industry

The evolution process involving the banks started with the birth of Euro market for financial activities. It also profoundly modified not only the ownership structure and its relative size but the production logic and the relationship with the demand system. The starting point, constituted by the combination of loans and deposits, underwent a profound evolution in the relationship with the market both for the families and for the companies and institutional operators. With reference to the first, the overlapping between the "deposit" element and the "management of savings" requirement was progressively loosened and then broken in favour of a wider, and therefore more contested, structure of the product–market combination. This new structure comprises the entire system of financial, insurance and pension requirements of specific families of users, sub-divided in relation to the size of the asset and the quality of the requirements signalled. With reference to the second, the banks' investment function as an exhaustive channel in satisfying the clientele has gradually changed its relative positioning compared to the growing demand of support services by the companies – advisoring and financial consulting – and towards a general widening in the range of financial products in order to satisfy the companies' requirements. Here, the growth in domestic and international stock markets and the greater mobility of the producers of specialised services has put in discussion the concession of loans as the only occasion for satisfying the companies' needs, thus provoking a process of deep change within the banks.

2.1.1 General Overview

The corporate banking development was born in the European banking system with reference to the pressure arising from the companies' market in terms of widening and qualitative growth in the range of the offer and available expertise. This is not so important for the large companies, whose ability to deal with the financial system is autonomous, equal and some times open to the international dimension, but especially for the large quantity of SME present in our economic system.

Although the main stimuli and objectives are clear, there still does not exist today not only an unequivocal definition of corporate banking but especially a basic application reference model. The situation seems much more complex if the offer is extended to the venture capital activity. The common factor to the different experiences is no doubt represented by the fact that corporate banking is defined as "a complex system of the offer of financial services and non, to a group of pre-defined customer companies". But it is the definition itself that opens the debate on the implementation process of the corporate banking, which requires (ABI-Prometeia 2000):

1. an ex ante choice on the typology of services that must be produced and sold within "the offer system", in order to ensure the satisfaction of the counterpart and the economic return for the bank;

2. a definition of the integration model of the sales processes for the individual products, in order to ensure the "completeness" of the service;

3. a specification of the criteria that allow to identify the segment of corporate customers.

The first interrogative is not limited to a simple choice from a catalogue of products but a wider selection of the bank's strategic positioning. The prevalent maintaining of credit products consents the bank to continue operating within the classical logic shared by the commercial banks. Vice versa, the widening of the products towards the corporate finance or financial risk management areas requires a strategic choice in the direction of merchant and investment banking, resulting in a clear transformation of the bank's business system (Kuhn 1990; Liaw 1999). The second interrogative entails defining the processes, structures and organisational roles that aim at increasing the satisfaction of the customer company and permits the bank to transmit the range of products available in a co-ordinated and profitable manner. The third issue requires the choice of existence criteria for groups of companies – the corporates – which must underline a different logic approach one from the other. This results in the bank declaring which companies represent its target market. In this direction, the simple size criterion, apart from not been unequivocally and easily determined (turnover, number of employees, overall activity, inclusion of participations, etc.), can be misleading due to the typology of territory and market served by the bank. Moreover, the concept of SME presents a significant quantitative extension and a strong internal differentiation.

2.1.2 The Evolution Path of the Corporate Banking Models: a Four Steps Approach

In relation to the aforesaid interrogatives, in the Euro banking system it today appears possible to identify a medium term development path in which the successive phases – partly taken and partly still to be undergone – are characterised by: a growing degree of the bank's productive diversification; a stronger autonomy (formally and substantially) by the organisational division that governs the market relationships with the corporate clientele (see Fig. 2.1). The two mentioned parameters therefore represent the essential criteria that allow a formalisation in logical categories, whose complexity is explained by a much higher and articulated number of relationships, conditioning and organisational characteristics. However, indicating a development path does not force to an obligatory direction, not only in the formal sequence of the phases illustrated, but in particular in entrering the path itself. This means that the choice of progressive growth and internal organisation of the bank's corporate system may also result in the deliberate exit of the business division. This position is inherent in the concept of the general bank, where the possibility to operate in a diversified manner in the different product–market combinations results in strategies of active and conscious selection rather than exit ones[1].

The first development phase characterises all the banks embarked on the divisional path by business area. Today, it is still common to the small and partly medium size ones that have given less attention over time to the relationships with the companies. This was usually due to a critical mass of contained resources, or because the banks attributed a low priority to the system of relations with the companies, as a direct consequence of explicit strategic choice in the medium term. In these cases, i.e. the phase of the "movement of opinion and of the growing organisational state" (Caselli 2001), the corporate banking is described not so much as a phenomenon and production process but rather as a "psychological tension" of the structure. Here, the management can also get to the formal creation of a corporate banking division with the general objective of window dressing and of developing a greater proactive – or aggressive – attitude towards the medium-large sized companies by adopting equal production processes. It follows that the credit function retains its organisational and product centrality, and the sales co-ordination tends to be managed by the commercial ability of the available resources.

---

[1] See Schwizer 1996; Mottura 1994; Forestieri 1993.

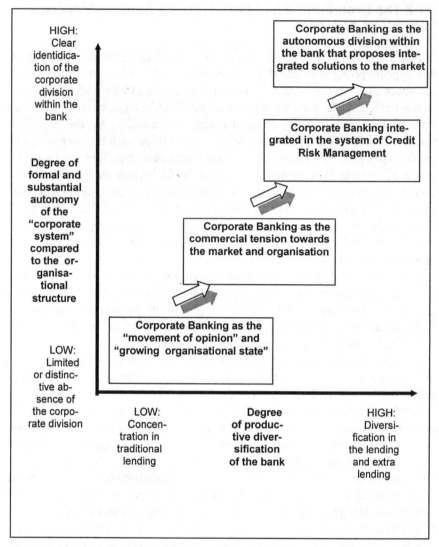

Fig. 2.1. The evolution path of the corporate banking models and the map of the banks' choices

The second phase enriches the previous one as the "growing state" originates a tangible organisational response due to the awareness of the need for a significant commercial drive for the products available. Hence, the "commercial tension" phase combines identifying a group of target customers and formalising an ad hoc organisational structure – the corporate division – that co-ordinates the commercial efforts with the objective of maximising the business volume. In this case, if the organisational response gives rise to a legitimate and proper existence, in the medium term the basic assumptions on which this response was founded appear weak. This can be analysed with reference to two distinct aspects.

First of all, the logic of the general approach is incremental and not discontinuous compared to the market policy of the credit department, in terms of expertise, structural organisation and operating choices. As a matter of fact, the resources largely, and often entirely, derive from the credit department itself, according to a process of transformation of the professional objectives and enrichment of the positions through an education service prevalently related to the commercial and sales activities. The organisational positioning in the corporate division is however ambiguous, because if addressing a wide segment of companies legitimates its proactive value towards the market, the limited definition of the co-ordination logic with the functional structures may not only generate problems in the organisation functioning but also deprive it of market credibility. Hence, the case of banks in which the responsibility of the customer companies is attributed to the corporate roles but the decision making powers remain within the credit department, appears significant. Moreover, the problem of the operating choices must also be considered, as the process of annual budgeting and the objectives establisment privilege the lending volume compared to other product areas, thus precluding the growth stimulus in the contact and relationships roles of a real sensibility in analysing and completely satisfying the companies' needs.

Second, the diversification process is partial and episodic as the sale of credit products remains preponderant compared to the other products offered, often considered as experimental or exploratory by the bank. Typical of this situation is the choice of numerous banks to create in staff in the corporate division, or within the commercial area as a whole, a "corporate finance department" or a "merchant & investment banking division". The common traits to these organisational solutions are, generally, the limited number of resources involved, the experimental character of the initiatives and the unclear specification of the relationship with the other bank's organisational structures. It follows that, even in the presence of a clear commitment by the bank's top management, it is difficult for the aforesaid structures to succeed in creating a substantial space for intervention in a significant number of companies and in terms of visibility and reputation towards the market.

The third phase introduces in the evolution process of the corporate division the system of values matured in the context of the overall credit risk management planning. This does not necessarily constitute in itself an evolution third step compared to the previous, as even the presence of a growing state can be coherent with the maturation of a complete system of credit risk management. However, the empirical experience that emerges from the Italian banks illustrates that, especially due to temporal order and organisational and cultural maturation motivations, the integration between the corporate and credit risk management functions occurs after the organisation of a functioning corporate division (Cairns et al. 2002).

The merge between the system of commercial values in the corporate division and financial ones in the credit risk management generates important production consequences for the bank. This can be observed with reference to the strengthening of the ability to control the lending activity, and especially to the growth in the potential capacity of originating those operations that are not necessarily lending, due to a more precise and incisive evaluation of the company risk. However, the stable and medium term valorisation of the aforesaid strong points relates to the presence of two factors of success. The first inevitably refers to the degree of effective diversification achieved in the bank's offer, so that its origination potential is transformed into a substantial and continuous deal flow. The eventual low ability of satisfying the requests originating from the credit risk evaluation activity precludes not only the business development but may generate dangerous side effects, such that the scarse company's satisfaction also affects the lending activity. Second, the refinement of the ability to evaluate the credit risk must be coherent with the utilisation, in the customer relationship, of the aforesaid judgement, and with the degree of acceptance within the bank's organisation structure. Therefore, only in the production logic of an internal rating, founded on the development of a bottom up schema compared to the centralising the external rating, there exists a significant and responsible notching, which makes the creation of a commitment between corporate roles and companies possible. Vice versa, the recourse to prevalently top down schemes generates dangerous opportunities for a conflict at the commercial negotiating level and reduces, with all probability, the degree of commitment of the corporate roles to the organisational structure. This second choice can be compatible with the general bank's orientation towards addressing a low priority to the entire corporate area, to the advantage of other market sectors.

The fourth phase finally constitutes the target of the corporate division evolution path, as the divisionalisation of the corporate system within the context of the bank's production structure is completed both in strategic and operative terms. First, the corporate division – or Business Unit – is described as the centre of the autonomous and recognised management in the relations with a segment of customer companies, considered as a priority for the bank. This results in a precise choice with reference to the weakness points in the previous phases and a consequential modification "for discontinuation" of the organisational structure, the roles involved and the management models. In support of this, the explicit declaration to enter in the market of relationships with the SME as a priority production centre in the bank's performance. Hence, this phase is the final point only for those banks whose top management considers the business corporate division critical for the overall performance.

Analytically, it is possible to observe that those banks oriented towards reaching the aforesaid phase, build their corporate banking activity in relation to four management themes, or development vehicles. These represent the most effective key to reading and interpretating the changes and evolution in course in the system of relations between bank and company (Geisst 1995). In other terms, they are the planning elements that should drive the investment activity of reorganisation and substantial intervention in defining the offer operating mechanisms. These four themes are:

1. the organisational and strategic structure of the corporate banking division;

2. the organisational roles and the expertise involved;

3. the credit risk management and company valuation models in a broad sense;

4. the market positioning.

The organisational structure represents the cornerstone of the macro choices related to placing the corporate banking within the bank, and the models of interaction between the offer (which products lines?) and demand (which segment of customers?) system. The roles and expertise theme defines the intensity and effectiveness with which the process of understanding and satisfying the needs must be developed, resulting from the co-ordination in the products sales. The credit risk management and company valuation models represent the base of knowing a company, and therefore the activating factor of any other relationship process with the company. Finally, the market positioning theme is the assessment and validation of the previous steps as it defines, with reference to both the choices of the medium term period and operating strategies for managing the client relationships, profitable results coherent with its production structure. It follows that the quality of the company's credit diagnosis affects not only the quality of the financial services but also of the performance obtained by the bank from the company.

2.2 The Design of the Corporate Banking Structure

The development path of the Euro banking system towards a relationship with the companies organised and managed on the basis of a corporate banking oriented approach appears today strongly lacking internal homogeneity, Moreover, it is still largely based on the credit component. However, from the second part of the '90, we assisted to a significan grow in the interest and concrete implementation of the offer processes and of organisational solutions in order to widen the portfolio of the financial services proposed to the companies. Most of the interest in the corporate banking division was developed within the banking system and much less at the level of specialised and dedicated operators. This is due to the fact that the orientation towards the banks, typical of the continental financial system, combined with the presence of a universal model and of the marginal role covered by specialised intermediaries, in contrast with the Anglo-Saxon experience, places the banks in the exclusive position for developing the Italian corporate banking for the SME[2].

2.2.1 Organisational Pattern

The theme of defining the organisational structure represents, as has been previously analysed, the first development vehicle of the corporate banking for all the banks that embark on an effective planning of the production processes and of the customer relations based on the presence of a Business Corporate Unit. The emerging tendencies highlight, as a fundamental solution to the organisation of solid relationships, and also valid in the industrial system, the implementation of a process of divisionalisation of the organisational structure. This results in the preliminary distinction of the segments of clients served and the consequential sub-division of the organisational structure in order to define productive and managerial solutions in line with the needs expressed. It follows that the presence of at least three macro areas of customers, i.e. private, retail - either private or companies - and companies, allows the bank to build three different business divisions in order to maintain each market's effectiveness. In general, they are called Corporate, Retail and Private division. This process belongs to the logic of strategic planning that starts from the "geographic area, typology of client, product range" scheme, as its discriminating element for organising the market strategies is the "typology of client" parameter. It represents therefore one of the possible choices for the European banks that, operating as

---

[2] Saunders and Walter 1994; Walter 1988; Cesarini 1994.

de-specialised intermediaries, and however focused on the entire market of privates and companies, must necessarily distinguish and organise the retail from the corporate oriented production processes. If, in fact, a mixture of them is maintained, it becomes difficult for the bank to obtain the minimum objectives of proper functioning of the production structure, i.e. the exploitation of productive capacities, the production of economies of scale, the control of the resources employed. Quite similar is the path undertaken by the German and French banks that, with the same starting fundamental characteristics of the Italian financial system, develop an intense process of divisionalisation by clients segment, on which is engaged a greater focalisation of the offer of the corporate banking services.

The realisation of a Business Corporate Division is the best operating solution coherent with the development of an offer oriented towards a partnership and problem solving logic of the companies' financial needs. Hence, the more the bank is able to propose itself as a reference point for assisting and supporting the corporates' financial choices, the greater will be its image, perception and effective positioning as a "relationship" and "home bank". The aforesaid competitive model is based on four main aspects, that must necessarily happen simultaneously and in a structural way:

1. the repeated matching and satisfaction between the company's system of the needs and the bank's system of services offered;

2. the high degree of co-ordination of the services offered, due to the presence of dedicated organisational structures;

3. the high degree of continuity of the exchange process between bank and company over time;

4. reciprocal, although not formalised, commitment to consolidating the relationship in the medium term as the added value element.

In relation to them, it is in fact possible to identify different approaches: the traditional commercial bank model, strongly transaction oriented; and the home bank corporate oriented one, centred on the competitive importance of the relationships, i.e. the ability of assisting the companies. In the first case, the limited broadness in the range of services offered depends on the ex ante small degree of co-ordination between the offer and the exchanges continuance. Vice versa, in the second one the high diversification of the bank's portfolio services is due to the strong operative correlation and high degree of offer co-ordination and continuance[3].

---

[3] See Canals 1997; Walter 1997.

The bank's affirmation of a partner model requires its simultaneous satistaction of the four aspects mentioned, as eventual unbalanced development paths have a negative impact on the efficiency and effectiveness of its competitive positioning. For example, solutions characterised by a strong diversification of the bank's product portfolio and a low level of co-ordination do not increase the added value of the customer relationship, thus limiting the possibility to provide a fully personalised service. Moreover, a low level of continuance of the exchange process, associated with a high diversification of the products and a significant degree of co-ordination of the offer, reduces the opportunities for the bank to intervene in the transformation phases of the company's business life, thus compromising in the medium term the stability and profitability of the customer relationship.

Realising a complete company approach, in terms of developing simultaneously a matching with the company's needs, the offer co-ordination, exchange continuance and medium term commitment to the relationship, must however be supported by the control of significantly different management expertise and technologies.

With reference to matching the company's needs, the wider the range of the bank's offer requires the availability of managerial and sophisticated technical-productive competencies, that are completely different from the traditional credit intermediaries' ones. Hence, the offer of products containing consulting, special financing or capital market services can be developed only in conjunction with the use of specialised resources, that at the same time possess a high knowledge of the particular product, and allow the offer to be in line with the client's needs. However, the widening offer range does not necessarily produce a symmetrical increase of the production capacity: the specialised products can in fact be realised in specific product companies and afterwards be distributed by the bank, that controls the sales process to the customer.

With relation to the co-ordination of the offer, the possibility to intervene in the market in terms of providing systems of services that are not overlapping among themselves and which are in line with the company's requirements is undoubtedly important. However, this must correspond to an intense development of the interface and management resources of the customer portfolio and an effective planning of the information systems in order to follow the continuous evolution of the companies' needs. The co-ordination of the offer is also directly correlated to the continuance of the exchanges over time, i.e. the bank's ability to satisfy the quantitative growth of the needs, without an evident discontinuity in the entire financial flows circuit generated by the company's real activity. This is strictly related to the availability of timely and flexible instruments of intervention and the capacity of the contact and management roles to strengthen their participation within the company.

Finally, with reference to consolidating the relations in the medium term, the prospective of giving rise to continuous dealings, rich of commercial opportunities, is founded on the counterparts' loyalty. This element has separated and characterised for long periods of time the concept of "relationships" orientation as conceptually opposed to the "transactions" one, typical of the traditional commercial banks. However, in reality the aforesaid opposition is less applicable to the concrete banking activity as the relationship content of the deals and the consolidation tension of the relations must refer to all clients' segments, as the minimum condition of existence in the market. The differentiation between segments refers to three different parameters that specifically characterise and distinguish the corporate from the retail division approach:

1. humanistic requisites;

2. professional requisites;

3. contractual requisites.

The first elements in the value creation process refer to the human profile, the standing and availability of the resources involved in managing the relations. This means that the corporate division's contact roles must be covered by people who are oustanding not only for their strong attitude towards communication and the construction of a trusting environment, but must also possess the indispensable qualities for realising complex negotiations, e.g. discretion, confidentiality, assertiveness, timeliness in the search of solutions and the ability to place the customer's needs at the centre of the production process. It follows that the bank's recruitment process must stress these parameters employ the best and identifying the resources with the highest potential.

The professional requisites underline the market expertise element of the corporate resources. This factor is frequently mistaken with the not well defined "consulting orientation", that should characterise the attitude of the bank's relations towards its corporate clients. The vagueness of this statement must indeed be solved with a specific analysis of the professional content of the contact roles in relation to the customer. This must be connected to the entire corporate production process, as maximum coherence must be sought between the typologies of diagnosis of the client's manager and the available solutions in relation to the bank's offer system. The gap between the capacity of diagnosis and the offer possibilities causes not only frustration but also undermines the bank's credibility. Vice versa, when the bank's possible solutions are more than the ability of diagnosis the client manager's position is undermined as it does not allow the full exploitation of the bank's available production capacity. This may be largely penalising in the start up phase of new product industries and when developing product areas as it slows down or precludes the possibility to achieve the break even point in a sufficient time period.

Finally, the contractual requisites refer to the product typology proposed by the bank as the contractual specificity of different financial services significantly affects the possibility of increasing the exchange commitment, loyalty and continuance. This can be verified under two different aspects. First, the inner characteristics of each product affect in different ways the degree of interaction and interdependence between bank and customer company in the medium term: the participatory intervention, medium and long term credit for big ticket operations and global insurance risk management services are complex – as the bank's professionalism is verified – but also binding contracts over time due to the nature of the rights incorporated and the importance of the financial services within the company's management system. Second, if the products share equal contractual specificity, the characteristics of the collateral and packaging proposed by the bank distinguishes its attitude towards the development of a trusting environment. Here, decisive indicators are the indiscriminate or calibrate use of guarantees, the style of imposition of covenants, the transparency of the services pricing and the rigidity in the contractual conditions proposed.

2.2.2 Positioning and Functioning

The realisation of a distinct corporate banking division within the bank responds to the challenge examined under three different profiles as the division:

1. represents the core resources managing the entire company relationship according to a unitary and global approach;

2. constitutes the company's clear reference point for formulating its complete range of financial needs;

3. has the necessary authority and organisational expertise in order to formalize specialised products - internal and external to the division - able of satisfying the customer's needs.

In summary, the corporate banking division performs those functions: external interface for the bank, as it is able to manage the complete commercial, information and personal relations with the company, proposing coherent solutions to its entire potential "financial problems"; internal interface within the bank, because it is able to acquire and co-ordinate the necessary expertise and products (see Fig 2.2). Figure 2.2 shows the direction and the intensity of the banking production process: it starts with the diagnosis of the company's needs in the corporate area, follows with the shopping phase developed within the bank to the corporate division, and then ends with an assembly of the products and expertise, in order to finalise the proposal and implementation of the appropriate solutions.

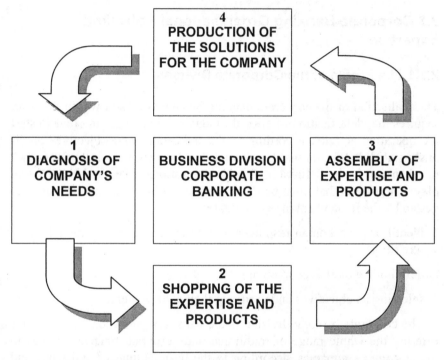

Fig. 2.2. The positioning and the functioning of the corporate banking division within the system of providing services of the bank

During the implementation phase, the corporate banking division model can assume different organisational schemes due to the overall bank's organisation–company structure and the management's decisions. In particular, the banks following the universal bank and centralisation model within their production functions tend to develop the corporate banking division as an internal business unit. Examples are Unicredito Italiano, San Paolo IMI, Credit Agricole, BNP Paribas, Deutsche Bank or Dresdner Bank, who develop their activity through a customer segmentation process and corresponding internal divisionalisation by business area. Vice versa, the banks oriented towards a group structure and whose production functions are externally arranged tend to develop theit corporate banking division as a centre of services – a link which is "external" to the bank but "internal" to the group. Significant are the cases of Banca Intesa BCI, MSBC and Santander, where the presence of production centres for group companies requires a "company centre", able to co-ordinate a corporate oriented offer system.

2.3 Corporate Banking Organisational Roles and Expertise

2.3.1 The Profile of the Corporate Division

The realisation of the corporate banking business division within the bank requires the identification of critical fundamental passages in order to start the operating activity, according to the indications and objectives previously stated. It follows that the second development vehicle of the corporate banking is the definition of the organisational roles and expertise in play. The phases that must be developed in the business division, as suggested by the Italian banking system, are:

1. identifying the human resources and defining the organisational structure of the;

2. defining the market positioning;

3. selecting its objectives and verifying their achievement.

The corporate banking division is the bank's organisational solution for offering the whole range of traditional and corporate finance services to the customer companies, according to the basic culture of assistance and partnership. The first significant step in building the division is identifying those criteria in order to distinguish its profile within the whole the bank's organisational structure - and thus also the retail's and, whenever present, large corporate and institutions division's one. In particular, the indispensable factors for creating and maintaining the corporate division are as follow (Rappaport 1997):

1. the division operates within a clearly defined market under the geographic scale or typology of company profile;

2. the division operates exclusively in the chosen market and thus does not represent a "supplier" within the bank but rather its external "competitor";

3. the division is a distinct, separate and clearly identifiable unit within the bank. It must be possible to define a strategy, market policies and independent management of the material and immaterial resources. These, in general, must not depend on the existence of other centres or units within the bank;

4. the management of the division has the full control on most the aspects that affect its success, including marketing, production and management of the resources.

It derives that the optimal and necessary placing for the business division within the bank's organisational structure is that of direct dependence from the general management. This appears to be necessary in order to legitimate its structure, guarantee an autonomous strategy and management control, identify an appropriate offer system together with the customer company, that is reliable and considerably differentiated from the other available channels, e.g. the single subsidiary. Vice versa, the bank's choice to position its corporate division under the control of the commercial management weakens its functions as the credit risk management probably remains a domain of the credit department.

The preliminary identification of the necessary organisational roles within the corporate division may start from an evaluation of the unit's production cycle in offering a service. The diagnosis of the company's needs and of the management of the services offered requires the presence of an organisational role in the customer management as a linkage between demand and offer, thus guaranteeing the maintenance and effectiveness of the exchange. The activity of organising and internally assembling specialised products with differentiated expertise requires a coaching role that, given the objectives of satisfying the client's needs, knows how to manage and co-ordinate the division's production activity. The activity of providing corporate banking financial services finally requires specialised products, placed internally within the division or externally in specific company-products, that characterise their intervention in relation to the know how matured in managing specific financial risk management, asset management or corporate finance environments.

2.3.2 The Microstructure of the Corporate Division: the Organisational Roles and Expertise

The valuation of the organisational tasks in relation to the production process of the corporate banking business division allows to highlight three important roles (see Table 2.1):

1. the client manager (or corporate banker or company manager), i.e. the relationship and customer manager;

2. the category manager, representing the specialistic product;

3. the professional, i.e. the consultant–professional in the environment of cross expertise.

The client manager is the key organisational role in carrying out the customer relation, coach and internal shopper functions, i.e. the manager of the assembly process of products and expertise, in order to guarantee the satisfaction of the customer's needs. He therefore covers a critical interface and connection position between the bank's production process and the company's financial function, assuming the responsibility and governing the performance of the entire relationship[4]. Therefore, interpreting the client manager's organisational role requires an expertise in managing relations, organising resources and analysing the company's needs. The client manager's expertise must be structured, although with different intensity, with reference to: the "industrial" knowledge of the company, for a correct evaluation of its credit risk; the knowledge and diagnosis ability of the company's financial needs; the knowledge of the corporate banking financial products; the ability of activating specialised interventions from the category manager and the professional; the ability of managing relations.

The bank's "industrial" knowledge refers to the ability if analysing the company's competitive position and evolution. This means developing analytical skills in the company's sector, market strategy and economic and financial structure in order to be able to appraise correctly its credit merits and competitive positioning. The aforesaid knowledge should therefore be "sound", as it is the necessary information ground for understanding the company globally, but not specialised, as the client manager must not get, for example, to a final evaluation of the company's worth. As a matter of fact, specialised expertise on company valuation must be performed by category managers or professionals.

---

[4] On the client manager's role and functions see: De Laurentis 1997; Caselli 2000.

Table 2.1. The tasks and the managerial objectives of the organisational roles of the corporate banking division

The organisational roles	The tasks and the managerial objectives
HEAD OF THE CORPORATE DIVISION	• Defining the strategies and directing the business division activities • Managing the marketing strategy of the Business division or rather the product, price, distribution and communication policy • Co-ordinating the client manager's activity and assigning the portfolios of customer companies • Link between the business division's and the entire bank's activity • Link between business division and general management • Supervisioning category manager's and professional's activity
CLIENT MANAGER	• Managing the portfolio of companies assigned by the bank and no longer governed by traditional roles in the subsidiary or credit analysis • Pricing the service based on the guide lines and autonomy fixed by the director of the Business division • Exclusive governing of the entire relationship with the single customer company over time • Developing the diagnosis activity of the customer's needs and of the complete credit risk in the relationship • Activating the category managers' and professionals' action based on the individual customers' needs • Managing and co-ordinating category managers' and professionals' action with the customer • Relations with the head of the Business division

Table 2.2. (cont.)

The organisational roles	The tasks and the managerial objectives
CATEGORY MANAGER	• Maintaining the technical expertise of the products attributed • Relations and support to the client manager's activity with reference to his specific expertise • Intervening with customers based on the client managers indications • Provisioning and managing the service or financial product requested by the client manager • Provisioning and managing the service or financial product towards the bank's entire organisational structure
PROFESSIONAL	• Intervening with customers based on the client manager's indications • Provisioning the specialised cross expertise for realising the financial product requested by the client manager

The knowledge and diagnosis ability of the needs is related to the client manager's know how, developed by area of the company's "financing problems". However, the needs extend in different graduations: from a purely financial to a more industrial context, such as the assistance for internationalisation; from a structural to a business character, e.g. the Euro needs, produced by the introduction of the new currency and extinguishing in the medium term. Vice versa, the knowledge of the financial products is the natural completion of the grid of financial needs as it allows the client manager to match solutions to problems. Therefore, the appropriate product knowledge requires the ability to combine the contractual aspects with the economic, fiscal and financial ones. This is why the profile of a client manager's expertise must not follow the product specialisation strategy but a "downstream" perspective, in order to maintain the general analysis and management ability. Hence, the recourse to specialist interventions is a natural step when correctly organising the expertise in the evaluation of the company's needs and financial products, as only a sound and systematic connection between problems and solutions allows the client manager to conceive the appropriate solutions. Moreover, the client manager's ability to "manage the work team" and to conduct and guide effectively the intervention of the resources of specialised product expertise is very important. From the client's point of view, the client manager mus present the specialist intervention not as a single transaction but as part of the whole relation with the bank.

Finally, the relationship ability with the client is a natural summary of the expertise mentioned, plus the specific and distinct skill to communicate or rather manage effectively the communication with the company. This is necessary whereas the client manager must have a "strong hold" on the client, functional to the correct credit risk valuation and fundamental for originating and developing the Corporate finance operations.

On the other side, the category manager is the organisational role presiding the technical expertise and product management. He therefore participates in formulating the client's solutions, interacts with the client manager and the client - if the kind of products need it, but does not manage the client relationship. This role may be internal to the bank or, where sufficient expertise are not present, external, positioned within group companies or a network belonging to specific companies-products. In this latter case, the category manager covers the gap in expertise that the bank is not able to insource. It follows that, independently of any organisational positioning, the number of category managers depends on the product industries the bank uses for organising the offer process. Running the corporate finance, asset management or financial risk management areas results in identifying the category managers or teams of category managers focused on managing single product industries.

The professional brings transversal expertise for realising the corporate products. In particular, he is specialised in legal and fiscal advisory and consulting, and his intervention is activated by the client manager in relation to the client's needs. Therefore, he brings only his consulting and professional expertise and, as for the category manager, may also have an external position to the bank. The creation of a further organisational category after the category manager one seems to be necessary due to the nature of the service offered, as the consulting activity is seen as the connection between the specificities of the financial products and the strengthening relationship and commitment between bank and client.

The marked differentiation of the roles and the requirement to distinguish among the operating profiles in terms of expertise, attributes particular significance to the process of recruiting and educating the human resources (Fabrizi 1998). This is not so much critical for the category manager and professional, where the direct belonging to a technical product or consulting area clearly determines the necessary professional profile, but for the client manager, who combines specific relations, products and origination characteristics. In this latter case, the selection criterion must be based, in the start up phase of the division, on researching resources showing a high know how of company valuation and significant relational and commercial abilities. As a matter of fact, it is the client manager who permits the realisation of the whole process within the corporate division: the diagnosis of the company's needs produces the organisation of the financial services and operating solutions. Therefore, he must respond as much as possible to the following requisites:

1. significant experience matured within a bank, i.e. for a sufficient period of time;

2. perspective of sufficient permanence within the bank;

3. highly accordance with the project;

4. prestige within the bank and by the management;

5. prior "field" experience in the credit department, especially in the management of loans and companies' relations;

6. high expertise in the company valuation and management of the financing process.

These requisites highlight those human resources with a significant experience at the management or operational level, with particular reference to customer companies, a skilled knowledge of the entire management process of the bank's credit, and an understanding of the companies' territory, needs and characteristics. However, the selection process of the resources, once the corporate division is in place, must be based on the employment of young people with high growth potential who, working together with the client manager, gradually acquire the relevant expertise for managing client relations.

The number of client managers must be determined in relation to the total number of target companies of the corporate division. As a matter of fact, each portfolio must take into account:

1. an appropriate number of visiting days and contacts with each company;

2. the maintenance of the "proximity" and strong ability to intervene in the company;

3. a sufficient number of days for managing the division's back office;

4. the management of the team of category managers and professionals for reaching the best solutions to the company's needs.

This means that the approximate number of companies per client manager can be fixed at around 70-90 units, depending on the geographic area concerned. Many European banks in the last years have shown scarce commercial effectiveness, in terms of problem solving issues, where the client manager portfolio comprehended more than 100 units.

2.3.3 The Plain of Activities

The presence of different organisational roles and figures within the business division provides a strong internal coherence and a high competitive effectiveness to the structure, so that: a portfolio of customer companies is assigned to the client manager; the presence of category managers, internal or external to the bank, guarantees the maintenance of the necessary product expertise in order to satisfy the clients' needs; the possible recourse to the professionals allows to complete the set of technical expertise for realising the corporate services. Therefore, the corporate division acquires a unitary direction under the guide of the head of the client managers, who defines the market strategy and management policies and responds of its decisions directly to the general management,.

The actions of the corporate division in the market must tend to the realisation of the base strategic objectives, i.e. affirming the bank as the reference partner for the solutions of the target company's financial needs. The development of a partnership model with the companies is a medium-long term objective, that permits the bank to modify the the production process of its economic result. This is verified where the client relationship based on a corporate relationship logic produces for the bank:

1. the expansion of the intermediary margin, as the range of services offered is widened beyond the boundaries of the traditional lending towards fee based products. Moreover, the bank's function goes from a simple provider of financial resources to the manager of financial products;

2. the consolidation of the interest margin, as the corporate lending activity for the company is entirely performed by the bank;

3. the stabilisation of the operating result, as the closer proximity to the company and more intense intervention ability allows to forewarn and manage pathology situations and possible crisis well in advance.

STRENGTHENING OF THE CUSTOMER RETENTION		
COMPLETE DEVELOPMENT OF THE CORPORATE AND INVESTMENT BANKING SERVICES	LONG TERM DEVELOPMENT OF THE DISTINCTIVE CORPORATE RELATIONSHIP ABILITY	REDUCTION OF THE FINANCIAL AND CREDIT RISKS
	INCREASE IN THE BUSINESS DIVISIONS' VOLUME AND MARGIN	

Fig. 2.3. The structural objectives of the Corporate Banking division within the bank

Therefore, the creation of the business division is justified and sustainable only if its actions in the market produce an important impact on the quality of the bank's income statement, i.e. if it produces incremental value compared to the traditional management organisational structures of the companies' relations. Hence, the basic objective must be realised through pursuing short-medium term strategic and operational targets, that allow to orientate and verify the effectiveness of the business division's actions over time. In particular, the aforesaid objectives are as follow (see Fig. 2.4):

1. complete strengthening of the customer retention;

2. development of product industries for the SME (traditional and innovative lending, corporate finance, asset management, etc...);

3. growth in the volumes and margins of the different business divisions served;

4. reduction of the assets's risk through the control of anomalous receivables, delays in payment and losses.

The objectives' evaluation identifies the economic significance and competitive importance of the customer relationship. The development of the relations activity means involving the company in the exchange process, thus limiting the possibility of exit. Therefore, it allows to carry out a fundamental activity in maintaining the existing customers, because they are considered as a valid and strategic resource for the bank's success. It must also be directed towards developing product areas which are not the bank's domain, thus searching new profit sources. This is verified if the bank's offer is a problem solver, able to activate appropriate financial solutions to the customer's needs, independent of its internal production ability. Finally, the involvement between bank and company in the client relations context permits the bank to increase the quantity and quality of information available on the counterpart. This is the fundamental starting point for improving the know how in terms of credit evaluation and thus complete efficiency in the corporate banking management activity.

2.4 The Market Positioning of the Corporate Banking Division

2.4.1 The Importance of the Bank's Size Profile

The choice of the corporate division positioning requires a complete evaluation with reference to analysing the market structure and verifying the available expertise for the strategies' implementation. This means proceeding to a study of the competitive environment for each product–market combination, typical of the segmentation process. Moreover, it is important to focus on the critical organisational factors, resources and procedures for the success of the target product–market combination for the offer policies. It follows that the evaluation process of the alternative strategies must lead the bank to a high level of internal coherence - in terms of the relations linking the environment to the business strategy and organisational structure - and of competitive effectiveness, i.e. the ability to produce a good market performance and maintain the competitive advantage obtained with reference to the competitors.

The valuation of the alternative strategies and successive portfolio composition of the strategic business areas for achieving the best positioning of the bank's offer depends on two parameters: the typology of clients served and the financial services offered[5]. Therefore, the process of market segmentation is significantly affected by the bank's size and market. This is due to the fact that depending on the type of operations, not only the boundaries and content of the large corporate & institutions, corporate, retail and small business divisions are different, but also the relative advantage of maintaining the divisions themselves and the underlying economic mix for achieving the expected performance.

The analysis of the main international experiences in the financial services sector identifies different organisational models in the portfolio of client companies' segments, and thus several choices for the bank's positioning (Forestieri 2000). Although it appears difficult to categorize the behaviour of the financial intermediaries, given the variety of economic

---

[5] This recalls the model "Client – Arena – Product (CAP) Matrix" developed by Walter 1988.

environments, institutional structures and characteristics of the demand, it is useful to point out the basic tendencies in place and the most common solutions, highlighting their distinctive traits (Hunts 1995; Walter 1997).

The first trend is the developing global strategies. The financial intermediaries widen the offer function outside their original market's boundaries proposing financial services to companies in the entire international market. This leads them to compete in a broader environment with financial intermediaries, if present, operating locally in the different countries. The development of global strategies requires (and produces) a significant mass of financial and non-financial expertise as well as the achievement of economies of scale from the simultaneous maintenance of the entire spectrum of relevant expertise and the demand led stimulus of the international profile of the clients' segments. Moreover, the international importance of the financial operations made is an important factor: for example, most quotations on the stock exchange, privatisations and capital market activities, although could be referred to companies operating in a domestic environment, usually have an international size due to the integration with the market circuits. It should also be underlined that the global strategies tend to generate self enforcing effects, so that the progressive market growth causes further benefits to the financial intermediary in terms of improving the portfolio diversification, increasing the available information and market reputation, reducing the obligations and protectionist barriers in certain countries. These advantages nourish a new growth of the intermediary's market share, therefore justifying the basic strategy adopted.

However, the global strategies refer to the larger segments of clients showing requisites of inherent internationalisation and importance in relation to the average size of the operations (big ticket operations). The adaptation of the global strategies to the SME seems to be more complex, as the intermediaries' relational and contact abilities and the detailed knowledge of the territory are more important than the financial and non financial expertise. Therefore, the size of the bank oriented towards global strategies imposes an effective control only over the large corporate & institutions clients[6].

---

[6] In the case of large banks following global strategies, the large corporate & institutions clients are generally segmented on the basis of products and, sometimes, dynamics. The commodity segmentation results in classifying the companies served on the basis of the sector (industry) they belong to. This allows the financial intermediary not only to orient its internal research activity, thus gaining significant expertise within the specific sectors (telecommunications, transport, chemicals, etc...), but especially to measure the offer function on the characteristics of the company's industry. The dynamic segmentation classifies the compa-

Similar conclusions are reached for second tendency emerging from the market, i.e. the international developing strategies, that are differentiated from the global ones due to their reduced geographical area. Also in this case, the financial intermediaries widen the offer function outside their original market's boundaries, but limiting the expansion to specific macro areas (e.g. Europe, Asia). This allows, as for the global strategies, to mature a high mass criteria of expertise and achieve significant economics of scale. However, the financial intermediaries adapt the characteristics of their products portfolio of products to a smaller number of clients segments and limit their competitive arena.

The third market trend is developing domestic strategies, where the intermediary competes mainly within its market. This strategy, compared to the global and international ones, may turn out to be an obligatory choice where the financial intermediary does not have a strong visibility outside the national market and a real ability to access the international or global (financial and not financial) networks of expertise. Therefore, the financial intermediary is exposed to competitive attacks from the global and international intermediaries, capable of exploiting not only a high reputation level but especially a greater expertise. As a matter of fact, a domestic strategy, especially in the large corporate & institutions segment, is sustainable only if combined with a consolidated tradition and reputation over time, a strong ability to communicate with the entrepreneurial system and a sufficient volume of operations. Vice versa, the natural market refers to the SME segments of clients, i.e. the corporate, retail and small business segments.

The domestic strategy appears therefore more effective in the SME segment, that is clearly less exposed to the attacks from external competitors. However, the focus on the "SME – domestic market" productive combination is sustainable for the intermediary if the business volume is compatible with the costs of the structure of financial and non-financial expertise necessary for producing the services to the companies. Therefore, the domestic strategies towards the SME must anticipate the risks related to:

1. the operations rarefaction;

2. the reduction in the average size of the individual operations;

---

nies on the basis of the position reached in their development life cycle. In many cases the "growing" companies are distinguished from the "mature" and "declining" ones, in order to adapt the characteristics of the financial products offered to the specific financial needs of the different phases in the company's life cycle.

3. the increase in the risk associated to an almost exclusive maintenance of the lending product area.

First, the reduction in the volume of operations, apart from reducing the overall profitability of the corporate banking activity, generates dangerous effects on the expertise: If present, it weakened and progressively exposed to depletion; if still to be constructed, the low profitability does not push the bank to invest in this direction. Second, the reduction in the average size of the operations does not only affect the overall volumes of activity, with similar consequences to the previous case, but also exposes to competitive attacks from the professional agencies – i.e business consultants and private practices – that are generally active in lower value operations based on fiscal, legal, financial, and partly organisational analysis. Last, the eventual absence of a complete development of the different product industries leads the bank to concentrate its market operations in the credit area. This generates, with reference to equal volumes in the case of diversified offer, an increase in the risks and a consequential decrease in the corporate division's attractiveness for the bank.

The "domestic strategies" reference model well adapts to the current Italian market. The essential traits that emerged in the first chapter are in fact as follows:

1. at present no intermediary concretely follows a global strategy;

2. few intermediaries follow a domestic strategy focused on producing services for large corporates & institutions, due to their consolidated reputation over time, the strong relationship ability with the entrepreneurial system and an appropriate volume of operations;

3. some intermediaries tend to specialise in the medium sized companies (e.g. Sopaf and Arca Merchant), with reference to extraordinary financing services, and holdings of industrial participations;

4. the larger banks and numerous medium sized banks are widening their offer function from the traditional credit division to corporate banking services for the SME. This process, in most cases, does not assume precise and definitive connotations and is not generalised and related to a high number of banks.

Therefore, the Italian environment, oriented to local strategies, usually exposes its banks to the risks previously mentioned, becuase entering an area of activity different from the traditional credit one requires significant investments - not only monetary - for achieving the necessary expertise. This is justified and profitable if the choice of production diversification in the corporate division has a priority and is not accessory to the traditional credit activity. In this case, the likely limited number of exchanges not only undermines the diversification orientation of the banks, but does also not permit to develop a process of significant learning, that is the starting point for creating visible, recognised and concretely expoitable expertise in the market.

2.4.2 The Corporate Division Positioning: the Medium-Long Term Choices

The market positioning choices in the corporate division are strictly related to the segmentation process selected and the bank's analysis of the combinations of product industries–group of SME served. This means that the segmentation methodology selected by the management strongly impacts on the prescription obtained by the bank at the end of the process of market mapping and understanding. The identification of the bank's product industries and of the homogenous groups of companies, in fact, allows the bank to be supported by strong operative characteristics in defining both the market strategies adopt in relation to the different combination of products-markets, and the necessary resources and expertise in order to strengthen the offer processes. In this perspective, the more the selected segmentation criteria are sensitive to the market dynamics and proactive towards the environment interpretation and the anticipation of the exchange behaviours, the more the bank can progressively consolidate a coherent differentiation policy of the service and organisational functions.

Therefore, the identification of the characteristics of the competitive system describing the product industries and the choice of the model for aggregating the SME in categories can originate different uses for the bank, as the positioning issue includes a variety of actions with different significance and application. Analytically, the models used are as follow:

1. definition of a complete map of the product–market combinations functional to identifting the bank's strategic business divisions and preparing the organisational structure and the most appropriate production processes for the satisfaction of the clients' needs;

2. implementation of an effective methodology for understanding the market dynamics, in order to create a tableau de bord for the corporate division functional to organising the overall market strategies and the single client managers;

3. definition of an effective methodology for understanding the market dynamics, with the objective of focalising the attention on the appropriate positioning of a specific product division of the bank;

4. construction of an effective methodology for monitoring the market, with the objective of permitting the individual client managers to perform marketing operations or correct the more general market policies towards the clients.

The first model has a greater width as it places the segmentation at the peak of the bank's strategic planning: not only the identification of the target market is an independent variable, but the organisational structure is considered changeable and adaptable in function of the composition and aggregation of the strategic business divisions. On the other hand, the other three models internalise the constraints of the organisational structure and, in relation to the utilisation plan, combine the market understanding with the most appropriate operative processes for the objectives' achievement. Therefore, if the bank starts the divisionalisation activity with the definition of the large corporate & institutions, corporate, retail and small business market divisions for credit risk management, the segmentation process is necessarily performed later and internally to the market divisions identified.

For this reason, the segmentation process assumes greater application significance in the corporate division, where the market importance, the variety of the SME models and the requirement of a strong hold by the client manager stimulates a deeper understanding of the client's behaviour. Vice versa, in the retail division, the reduced complexity and small size of the companies strongly reduces the possibility of aggregation in homogenous categories. The consequent positioning must therefore prevalently play on product strategies as the product industries' and companies' market analysis are studied together and compared.

2.5 Conclusions: the Quest for Success in the Corporate Banking Industry

The path traced up to now represents the "project" that the bank must design and implement in order to define in operational terms the Corporate Business Division, orientated to producing value for the bank in a consistent and lasting way. This is the effort necessary to overcome the corporate banking - intermediary and partial - solutions during the "movement of opinion", commercial tension and integration with the credit risk management division. Here, the logical path for the bank that knowingly chooses to enter the corporate sector, deals with the organisational structure and strategy of the corporate banking division, the organisational roles and expertise involved, credit risk management models, company valuation and market positioning with reference to the segmentation logic selected.

The internal coherence of these elements and the subsequent strategic choices are not alone sufficient to guarantee the success and effectiveness of the project but are the essential benchmark. Therefore, it is finally necessary to concentrate on the conditions for success of the "corporate banking project for the SME", i.e. the critical issues for the bank, that must be overcome in order to develop the correct sequence of management themes.

For a clear representation of the nature of these critical points, it is possible to start with a preliminary distinction between internal and external ones. The first refer to the bank's organisational aspects, i.e. strategic rather than productive management, and the typology of links that the bank must develop with the entire financial system for the SME, in order to formalise the most appropriate and effective solutions in terms of performance. The second, instead, is related to the relational and contact aspects with the clients and the models of interaction with the demand function of the SME, with the objective of increasing the problem solving and customer satisfaction ability.

2.5.1 The Internal Criteria

The themes of discussion and detailed management for the internal criteria are:

1. a clear segmentation processes and organisational divisionalisation;

2. the constancy and determination in the research of humanistic, professional and contractual requisites for the managerial roles;

3. the affirmation of the centrality of the education processes;

4. the management tension related to governing the financial network, let aside the organisational – institutional model selected by the bank.

The activity of market segmentation is crucial as it represents a determinant pivot for achieving a substantial coherence between the organisational structure and the specificities of the demand, and an effective interaction between the credit productive system and the entire production and commercial system of the bank. For these reasons, the main choice must avoid standardised of systematic solutions that repeat the same peculiarities for most Italian banks, therefore generating a substantial levelling of the expertise and consequent low added value, inadequate frequency and disproportion of the solutions compared to the bank's characteristics. This also affetcs the efficiency and competitive effectiveness. Vice versa, the adoption of a personal market vision, fruit of explicit and radical management choices, represents a potentially strong competitive advantage and constitutes the correct fine tuning towards the reference market. In this context, the strongest attention must be given to the segmentation representing the basic infrastructure for the bank, i.e. the credit structure. This is due to the low reconfiguration of the segmentation for credit purposes and the high impact that it has, in the medium term, on the production processes. Moreover, the lending activity is the core product towards the SME and an incorrect configuration of the credit system may compromise the effectiveness of the whole bank's commercial action. This aspect is therefore highlighted by the fact that the SME represents a category that potentially crosses the corporate, retail and small business segments; the tailor made solutions appear necessary in order to gain a strong internal coherence for the entire offer process.

The constant search for the appropriate humanistic, professional and contractual requisites for the corporate banking roles is strictly related to the segmentation model selected, as the roles themselves strengthen the effectiveness and guarantee the functioning over time. The connection between roles and segmentation depends on three distinct elements, whose simultaneous presence determines the effective functioning of the corporate banking division:

1. the activities' content;

2. the activities' process;

3. the process engineering of all the activities in the corporate banking division.

Under the content profile, the skills involved must lead the bank to a concrete, substantial and exhaustive control of the products, services and activities contents performed on behalf of the clients[7]. In constrast with the mass retail market approach, the wholesale attitude originating from the organisation of a specialised corporate division cannot result in maintaining the content expertise due to the productive structure effectiveness. Vice versa the human resources, as the differentiation element, transfer to the service system a continuous adaptation to the clients' needs.

With reference to the process, the offer system relies on the production mechanisms that perform with effectiveness realisable solutions for the

---

[7] The concept of "content" makes reference, in a wide sense, to all the financial, contractual, legal and fiscal services characteristics, as well as the specificities of the impact that these present on the functioning of the demand and the economic-financial equilibrium of the counterparts.

client. This is possible if both the procedural structure and the habit and frequency of the resources in defining the deal are appropriate and significant. Therefore, the significance of the process becomes gradually more critical as the offer function shifts from traditional corporate lending services towards operations from different business functions. The process presents a clear significance and autonomy in function of its contribution to the correct functioning of the corporate banking division. The bank, even if with an adequate content, e.g. in the M&A sector, does not have sufficient realisable concreteness due to or in the absence of clearly defined procedures or execution rapidity and consolidated experience in developing the business resources.

Under the process engineering profile, the complete offer system in the corporate banking division must be equipped not only with functional but especially qualitative expertise and teamworking abilities, in order to recognise the system of client's needs as the main source of value production. This means that if content and process are the organisation "mechanics" for the services and products formulation for the SME, the process engineering is the organisation "chemistry" that, from the sum of the mechanical processes, generates relevant solutions to the client. This trajectory indicates to the bank that not only the constitution of a corporate banking division requires the continuous investment in teambuilding and teamworking, but especially that the organisational structure evolves towards the logic of a team of professionals, quite far from the productive and cultural archetype of the traditional commercial banks.

The affirmation of the centrality of the education processes is the third internal criteria of the bank. It shows a strong correlation to the content, process and process engineering themes, due to the fact that the cultural and professional profile of the resources is the only connection available for the bank between the variety and effectiveness of the production processes and the complexity of the demand functions. The education process must be characterised by:

1. the importance of strategic investments in the corporate banking division;

2. the continuance of the aforesaid investments over time;

3. the search of an absolute coherence with the set of necessary content for the activation of the offer system;

4. the consequent ability to produce education towards the client, with the objective of increasing customer satisfaction and widening the potential market space for more complex products.

Therefore, the education activity must permeate the entire planning and functioning cycle of the corporate banking, commencing with the determination of the economy's break even and prospective performance, in order to maintain the map of the services offered with constantly adequate expertise in line with the objective (see Table 2.4). The high variety of the expertise in play also needs an effective system of time to market updating, and the bank faces make or buy production choices (Caselli 2001). Where the offer diversification tends to increase and separate from the affirmation of a net superiority of the corporate lending activity, the choice tends to orientate itself towards a buy logic, where the bank's professional and organisational growth is "by discontinuance". Although this approach directly provides the result and satisfaction necessary to overcome a deficit of expertise, exposes the bank to significant risks, related to the possible rejection of the structure and the emergence of substantial differences in reaction and communication, that can lead in the end to a substantial paralysis of the production.

The tension towards governing the financial network is the fourth and final internal criterion in the corporate banking division planning. First, if compared to the previous ones, it appears to be more general and to involve the traditional problem of linking the institutional model selected to the strategic-organisational model adopted (Forestieri 2000). Given that each bank's framework of choices for connecing internalisation and externalisation of the production activity remain unchanged, the presence of a financial system for the SME leads to planning and governing a network of relations and alliances that can assume varied contractual and content degrees. This is due to the existence, in relation to the internalisation and externalisation process, of activities whose distinctive characteristics can be repeated with difficulty by the bank. For example, the activity performed by the professional agencies, the mutuals or even the venture capitalists, where the condition of success is often in the separation and in the conflict of interest with the bank's objectives[8]. The themes of independence, confidentiality and secrecy constitute, together, the physiological limits to the

---

[8] The concept of conflict of interest does not appear to be classifiable in rigid categories. It is however possible to note several situations in which traits of conflicts emerge, such as: i) the conuslting&advisoring activity is related to personal and reserved elements, whose divulgation can impact on the credit judgement; ii) the consulting&advisoring activity is not coherent with the appropriate transparency level for a credit transaction; iii) the activity developed is necessarily "external" to the bank for technical-production reasons (e.g. in the case of providing guarantees); iv) the entrepreneur wishes to distinguish the capital management from the company funding, with the objective of avoiding a cross collateral activity by the bank.

concept of the universal, but also corporate - if intended as auto-sufficient solution of the bank – banking, and the principles of internal diversification of the financial system in relation to the SME's needs. With reference to the bank's planning, a strategy towards internally repeating external activity is not possible. The bank had better undertake a policy of networking and selective alliances, that stresses the correct mapping of the value chain connectings bank and SME system, with the objective of communicating to the market, in the presence of conflicts, the separation of the unity with a strong and explicit business idea. This challenge will likely lead in the next few years not only – at the research and operating level – towards the networking and "bank network" direction, but also towards the implementation of stronger solutions, typical of a financial district.

2.5.2 The External Criteria

The external criteria, although strictly related to the internal ones, are:

1. the development of the family office policy;

2. the specific definition of the packaging strategy

With reference to the family office policy, it must be noted that the visual "family" of market relations between bank and SME is an element of substantial rupture compared to the traditional logic of relationship with the companies. As a matter of fact, it widens the size of the market space available and multiplies the relevant variables in order to achieve a growth in the profitable relationships. However, the risks, and not only the opportunities, are pretty high: the full consideration of the ownership family as the production centre of the financial needs and the element conditioning the companies's choice generates an overlapping and conflict of attribution between corporate and private division. Moreover, the family requires a partner characterised by professional independence, confidentiality and secrecy in the market relations.

Table 2.3 The expertise mapping within the corporate banking business division

Legend: ■ = Relevant expertise; ▦ = Expertise requested; □ = Expertise not requested

THE AREAS OF EXPERTISE	THE PRODUCT INDUSTRIES					
	PS	FS	FRHS	IRHS	AMS	CFS
1. Corporate finance						
Evaluation of credit risk	□	■	▦	□	□	■
Product speciality	□	■	□	□	□	■
Corporate evaluation	□	▦	▦	▦	□	■
2. Financial risk						
Risk evaluation	□	▦	■	▦	▦	▦
Speciality of product	□	□	■	▦	□	□
3. Insurance risk						
Risk evaluation	□	□	□	■	□	□
Speciality of product	□	□	□	■	□	□
4. Financial asset management						
Asset allocation strategies	□	□	□	▦	■	□
Speciality of product	□	□	□	▦	■	□
	□	□	□	□	□	□
5. Advisoring & consulting						
Financial	▦	■	■	■	■	■
Business-managerial	□	▦	▦	▦	□	■
Legal domestic	□	▦	▦	▦	▦	■
Legal non domestic	▦	▦	▦	▦	▦	■
Fiscal domestic	□	▦	▦	▦	▦	■
Fiscal non domestic	□	▦	▦	▦	▦	■
Project management	■	□	□	□	□	■

Relevant expertise ■
Expertise requested ▦
Expertise not requested □

- PS: Payment Services;
- FS: Financial Services;
- FRHS: Financial Risk Hedging Services;
- IRHS: Insurance Risk Hedging Services;
- AMS: Asset Management Services;
- CFS: Corporate Financial Services.

The solution to these criteria cannot be represented by an arbitrary attribution of the client's control to one of the two divisions as the risks and gaps of effectiveness tend to distribute and manifest themselves equally. More likely the path to be adopted, given the attribution of the SME to the corporate division, is that of defining contractual and production "environments", dedited to managing the relations with the family of entrepreneurs. The family office solution refers here to a production – a specialised production centre – and contractual solution – the stipulation of the "family office" contract – that commits the bank to manage in the medium and long term people, risks and family's wealth[9]. This form of intervention positions the proponent bank as the structure of support, trust and operational exclusivity, aimed at satisfying the whole range of the family's needs. In fact, due to the "delicacy" of the deals, the relations between bank and client tend to become stronger, binding and full of exchange opportunities. The banks also receives benefits in terms of understanding the family dynamics and thus protecting the credit risk towards the company.

With reference to the explicit definition of the packaging strategy, the complexity of the segmentation models requires a re-equilibrium of the responsibility for the commercial success between client manager and production structure. The back-office activity must therefore go in the direction of a "marketing laboratory", i.e. the creation of centres for the innovation that not only formalise the specificities of new and old products, but in particular defines the combination or packaging logic. This is

---

[9] The theme of family office represents an area of reflection and emerging action in the international financial system. Significant is the examples of Pictet and of Northern Trust. For a preliminary analysis of the phenomenon see Corbetta et al. 2001. With the term family office we mean a package of personalised services that several foreign banks have offered for some time to the high end of family companies. These services are aggregated on: asset allocation/manager selection/ asset management; venture capital/private equity consulting and management; global custody/reporting/monitoring; other services. For example, this latter can relate to: fiduciary services; legal, fiscal, company structure, real estate, insurance and pension positions consulting and management; personnel insurance management; management of any family member's personnel; consulting and training of the younger family members; managerial consulting for the activity of the single family members; planning and management of the family museum; planning and management of family archives; planning and management of the family's philanthropic initiatives. The family office is therefore a service adapted to the family: with high financial resources, and a large number of members (actual and future), normally resident in different countries, relatively united and interested in passing the wealth on to the next generation.

relevant as the package approach constitutes the conjunction and matching element between the system of products and the client's needs, together with the client manager's efforts. For this reason, the products packaging must refer to the client's area or a specific context where the main element does not arise from the client manager's diagnosis ability, but from the correct functioning of the solutions, the rapidity of execution and the overall effectiveness. Examples are the packaging for real estate operations and financial risks management, or for the development of export activities. Nothing prohibits, however, to drive the packaging approach towards more complex situations, represented for example by financing a start up. In conclusion, the packaging logic appeals to two fundamental assumptions: first, the recognition of the client manager's impossibility to perform effectively a global player role of the bank's offer towards the SME, where the package availability frees time and space of action; second, the bank's strategic choice to complete the map of product– market combinations, supported by the expectations of success with the demand.

3 Opportunities in the Quotation of Private Equity Companies

Marina Maddaloni and Maria Pierdicchi

3.1 The Advantages and Risks in the Quotation of Private Equity Operators

An interesting opportunity for the Private Equity operators can be represented by a Stock Exchange quotation and in this case, given the high correlated risk return profile, on a market dedicated to innovative companies and high growth. The international scenario today underlines the presence of numerous Private Equity companies quoted on the financial markets and the concentration of this phenomenon appears higher where the financial markets in the different countries have been established for a longer time period: evidence is the fact that numerous VCs are today quoted on the Nasdaq, the London Stock Exchange and the Swiss Market. In particular, the Swiss Stock Exchange gave life to an ad-hoc segment for the Investment Companies providing a more strict regulation compared to those of the ordinary segments in terms of admission procedures and requisites for permanence (for further details see paragraph 3.1.2).

In recent years this was also made possible by a particular inclination of the market to accept higher risk levels compared to the past: this phenomenon is evident in light of the many quotations of companies active in technology sectors and with high risk that were introduced to the market in the course of 1999 and 2000 with a significant presence of retail investors in IPO.

If the returns volatility of the securities is accepted as the indicator of the risk of a share/index, it is evident from an analysis in Fig. 3.1 that the Private Equity shares most representative on the Nasdaq have shown a volatility much higher than the market they belong to (on average 3 times greater); index that has in turn over the three year period analysed a higher volatility, if compared with indexes representative of more mature/traditional sectors.

Volatility during the period 1999-2001

	on annual basis	referred to Nasdaq Composite
ACACIA RESEARCH	112.1%	2.7 x
C M G I	128.6%	3.2 x
D V IN E IN T E R V.	129.6%	3.2 x
FRONTLINE C G	141.1%	3.5 x
INTERNET C G	147.9%	3.6 x
SAFEGUARD SCIENTIFIC	104.4%	2.6 x
Indice Nasdaq Composite	40.8%	
Indice S & P 500	20.7%	
Indice Dow Jones	19.6%	

Fig. 3.1. The volatility of the principal Private Equity companies quoted on the Nasdaq (Source: New Market calculations on Bloomberg data)

The quotation on the Stock Exchange for Private Equity companies thus represents an important lever for growth, recognised by the market, and allows to obtain several advantages and benefits, among which:

– raising of capital, recourse to a widened investor base, at international and national level;

– access also to the public of small investors interested in the venture capital activity and the results achieved by the investors in previous years;

– acquire an international visibility, with benefits in terms of number of projects to view and invest in;

– realise with greater ease partnerships with technological suppliers and operators connected to the high technology sector;

– have the possibility to manage a wider and more diversified portfolio of companies, guaranteeing the possibility to support projects requiring large investments (consequence of higher fundraising).

3.1.1 The Segmentation of the Private Equity Operators. Incubator, Venture Capitalist and Investment Companies: Distinctive Characteristics

The analysis of the advantages and risks in the quotation of PE operators in the Stock Exchange necessarily passes through a segmentation of the specific market which evidences the peculiarities of the risk/return correlated to the investment in different phases of the development of the company financed (see Fig. 3.2). In particular, for this purpose it may be useful to sub-divide the universe of private equity in three macro-categories, Incubators, Venture Capitalists and Investment Companies, in relation to several distinctive factors:

– financial commitment and timeliness of the intervention;

– typology of risk;

– typology of support.

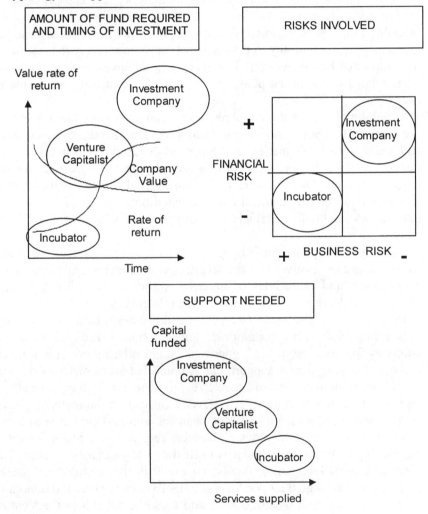

Fig. 3.2. The distinctive factors of Private Equity companies

Financial commitment and timeliness of the intervention: given the growth curve of a company over time, naturally the greater the intervention by an institutional investor is made in the initial start-up phase of the company, the more contained is the financial commitment requested and greater the correlated risk (high mortality of start-up companies). This consideration seems greater the more the time horizon of the investment in the start-up company is considered as typically a medium-long term one (4/7 years) and discounts greater difficulty in liquidating the participation; on the contrary, the greater the intervention relates to an advanced phases in the development of the company and thus maturity and consolidation of the business, the greater the financial effort and the lower the company's mortality risk. The investment in addition discounts the reduced time period and a greater liquidity in the participation. Consequently the rate of return requested by the investors is higher when the investment in risk capital is related to the initial phases in the life of a company and decreases over time.

Typology of risk: in the initial phase of a company's life the risk of failure is typically a business risk embodied in the fact that the company must still build a business model, develop a product/service desirable to the market, set up a business structure, source managerial expertise, acquire market share, etc. On the contrary, for consolidated companies the greatest risk appears to be the financial risk correlated to the relationship between indebtedness and risk capital and thus the optimisation of the financial leverage.

Typology of support: initially the start-ups require a series of operating support that can be offered by the investors who, therefore, do not perform a role solely and exclusively of financier to the initiative, but contribute through added value services to the company's growth.

In general, the Incubator intervenes in the initial phase of a start-up, with a limited financial commitment and a greater risk of failure of the company financed, requiring higher returns and offering not only a financial but also an operative support, that is translated into a series of services of different nature (rental of space, recruitment, accounting, consulting, etc). The Investment Companies intervene, instead, in the maturity phases of the companies, with a significant financial commitment but with lower mortality risk of the company and therefore require lower remuneration of capital; typically the investment is exclusively of a financial nature. The Venture Capital companies, finally, intervene in the intermediary phases and therefore arrange their portfolio in order to balance risks and returns.

In conformity with what has been said it can be noted how the Venture Capital activity can be orientated in terms of business model alternatively towards the Incubators, or the Investment Companies, assuming each time only some of their characteristics.

This sub-division, as is necessary, evidently takes into account the difficulty sometimes met in outlining a precise boundary line between the different typologies of operators.

The business model and value creation of the Private Equity companies varies significantly depending on whether they relate to Incubators, Venture Capitalists or Investment Companies, and this consequently modifies the considerations on the greater or lower risk for the quotations on the Stock Exchange of these types of companies and their "quotability".

The process of the Incubators's value creation develops through five fundamental phases, that can be generally summarised as follows:

1. identification of the investment opportunities through a shrewd selection of the business plans and analysis of the management team that is characterised by its successful track record;

2. investment of financial resources: it typically relates to seed capital (therefore the most embryonic phase of the financing) and thus contained amounts for starting the entrepreneurial initiative. This investment is normally made taking into consideration the possibility of successive financing rounds;

3. operating support to the company in order to allow the management to concentrate on the business development: this support goes from the rent of physical space for offices, to maintaining the accounting records, to a support in terms of recruiting human resources with adequate skills, etc.;

4. access to a network of contacts that assist the management in developing the business: supply of technology, partnership, customers;

5. exit of the investment at a higher value: it can be represented by trade sale operations or if the Incubator has made different financing rounds and thus accompanied the company in its growth path, also with an IPO.

The mission of the Investment Companies is to create a portfolio of companies diversified with a view to stock picking and risk diversification. The business model goes through the following phases:

1. identification of the investment opportunities with reference to companies not yet quoted but in advanced development phases, or quoted ones, or finally funds that in turn invest in different participations ("fund of funds"); always with a view to stock picking;

2. investment in financial resources for the expansion phase or late financing rounds;

3. exit of the investment at a higher value using an exit compatible with the diversification choices adopted and orientated to a time horizon even in the short/medium term.

Again the business model and the creation path of Venture Capital companies lies within the intermediary boundary line between the activity of the Incubators and those of the Investment Companies and may relate to a model that simultaneously includes the two.

The quotation on the Stock Exchange of Incubators, Venture Capitalists and Investment Companies, thus responds to the different requirements expressed by the market: higher risk and consequently higher returns in the case of the Incubator, greater diversification of the risks in the case of the Investment Companies, that invest in mature companies and that at times may specialise in several specific sectors (see the case of BB Biotech, that invests mostly in quoted companies and specialised funds in biotechnology) and, therefore, can be seen by the market as an investment choice as a kind of "sector index".

3.1.2 Advantages and Risks in the Quotation

Compared to an industrial or service company, the venture capital and investment companies present significant characteristics in relation to a quotation. As described in the previous paragraphs, they relate to companies with a high risk profile in line with the nature of the activity undertaken, characterised by industrial and financial risks typical of the risk capital investment.

The degree of risk varies however according to the typology of operator, and it is higher where the investment is focused on the company's initial development phases (from start-up to early development), technological and innovative sectors and on a lower degree of diversification. In general, the more one moves towards investment companies the lower the risk profile, to arrive finally at the investment companies in funds and quoted companies that represent the type historically present in the different stock markets (see Fig. 3.3).

The limited liquidity of the assets in portfolio, that represents one of the critical elements in the quotability of a company, reflects the uncertainty of the value that can be created and the ability of the investors to monitor it. In fact a company with a portfolio of non-quoted participations derives its value from the ability to generate capital gains that are realised somehow only at the moment of exiting from the investments. In several cases, and in particular where the exit arises through an IPO, the realisation can also be dissolved over time due to the effect of lock-conditions obliging the shareholders to maintain the shares for a period after the IPO (on the main stock markets between 6 and 12 months). In this case part of the exit may occur through sale of the shares in IPO at a price defined at the closing of the offer and a part can be requested at a later sale on the market or with a private arrangement at the expiry of the lock up. It should however be underlined that for the high growth companies that have been placed on the market in recent years, the quotation does not represent even a partially way out, as the presence of the institutional investor in the shareholding structure is an unconditional element in order to accept the already high degree of risk. In this case the uncertainty on the value created is extended over time, as the effective time and value of the exit for the venture capitalist depend on the market conditions, on the trend of the quoted security after the quotation, and many other specific and market factors.

This is one of the aspects that may induce many institutional investors, particularly those operating in the most advanced phases of the company's life cycle, to prefer an exit through "trade sale" compared to the IPO, in order to guarantee a sure and immediate value through the capital gain realised with the sale to third parties. This relates to a choice not always available: the full sale of the shares in an IPO is more frequent for the companies that have been subject to LBO or turn-around operations and that present a consolidated industrial history; the younger companies and with a higher risk return profile in fact normally receive new capital to finance growth and the sale of existing shares is not compatible with the objectives of the quotation, at least in the initial phase.

Venture Capital

INCUBATORS

Opportunities
* Low amount of funds provided
* Potentially, very high capital gains
* Focus on specific segments: scale economies (- costs)
* High managerial skills available to firms in the investment portfolio
* Positive Network effects for the companies included in the portfolio

Risks
* Portfolio composition: start-ups
* Focus on a specific segment: influence of the market trend of the same industry securities

INVESTMENT COMPANIES

Opportunities
* Portfolio diversification, hence risk diversification
* Funding at later stages in the firm life cycle

Risks
* High amount of funds provided
* Limited capital gains
* No involvement in the management of the participated firms
* Likely network diseconomies

Fig. 3.3. Opportunities and risks related to private equity operators

   In absence of exiting the value of the participations in portfolio must be periodically estimated in order to calculate the net asset value of the fund or of the assets in the financial statements. This concerns a practice that although being based on the best techniques and specialised operators in valuation present areas of uncertainty related to the structural difficulty in defining values for companies still immature, subject to market, technical, commercial and operating uncertainties. The operators avail of the most well-known techniques; for example, in case of a transaction on investments with third parties, for the entry and exit of new shareholders, the price paid represents an important reference for valuation but does not obviously constitute an effective value of immediate potential liquidity, as can be theoretically a price on the stock exchange for a quoted participation. The uncertainty and the possible scarce significance of the net asset value of a pool of investments represents one of the most critical elements in the quotability of a venture capital company: the greater the expected exiting is extended in time, for example due to the concentration of the portfolio in the company's initial development phases, the greater the investors of the VC companies must count on the net asset values made available by the company and the analysts in order to evaluate the shares and elaborate their consequential investment strategies. In addition the difficult, if not impossible, comparability of an investment company with other similar quoted companies further complicates the investment choices, strongly reducing the value of comparative analysis based on multiples. This is reflected in two distinct aspects of the companies in question: they do not generally provide an alternative diversification purely by sector, except where there is a high specialisation and a very wide portfolio able to strongly mitigate the risks of concentration (investment companies); and given the difficulty in the short and medium term of forecasting the results, due to the inherent illiquidity of the assets generating cash and profits, are subject to a very high volatility, normally higher than those of the shares the investments are held in.

The difficulty in forecasting the annual results of a private equity are thus evident, taking into account that the quotation imposes quarterly reporting and analysis. Therefore, the quality and the quantity of research available for the operators in this sector tends to be limited, particularly in the markets that do not have a specific tradition in the sector.

Given all the previous mentioned difficulties, it can be legitimatedly asked why a fund/ VC company can be considered suitable for quotation.

The advantages should be considered from two viewpoints: the issuer and the market, i.e. the investors.

From the issuer's viewpoint the following aspects are considered:

1. the quotation widens the ability of raising funds for the operator, providing a retail "fund raising" channel which is normally accessed by institutional operators in risk capital. Particularly during very favourable market conditions, as in the boom period of technological shares between 1997 and 2000, the quotation permits the quick raising of funds and diversification of the fundraising channels;

2. on quotation, an operator provides to his original investors an indirect liquidity of the investment, repairing, in part, to the low portfolio liquidity. However, the effective liquidity depends on the shares on the secondary market, an important factor considering the sector in question that may limit the base of institutional and retail investors;

3. as for other companies, the quotation permits a broader imagine and visibility, also influencing the companies financed, that are thus in the public domain and can take advantage of new business opportunities;

4. in some cases the quotation allows a better valuation of the assets held by the group, by an industrial company or diversified holding, separating the investment activity and VC from other businesses that prevents the correct valuation in the financial statements. For example, the VC activities and incubation of large technological groups, often subject to spin off or quotation for the above mentioned reasons.

Passing to the desirability for the market the following aspects can be noted:

1. the Private Equity companies offer an investment opportunity with one security in several companies, giving therefore access to retail investors in risk capital not otherwise accessible. This is the basis for the success of many quoted VCs in the USA during the boom of the technological shares in which investors were searching for new economy companies held in the portfolio of VCs with expectations of high returns. The desirability of the investment in the fund, or in the investment company, is greater when the participations held are close to quotation (pre IPO), in the expectation that the exit value will be much higher than the Nav, a factor that the market should reflect on quotation. In a favourable market this aspect favours the investment company, but in adverse conditions, where the risk perception increases due to a greater rate of expected default of the underlying portfolio, the contrary happens: the investment companies tend to be penalised more than the companies financed, for the greater liquidity costs and uncertainty on the results obtainable;

2. the inherent reduced transparency of the results (even the best accounting criteria tend to underestimate the risks), together with the illiquidity of the assets can represent an obstacle to placement with the retail public, while it is a situation that favours the very specialised investors and in particular funds of funds, that find in the quoted fund a guarantee that provides a certain degree of transparency and liquidity;

3. the composition, size and diversification of the portfolio are variables that can strongly influence the placement of VCs on the stock exchange. The sectorial specialisation renders more desirable the instrument to strongly specialised institutional investors (industry related funds). With equality of specialisation, however, the preference is to quote funds with a significant part in quoted investments, for the greater transparency of

the nav and the greater assets liquidity. The size is generally strictly related to the companies diversification (or by companies and funds) and can therefore represent an additional point;

4. required return: this is the return compared to the risks, higher compared to traditional equity and thus potentially useful for increasing the portfolio return;

5. the element that strongly characterises a VC operator is the management and its track record, i.e. the successful investment operations and the relative IRR generated. The fund management makes the difference, in its capacity of choice and investment management and all the related activities. The qualities of possible co-investors in strategic participations or in other funds may also be important: there are funds with high reputations that represent a kind of guarantee and a co-investment with them can provide an important message to the market on the quality of the relationship and the sector "entry".

The aforesaid considerations provide several elements in understanding why the phenomenon of the VCs quotation is relatively recent and not insignificant. The trend, which began in the USA with the boom of the new economy and technological shares, has seen a moderate development in Europe, particularly in the markets where the private equity industry is historically more developed and timid development in some of the other markets, but has never taken off. The operators with a broad activity range rather than those highly specialised tend to be prevalent, while the incubators are almost entirely absent, due to the high risk and the scarce development in Europe. The industrial incubators, typically related to industrial groups have rarely been the subject of spin-offs or quotations and there are no reasons to believe that this situation may change.

3.1.3 The Due Diligence Process for the Quotation: Some Issues

Where a VC operator goes for a quotation he must chose the most appropriate market and structure the offer in the best possible way. The company managing the market and the relevant authority must verify quotation and information requisites, taking into consideration the peculiarities of the potential issuer.

In relation to the market choice, this may be influenced by several factors and specifically by the presence of "comparables", on specific segments or regulatory markets, from the market in which the funds prevalently operate and so forth. Typically the greater number of quoted operators is found in the more developed markets both from the offer side, number and variety of issuers, and from the demand side, i.e. number, specialisation and culture of the final institutional and retail investors. In particular in the USA and Great Britain there are the most and highly significant quoted VCs, while the Swiss stock exchange has a specialised segment of "investment companies" that includes different types of operators among which VCs and private equity funds. The creation of an ad-hoc segment for investment companies effectively responds to the requirement of considering the peculiarities of these companies in both the listing phase and the post listing phase, separating them from the production companies.

In relation to the admission procedures for quotation, the company's quotability profile must be valued with methodologies and criteria rather different from those traditionally used.

Among the most important documents that must accompany a company's quotation is the three year business plan, a document that must provide to the admission body a detailed description of the issuer's activity, development plans, use of funds raised from the quotation and the strategic and commercial direction that it is going to follow for reaching the growth objectives.

For a company whose principal activity consists of the investment in different participations the "business plan" cannot be formulated in the same terms. The fund's or company's ability to generate value must be evaluated by taking into consideration a series of elements, some of which are difficult to quantify:

- the "track record" of the previous investments in portfolio, i.e. the internal rate of return obtained and the nav (net asset value) maturated on the investments in portfolio. This relates to elements that provide indications on the ability to make valid investment choices, that however does not guarantee the future ability to generate value;

- the rate of failure of the portfolio: the greater the investment activity is oriented towards the company's initial development phases the greater the industrial and financial risk of the portfolio. Hence the importance to verify and compare the rate of failure, i.e. of investments that have been written-down or have been liquidated in losses with the best operators in the sector. The comparison can be very difficult due to the scarcity of available information as well as the difficulty to make "homogenous" comparisons given that each operator can distinguish itself for specialisation, diversification, size of the fund managed, development phase of the companies financed etc. Just as important is the evaluation whether the recording and monitoring instruments of the risks are adequate in providing the public with timely information on the trend of the participations' value;

- the professional profile of the management team. This is a critical aspect of due diligence, an indicator of the operator's skills. The evaluation of the expertise and reliability of a team refers especially to the experience and track record of the individual partners and managers, the investments made in previous experiences or funds managed, the specific expertise in the sector possibly matured, the professional integrity shown and so forth;

- the quality of the original investors or shareholders of the company or fund is another important aspect as this tends to characterise the quality of the operator and of its management team. In particular the presence of prestigous industrial investors, or with strong positioning in the sector in case of specialised funds, may indicate that the management team has the ability to create attractive returns. In some cases the investors are also commited to provide know how and support to the management team in order to facilitate or better accelerate the growth of the investments, thus as "enhancer" of value and therefore representing a competitive lever important for an operator;

- the company's "governance" is another critical aspect as it establishes the underlying rules to the investment decisions and the later control and valuation of the activities in portfolio. It is therefore important to verify that the company has adequate resources and procedures to ensure a correct valuation of the potential investments, a transparent mechanism for their approval, a monitoring system for the investment performance and a periodic verification of the net value of the assets based on the best international practices. The remueration polices of the management are very important in order to protect the shareholders in coherently pursuing the objectives of the strategic investments. Connecting the remuneration of the management team as much as possible to the returns generated on each individual investment even through direct participations is an effort to reduce possible agency problems;

- the risk profile of the specific investment activity must be valued with particular reference in generating value over a time period coherent with the investors' expectations, an element that is obviously very difficult to estimate ex-ante. The explicit investment policies represent the "philosophy" of the fund and provide limits on the managers' discretion. It is important for the final investors to establish whether the resources of the fund are adequate in achieving the objectives, e.g. in relation to the monitoring capacities, managers' expertise in the sector, portfolio synergies, risk diversification etc.;

– reporting. The transparency in reporting and clarifying the valuation and accounting criteria of the assets are indispensable instruments in protecting the investors and therefore should be carefully verified before the quotation and supported by the auditors' opinion. The requirement to supply quarterly data, as a practice by all the regulatory markets for the quoted companies, can be met with the impossibility and the burden of the processes needed to define the carrying value of an investment company. In some cases the net asset value is calculated twice yearly and the quarterly report is limited to providing more general information on the trend of the investments. The possibility of a less frequent reporting and standardisation of the criteria in calculating the NAV are among the reasons why the creation of an ad hoc segment for the investment companies is desirable, with specific information requests. In general the more the company invests in "mature" or quoted companies the more immediate, transparent and reliable is the monitoring of the portfolio value and the progressive results. However, in case of funds of funds, the verification of the valuation criteria is more complex as requests are made to other parties, and not necessarily the management and accounting criteria are in line with the quoted company and are subject sometimes to a different legislation. If thus the fund of funds reduces the risks in the portfolio thanks to the indirect diversification of the investments and management of the operations, it also shows a reduced transparency.

The process of verifying the quotability of an investment company is limited however to the analysis of the ability to follow the announced investment choices and to assure the market of the the risk and information monitoring. The market will evaluate whether the overall industrial and financial risks in the portoflio participations, together with the skills of the management team, will provide sufficient indications to accept the risks of future investments, not determined ex ante if not for the investment policies pre-fixed and for any returns realised prior to the quotation.

Considering the high uncertainty that remains even after the aforesaid verifications, it is not surprising that the quotation experiences are effectively limited. In Europe, where venture capital is a relatively new and not widely developed sector, the quoted investment companies are almost all relating to private equity, funds of funds and investment companies in quoted companies (very similar to open investment funds). In these cases as well the experience has not been positive. With few exceptions these issuers have suffered from high volatility, negative performance and scarce liquidity, phenomena accentuated in negative market periods or in the underlying markets.

During the boom of the high growth markets the public began, especially in the USA, to be interested in this type of investment, driven by the attractiveness of the "upside" that some operators promised in the innovative industrial sectors with high and relatively "quick" returns, thanks to favourable IPO market conditions, or exiting. In fact the more the stock market appeared open to high risk return profile the faster the investment and exiting process became through quotation of venture capital. Being able to participate in the "first levels" of the chain for the retail public, typically excluded from the high entry levels to the VCs, meant being able indirectly to share in the valuations in the IPO, in some cases in the order of three figures. With the drastic inversion of the markets which began at the end of 2000 and worsened in 2001, the dramatic increase in the risk of failure of start-ups and young companies was significantly reversed on to the most exposed VCs in the high tech sectors and the early stage phases, widening the risk perception of the portfolio.

Effectively the "closing" of the exit markets in itself accentuated the risks of the most exposed operators in the sectors in question, lengthening the investment cycle, requiring new investments to support the companies in difficulty, lowering the valuation prospects of the exit values on the more mature investments and increasing the losses in the portfolio. Not only did the markets take the risk threshold to more "acceptable" levels more coherent with those traditionally registered on the stock markets, lengthening or eliminating tout court the way out through IPO, but the institutional investors in risk capital saw the possibilities of quotation reduced due to the negative trend of the portfolio. To the industrial difficulties in the underlying sectors were added the financial problems related to the requirements of refinancing the companies financed, at valuations inferior to the initial ones (downrounds). With the drastic fall in primary fundraising, the activity was centered on refinancing rather than on new investments, thus representing a check on diversification.

In light of the experiences in the last years, it is reasonable to expect a streamlining of the sector, with a selection of the players and a progressive concentration, particularly in the most popular markets (USA, UK ,Israel), and a reduced recourse to quotation for the aforesaid reasons.

For the industrial operators with their own investment activity or incubators too the losses registered were high, increasing the loss of value on the securities of the parent company and contributing to increasing the transmission of the perception of risk in entire sectors.

Below is a clearer overview of the most recent international experiences.

3.2 The International Experiences

3.2.1 The International Scenario: Historical Overview

Historically the private equity was developed in the United States after the second world war, principally thanks to the wide availability of long term savings, deriving from pension funds and funds raised from insurance companies. Over the years, the practice of venture capital was consolidated allowing a significant remuneration on the investments that today can count on over one thousand United States risk capital funds, two thirds specialised in the financing of innovative and high tech projects. The reason for the greater maturity of the North American market compared to Europe in general is found not only in typically historical factors but also in some specific characteristics of the economic fabric in the United States, and Anglo-Saxon in general, that facilitated the development: positive attitude towards risk and consequently to failure, more favourable legislation environment to the creation of new companies, less discouraging and penalising laws relating to failure towards the entrepreneur, greater flexibility in the work place, that consents the start-ups an easier management in the employment of personnel for an indefinite period, according to the requirements of the company.

3.2.2 The International Scenario: a Benchmarking Comparison

At the end of 2000, the Nuovo Mercato carried out an analysis on the Private Equity companies quoted on the foreign markets in order to evaluate the international scenario and the tendencies in the financial markets. In the course of this analysis, 50 Private Equity companies were identified, quoted on financial markets in Western Europe and USA.

The methodology of the analysis: the analysis was performed on the principal United States and European financial markets: Nasdaq, LSE (AIM), OMX, Neuer Markt, SMAX, SWX, Nouveau Marché, AEX; the sample used was determined making reference to the segmentation of the sectors in each market (ex.: Nasdaq – General Financial Services; SWX - Investment Companies, etc.).

Included in the sample were financial holdings of groups having an internal structure exclusively dedicated to Private Equity (for example CMGI: CMG@Ventures). On the other hand the Private Equity companies with the portfolio oriented towards investments in companies typical of the "old economy" were not taken into consideration and are only by chance included in "hi-tech", as well as the mutual investment funds.

All of the companies belonging to the sample were analysed in terms of: type of private equity operator, investment sector, phase of the intervention, number of employees, reference market, information on the trend of the shares, financial and profit information, brief description of the business, companies invested in (see the example report on the company Acacia Research shown in the Fig. 3.4).

Revenues (M $)						
	dic-99	%	dic-98	%	dic-97	%
Tot. Revenue	0.100	100%	0.400	100%	0.500	100%
Oper. Inc.	-9.900	-9900%	-5.800	-1450%	-3.400	-680%
Pre-tax Inc.	-10.400	-10400%	-6.400	-1600%	-3.300	-660%
Net Inc.	-8.200	-8200%	-6.200	-1550%	-2.900	-580%

Revenues Variation (by year) %		
	98-99	97-98
Tot. Revenue	-75.00%	-20.00%
Oper. Inc.	n.a.	n.a.
Pre-tax Inc.	n.a.	n.a.
Net Inc.	n.a.	n.a.

Investm. Sector	%
IT/Infrastructure	55%
B2B	10%
B2C/e-Tail	0%
Internet Content	0%
TLC	10%
Biotechnology	15%
Media	10%

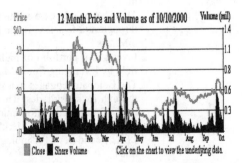

Fig. 3.4. Example report (Acacia Research)

Company Type:	Incubator, Accelerator				
Investment Sector:	Internet and technology companies (science, B2B, communications, e-commerce and infrastructure)				
Point of Intervention:	Seed, Early				
N° Employees:	93				
Listed:	Nasdaq, ACRI				
Web:	www.acaciaresearch.com				
IPO Date:	n.a.				
IPO Key Facts:	n.a.				
Equity Information:	Recent Price:	26.750 $	Mkt cap. at IPO:	n.a.	
	Max Price (12 months):	59.000 $	Mkt cap.:	427.68 M $	
	Min Price (12 months):	13.375 $	Float.(%):	n.a.	
	Performance (since IPO):	n.a.	Outstanding Shares:	15.988 M	
	Performance (12 months):	+25.17%	EPS:	-1.06 $	
Sponsor and Underwriter:	n.a.				
Business Area and Strategy:	Develops and operates life science and enabling technology companies The Company's core technology opportunity has been developed through its life sciences and enabling technologies subsidiary, CombiMatrix Invests in a portfolio of high growth companies in the life sciences field by leveraging CombiMatrix's technology The Company engages in a variety of technology-related businesses, through its subsidiaries: The EC Company, CombiMatrix Corporation, Greenwich Information Technologies LLC, Mediaconnex Communications, Inc., MerkWerks Corporation, Signature-mail.com LLC, Soundview Technologies Inc. and Soundbreak.com Incorporated.				
Listed Partners:	None				

Fig. 3.4 (cont.)

Detailed Company Profile:
By providing seed capital, critical management and technical advice, and ongoing operational support, the Company provides an infrastructure that allows early-stage companies to focus on their core strengths, which are creating new products and services. This support, in turn, allows developing businesses the opportunity to reduce their time to market. The Company also obtains ownership positions, through strategic investments, in businesses that fit well with its existing operating companies. Through its subsidiaries, the Company engages in a variety of technology-related businesses: The EC Company, CombiMatrix Corporation, Greenwich Information Technologies LLC, Mediaconnex Communications, Inc., MerkWerks Corporation, Signature-mail.com LLC, Soundview Technologies Inc. and Soundbreak.com Incorporated.

The Company provides its subsidiary companies numerous operational and management services, especially in early stages of their development. The Company's corporate staff provides hands-on assistance in the areas of marketing, strategic planning, business development, technology, accounting and finance, human resources, recruiting and legal. The Company also supports its subsidiaries by providing, locating and structuring financing.

Participations:
Acacia Research develops and operates a broad range of technology companies, including life sciences, business-to-business, communications, e-commerce and infrastructure. Current partner companies include:

CombiMatrix Corporation	Mediaconnex Communications Incorporated	Soundbreak.com Incorporated
The EC Company	MerkWerks Corporation	Soundview Technologies Incorporated
Greenwich Information Technologies LLC	Signature-mail.com LLC	

A first indication of the Private Equity companies quoted on the foreign financial markets is from the segmentation by country of the entire sample, as follows in Fig. 3.5.

Of the companies analysed 97.8% are quoted on technological markets (Nasdaq), specific segments (AIM, SWX- investment companies, SMAX), or on the New European Markets (Nouveau Marchè, Neuer Markt, AEX, OMX NM).

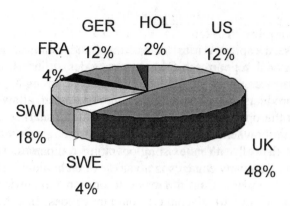

Fig. 3.5. Segmentation of the sample by country

At this point it is possible to gather information on the individual countries analysed (it is recalled that the analysis was made on data at December 31, 2000):

1. USA: in the USA there were 6 Private Equity companies identified, quoted on the Nasdaq. These companies took on the form of Incubators although in reality they provided financial support in terms of seed capital and in the successive phase to the development of the companies in portfolio. They often relate to large groups active in Private Equity for a number of years (CMGI since 1994); representing in this sense an interesting track record:

   – in possession of extremely diversified portfolio of companies (from companies active in infrastructure and TLC, to e-business, and the life sciences);

   – a part of these companies are in turn already quoted on the Nasdaq (ex. CMGI, with 9 quoted companies out of 70);

   – have significant volumes of sales even if frequently they demonstrate income statements in loss (already at operating income level).

The existence of negative margins is often attributable to the lack of economies of scale in the management of operating costs following the extreme diversification of the portfolio (ex. logistics support). In this context, the most recent tendencies to a greater focus of the investment activity demonstrated in the largest United States incubators can be interpreted:

- Safeguard Scientific: concentration of the investment activity in companies active in the internet infrastructure sector;

- ICG: activities related to the B2B marketplaces;

- CMGI: announcement of a complete reorganisation of the group in six different business lines business.

The investment activity is almost exclusively orientated to the domestic market, with at times a reduced diversification towards UK companies. A possible trend in the coming years will be a move towards the European market.

2. UK: in the UK there were 24 Private Equity companies identified, quoted on the London Stock Exchange, on the AIM segment, of which 7 were incubators and 17 investment companies. None of the quoted companies were included in the technological index; the quotation on the AIM does not appear particularly significant in terms of compatible with quotation on a growth market, given the extreme variety of the companies involved; in particular, the Investment Companies quoted on the AIM do not appear to be "innovative" companies, nor with high growth rates, limited to providing support in terms of capital to companies in the advanced phase of development and with consolidated business models (ex. Inflexion Plc.). For the Incubators of the old generation (1999), it is possible to note a general tendency to the focus on specific investment sectors (an exception is represented by the 3i Group), that already from 2000 consented the forecast of the realisation of positive margins (ex. New Media Spark, specialised in Internet). Some of the companies quoted in 1999 already had companies in portfolio that were quoted (ex. New Media Spark, 5 companies quoted out of 13). The investment activity is orientated towards the domestic market (prevalently), but also to other European countries (Scandinavia, Germany and France); only exception is the 3i Group present in different countries where they operate locally.

3. Switzerland: in Switzerland there were 10 Private Equity companies identified quoted on the Swiss Stock Exchange; they generally relate to investment companies of banking origin (ex. BB Biotech, BB Hitech, BB Medtech, all originating from Banque Bellevue). The Swiss stock exchange is characterised for the existence on the traditional listing of a segment dedicated to the investment companies equipped with an ad hoc regulation; within this, additional regulations are prepared that discipline:

- the clear clarification of the underlying principles to investment policy;

- the clear clarification of the valuation criteria of the companies invested in, especially if not quoted (periodic valuation - 6 months – of the asset not quoted);

- the clear clarification of the risk components of the business;

- a more precise publication of the annual and six-monthly financial statements, that in justified cases may be requested at shorter intervals;

- a minimum investment level requested at the moment of the quotation.

The investment activity is orientated to the following markets: Switzerland, USA, UK, Germany, Israel and Scandinavia.

4. Germany and France: the research has shown the presence on the Neuer Markt and on the Nouveau Marchè of a very small number of Private Equity companies quoted; in particular:

- Excluding the dual listing of BB Biotech (quoted since October 2000 also on the Nuovo Mercato and thus a triple listing), only one venture capital company is quoted on the Neuer Markt (Blue-C since August 28, 2000);

- 5 companies (all Venture Capital) are quoted on the Franckfort Stock Exchange in the SMAX segment (similar in some respects to the new STAR segment of the Italian Stock Exchange), dedicated to companies of medium capitalisation, defined as of quality, not necessarily high tech;

- Only 2 venture capital companies on the Nouveau Marchè (Altamir and Astra-Tech since 1997).

The activity of investment is orientated to the following markets: Germany, France, UK, Israel and Scandinavia.

5. Holland and Sweden: in Holland there is an Incubator (Newconomy) quoted on the New Holland Market on the AEX NM, and invests in the Benelux, UK and France. In Sweden there is an Incubator (Ledstiernan) and a Venture Capitalist on the Swedish New Market (Core ventures), which invests in Sweden, UK and USA.

3.3 Some Considerations on the Performances

Examining the trend of the shares of the quoted companies in the period October 1999 – October 2000, it is possible to note important differences in the results of the three macro-categories.

Return
Compared to the reference markets the quoted incubators were under-performed: an explanation could be that the investors perceive negatively the high business risks in the sectors operating in (high rate of failure of the start ups).

The investment companies quoted instead were overperforming; return-ing to what was previously said, this result could be due to the reduced risks of the projects in which the companies invest in (expansion or in-vestment in other quoted companies).

Volatility
The average volatility of the incubators within the sample (expressed in this case as the percentage difference between actual price and maximum price) is more contained compared to that in the investment companies (comparison Fig. 3.6 and Fig. 3.7).

This can depend on different factors:

– less visibility of the incubators quoted, reduced volumes traded, and thus variations in the price more contained, vice versa for the investment companies;

– the companies in portfolio of the investment companies are prevalently quoted, thus inherently sensitive to the trend of the financial market that it belongs to.

The analysis of the results of venture capitalists was more complex, as it was not possible to identify a common line, but rather a trend that at times is close to that of the incubators (lower returns and contained volatility) and at times that of the investment companies (higher returns and volatil-ity), probably in relation to the orientation and characteristics of the com-pany (number of companies in portfolio, sectorial diversification, know how, track record and perception of the investors etc.).

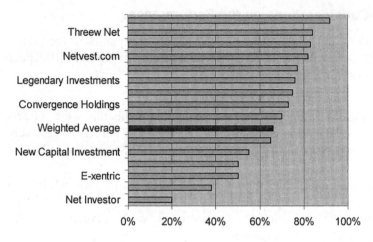

Fig. 3.6. % difference between actual price and maximum price for Incubators (indicator of volatility)

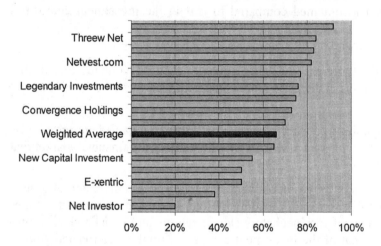

Fig. 3.7. % difference between actual price and maximum price for Investment Companies (indicator of volatility)

4 The Venture Capital Industry in Europe: Trends and Figures

Lucia Spotorno

4.1 Introduction

The term venture capital has different meanings depending on the context it is referred to.

In the US it comprises three types of investing – seed, start-up and expansion investments for companies previously funded by venture capitalists[1]. Therefore, neither buyouts financing nor middle market private companies funding is considered venture capital investing. These types of activities are labeled "non-venture private equity investments".

On the other hand, in Europe the boundaries between venture capital and non venture private equity are less well-defined. For example, the European Venture Capital Association (EVCA) collects data on the European private equity industry without differentiating between venture and non venture expansion financing. Moreover, due to national legislation and to different structures of the financial intermediaries, in some countries US venture type investments are made by institutions that do not meet the criteria of the EVCA definition of private equity and are thus left out of its data collection process[2]. Therefore, as EVCA is the main source of data about the European venture capital market, any comparison with the US is biased. However, comparing the two is essential in order to analyse the growth patterns of the European industry. The US model has been highly

---

[1] Seed capital is typically used to finance initial product research and development and to assess the commercial potential of ideas. Startup investments are targeted at companies that have moved past the idea stage and are starting to produce, market and sell their product. Investments in either seed or startup stage are also referred to as early stage investments. After a company has passed through the early stage, it becomes a potential candidate for expansion stage investing. In the expansion stage, a company that has already established its products in the marketplace often needs additional capital to fund the growth of its manufacturing and distribution capacity and to finance further R&D.

[2] This is the case of Germany, where banks provide small and medium sized enterprises with capital. No data are available to estimate the extent of this activity. See Ooghe et al. 1991.

successful in spurring innovation and economic growth[3]. Thus, only a better understanding of the differences between the US and European venture capital industries may help to shed some light on the ability of the Europeans to duplicate US results.

From this perspective, the present study examines the economic foundations of the venture capital market, describes its institutional structure and analyses its development in Europe and in the US. It emphasises two themes. The first is that the US model cannot be seamlessly transferred abroad. Different employment practices, regulatory policies, public market conditions and quality levels of companies seeking venture capital influence the performance of investing. The second topic is that even among European countries there are differences in venture capital investing patterns. Therefore it is not correct to address the European venture capital industry as a homogeneous whole.

4.2 The Venture Capital Market: an Overview

Venture capital encompasses three different markets: the organised venture capital market, the market for angel capital and the informal venture capital market (Fenn et al. 1995).

Angel capital, which is regarded as a critical source of seed capital, refers to investments in small, closely held companies by wealthy individuals, many of whom have experienced operating similar companies. Angels' investments range from $50.000 to $500.000, and there is evidence to suggest that angel capital represents the largest pool of risk capital in the US and the UK (Wetzel 1981; Harrison 2000). Angel capitalists may have substantial ownership stakes and may be active in advising the company. Their expectations in terms of investment results are often low when compared to those of professional venture capitalists as they are usually motivated by non financial rewards[4].

---

[3] The National Venture Capital Association Annual Study on the economic impact of venture capital in the US reveals that between 1991 and 1995 venture backed companies increased their staffs on average 34% per year. During the same period, Fortune 500 companies laid off staff by 4% per year. See: Jeng and Wells 2000.

[4] The list of nonfinancial considerations includes ventures creating jobs in areas of high unemployment, ventures developing socially useful technologies, ventures contributing to urban revitalization, ventures created by minority or female entrepreneurs, the personal satisfaction derived from assisting entrepreneurs to build successful ventures in a free enterprise economy (see Wetzel 1981).

Angel capitalists typically learn of investment opportunities through a network of friends and business associates. Tapping this informal network for an entrepreneur is not easy as no reliable source of references exists. By the same token, angels largely rely on random events to bring investment opportunities to their attention. Therefore, in spite of its considerable size, the market for angel capital is fragmented and highly inefficient.

In the informal venture capital market, securities are sold to institutional investors and accredited individuals. Investors do not have an active role neither in negotiating the terms of the investment nor in monitoring the company, as insiders typically remain the largest and only concentrated group of investors.

On the other hand, the organized venture capital market encompasses professionally managed investment in new or young ventures equity. Investment management is supplied by specialised intermediaries, who raise money from individuals and institutions, identify investment opportunities, structure and execute deals with entrepreneurial teams and monitor investments in order to achieve some return on their capital. Their stake in the company is usually large, even though not always a controlling one. They typically become members of the board of directors and retain important economic rights in addition to their ownership ones.

Through the services they offer, venture capitalists allow the funding of high risk ventures which would otherwise not be financed. Venture capitalists thus create a unique link among investors and issuers. The former are attracted to the venture capital market by the high expected returns and diversification benefits that this kind of investment provides to their portfolio, but lack the experience and the abilities needed for structuring, monitoring and exiting deals. The latter are firms that cannot raise money from cheaper sources. They are usually companies developing either innovative products or firms innovating the way an existing product is marketed. Both are projected to show very high growth rates in the future.

Besides independent venture capital firms, that raise their funds through the market and aim at extracting the maximum value from the portfolio companies they invested in, captive firms also operate in the market[5].

---

[5] Captive funds are prevalent in Europe, while almost absent in the US. Funds are labeled "captive" if more than 80% of their financing derives from one source. In Europe, new funds raised by captives were 18,7% of total new funds raised over the 1996-2000.
In 2000, new funds raised by captives were 23,5% of total in France, 40,3% in Germany, 23,9% in Italy, 20,5% in the UK. In the US, funds managed by subsidiaries of financial and industrial firms over the 1996-2000 period were 7,1% of total capital under management in 2000.

These are usually either banks, financial companies or industrial corporations subsidiaries. Their goals are not limited to the financial return on their investing, as captive firms activity also allows the backing institutions to acquire some strategic advantages. Once nurtured, portfolio companies may become either new customers or acquisition targets. Moreover, funding new businesses through specialized subsidiaries may attract corporations as a mean to learn about innovative and fast growing technologies whose internal development would be time consuming and risky.

A third kind of organized venture capital firms is formed by those partly funded by government agencies and private endowments. In this case, investments are usually focused on specific recipients in order to foster local development and social promotion. Return on investments, even if always a major interest, is thus coupled with considerations of non economic character.

4.3 How Venture Capital Works

The beginning of an organised venture capital industry in the US is usually dated to 1946. In 1946 the American Research and Development Corporation (ARD), a publicly traded, closed-end investment company was formed. ARD was created "(..)..to devise a private sector solution to the lack of financing for new enterprises and small businesses..(..).." (Fenn 1995), an issue of great concern in the early 1930s and 1940s. Besides the funding of new businesses, ARD aimed at providing managerial expertise to the companies it financed, as adequate management skills and experience were thought by ARD founders a critical ingredient for success in business venturing.

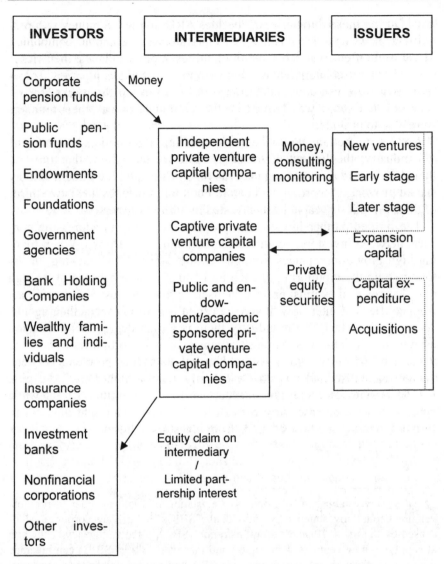

Fig. 4.1. Participants to the venture capital market (source Fenn 1995)

In spite of its unique features and persistent promotional efforts, ARD did not succeed in attracting much interest among investors. Moreover, no other publicly traded venture capital company was formed before 1959, when Small Business Investment Companies (SBICs) were established by law[6]. During those thirteen years, besides ARD, in the US money was provided to new high risk businesses by private venture capital companies. These were often created by wealthy families in order to invest their riches and did not act as intermediaries between investors and companies. Therefore, as private venture capital companies did not tap the market for resources, their scope was limited by the amount of money their founders were able to provide.

Not before the late 1960s the now prevailing organisational form in the US industry, the limited partnership, came to age[7]. From that time on though, it spread at increasing pace, as fund raising and investment activity started growing. According to Venture Economics data, by 1987 two thirds of the U.S. industry capital was invested in limited partnership.

The limited partnership success is due to its ability in addressing conflicts between investors, venture capitalists and funded firms. When funding high growth companies through venture capital firms backing, investors give up the monitoring of each individual venture because of its high cost. However, they must protect themselves from the losses that originate from conflicts of interest with venture capitalists, which act as their agents. As a matter of fact, in the industry there is a high degree of information asymmetry, which gives venture capitalists many opportunities to take advantage of their privileged position. Unresolved agency problems between venture capitalists and investors eventually mean a higher cost of capital for the businesses which are financed, as funds committed shrink while perceived risks increase. Adverse selection phenomena might thus occur, further hindering the flow of funds from investors to businesses.

---

[6] SBICs were established by Congress Legislation in 1958, in order to promote venture capital investments by individuals. SBICs are private corporations licensed by the Small Business Administration (SBA), a federal agency. They aim at supplying new ventures with capital and manageral skills. SBICs can leverage their capital through low cost loans from SBA and can take advantage of several tax benefits. On the other hand their investments are limited to non controlling stakes in companies whose minimum size is restricted by law. Besides that, due to the fact that SBICs are required to make interest payment over SBA loans, originally SBICs financing was mainly addressed to later stage, cash generating ventures. Despite the initial success of the program, by 1976 the number of SBICs was depleted to a third. This was mainly due to unsound management practices and self dealing, which made SBA revoke almost half of the licences it had granted.
[7] Tommy Davis and Arthur Rock are credited with developing the first limited partnership in 1961 (see Bygrave and Timmons 1992).

Moreover, venture capitalists are involved in an agency relationship with the entrepreneurs they finance and have to address the conflicts proceedings from it.

In order to increase the market efficiency, during the years the industry operating procedures and contracting practices adapted in order to limit agency costs (Sahlman 1990). In this respect, the limited partnership agreement provisions have proofed highly effective.

Morover limited partnerships are investment vehicles that enjoy a favourable tax treatment provided that certain conditions are fulfilled. Their income is taxable at the individual investor level only.

Therefore, double taxation problems are avoided. In addition, the gain on the securities possibly distributed by the partnerships is recognised as taxable income when the assets are sold and not at the distribution stage. This makes the limited partnerships an attractive way to address managerial compensation problems through stock options and other forms of performance based compensation.

Two different kind of partners are involved in it: general partners and limited partners. General partners manage the funds they raised from limited partners, investing them in promising ventures. General partners commit only a small part of the total amount of cash raised but bear unlimited liability[8]. On the other hand, limited partners' liability is limited to the funds they committed, provided that they not participate in active management. However, limited partners are almost always permitted to vote on key issues such as amendments of the limited partnership agreements, general partners removal and the partnership ending before its termination date.

The economic life of the partnership is usually set at ten years, even though most partnership agreements include an extension period provision. During the first three to five years the partnership's capital is invested. Investments are then managed and gradually liquidated. At the end of the partnership existence, all cash and securities are distributed.

The partnership managers usually raise a new and legally separated vehicle when the investment stage at an existing partnership has been completed. Thus, general partners tend to manage different funds at the same time.

If they are to raise new cash, general partners must establish a favourable track record. The importance of a favourable track record is twofold. On one hand it directly strengthens the general partners reputation for investing successfully. This reduces the costs and the amount of time needed

---

[8] On average, funds committed by general partners are 1% of the total amount raised. (Sahlman 1990).

in order to set up new partnerships. On the other hand, by making the general partners able to form new vehicles, a favourable track record builds up their ability to collect funds in the future, as experience itself is regarded as an asset.

In order to collect money, the general partners tend to turn first to investors that funded their previous vehicles. Funds are often raised in several closing stages. This allow the general partners to signal to the investment community that a fund is being successfully raised, implying a favourable evaluation by those who have already committed their funds. Investors who are familiar with the general partner ability are usually more prone to infuse cash, provided that they have a long-term commitment to venture capital investing. Therefore, as the process of forming new partnership can be very costly and lengthy, the stability of an experienced investors' base is crucial in order to limit the costs of business venturing[9].

However, a good reputation is not enough an incentive to behave properly, especially when the general partners are either at the beginning or at the end of their careers. Thus, the provisions of the partnership agreements work as incentive mechanisms to align general and limited partners' interests.

General partners' opportunistic behaviour is prevented through adequate performance incentives – which comprise the terms of the general partners compensation structure and the way fees and profits shares are calculated – and through direct control mechanisms. General partners receive a yearly management fee, which usually ranges from 1% to 3% of the total capital committed, depending on the size of the fund. They are also entitled to a share of the profits over the life of each partnership they manage (carried interests).

Carried interests are the bulk of the general partners' compensation and may be several times larger than the management fees. They are most often set at 20% of the partnership's net return (Gompers and Lerner 1998). This entails "….that the venture capitalists have incentives to engage in activities that increase the value of the carried interest, which is precisely what benefits the limited partners…." (Sahlman 1990).

However, this compensation structure (and especially the general partners' claim on profits) make venture capitalists apt to increase risk taking beyond reasonable levels. Their share of the capital initially committed to the partnership is small. Moreover, even if general partners bear unlimited liability, bankruptcy is a far off possibility, as venture capital partnerships typically do not borrow money. General partners' loss in case the partner-

---

[9] In the US, pension funds investing allows venture capitalists a low cost access to long term capital.

ship's investments yield negative results is thus restrained to a limited amount of money. On the other hand, if the partnership investments yield positive results their profits share is considerable.

Therefore, partnership agreements also protect the interests of the limited partners through covenants that place restrictions on investments and on certain other activities of the general partners. Partnership covenants thus usually limit the percentage of the capital that may be invested in a single venture or constrain the aggregate size of the partnership's two or three largest investments. Covenants may also forbid investments that deviate from the partnership primary focus (e.g. LBO investments), and restrain the use of debt. Partnership agreements may also call for mandatory distribution of realised gains, restrict co-investment with the general partners' earlier or later funds and restrain the ability of the general partners' or their associates to co-invest selectively in the partnership deals.

Once the partnership is created and funds are raised, venture capitalists start their investment activities.

Venture capital firms choose to manage a relatively non diversified portfolio of highly risky investments in the management of which they are intensely involved so that losses are limited. As expertise acquired with previous investments is crucial in this respect, venture capital partnerships tend to specialize by industry and by geographic area, thus forcefully giving up diversification benefits. Survival in the long haul is based on the ability to select and foster big winners.

The partnership investment activity is usually divided in four stages. First, investments are selected. Then deals are structured. Terms and conditions of the relationship between the venture capitalist and the entrepreneur are defined. At this point, the venture capitalist is entitled to actively participate in the management of the company. Finally, once her investment is ripe for cashing in, the venture capitalist liquidates it by either taking the company public or through the private sale of her stake.

The investment selection stage involves dealing with a severe sorting problem. In the US, only 1% of the investment proposals received by the general partners are accepted. Proposals are screened during an extensive due diligence process. As little information about the ventures is publicly available, venture capitalists must rely on information they collect by themselves during the sifting process.

Sometimes, the partnership judgement is validated through syndication. Syndication as a mean for the general partners to double check their thinking about a promising firm they are willing to fund is frequent in US high technology ventures at first-round investments[10]. Syndication is also due to the size and the location of the deal. Larger deals are financed via syndication as partnerships' provisions may prevent a single fund to acquire a controlling stake in the company. Syndication is even used when the company location is far away from the general partners'. In this case, extensive monitoring may be costly. Therefore, local monitoring is delegated to a geographically closer venture capitalist involved in the deal.

Finally, investments' sharing proposal informally oblige others to return the favour in the future. Therefore, syndication is a mechanism for limiting investing costs.

The structuring stage of the deal involves negotiating an investment agreement that settles the main facets of the relationship between the venture capitalist and the company. The amount of ownership the partnership is to acquire in the company is determined and managerial incentives are defined. The partnership ability to exert control over the firm if its performance suffers is established as well.

The investment agreement with the company is called the stock-purchase agreement. It fixes the amount and timing of the investment. Venture capitalists typically infuse cash in the company at different rounds. At each round, funding allows the company to reach the following stage of development only. At that time, further capital will be required to keep on growing.

On one hand, should the company prospects fade, investment staging makes venture capitalists able to abandon the project before serious losses are incurred. On the other hand, investment staging provides a mechanism to prevent the company's management misbehaviour. As venture capitalists restrict capital commitments, they can credibly threat to drop the venture before it accomplishes its business plan. Investment staging even allows venture capitalists to prevent the company from getting funds elsewhere as "...by denying capital, the venture capitalist signals to other capital suppliers that the company in question is a bad investment risk..." (Fenn 1995).

In the US, many venture capitalists fund companies purchasing convertible preferred stocks. Preferred convertible stocks offer two advantages to venture partnerships. First, preferred convertible stocks holders are paid before common stocks holders. Hence, venture partnerships' risks are re-

---

[10] "...venture capitalists, upon finding a promising firm, typically do not make a binding commitment to provide financing. Rather they send the proposal to other investors for their review. Another venture capitalist's willingness to invest in the firm may be a important factor in the lead investor's decision to invest....", cit. in Lerner 1994.

duced. Second, preferred convertible stocks make management intensify its effort to produce value, as management typically holds common stocks. If the company is only marginally successfully and payments go first to preferred stocks holders, common stocks value will be low. The implication is that management is rewarded only if the company is remarkably successful. Therefore, it will try hard to pursue the best results.

The agreement between the venture partnership and the company sets forth the conversion price for the stocks, which can vary according to the performance of the company. It also settle a liquidation preference. The convertible preferred stocks dividend payment is subject to the board of directors' approval. Nevertheless, some issues have provisions that call for accruing dividends.

In the stock purchase agreement, governance issues are also addressed. Provisions usually grant the venture capitalist: boards membership, the right to inspect the company facilities, its books and records and the right to receive timely financial reports and operating statements. Moreover, even if the partnership holds preferred stocks, voting rights are typically granted on a as-if-converted basis.

For early stage new ventures, the investment is often large enough to confer majority ownership. Alternatively, the partnership may obtain voting control even without being the majority shareholder. At any rate, the partnership can exert a large influence on matters that come to a shareholder vote.

Moral hazard risks are also reduced through performance incentives to the ventures' management. A common provision in the venture capital financing is the "equity earn-out". Should certain performance objectives be met, the equity earn-out provision grants management an increase in their ownership stake at the investors' expenses. On the other hand, misuse of capital is severely reproached. Thanks to the investment staging, venture capitalists may deny further cash. Even if they grant more money to the company, at any cash infusion, management equity share is diluted at increasing rates. As a result, the venture management is encouraged to save capital.

Besides monitoring and governing the company though, general partners furnish managerial assistance, helping companies arrange additional financing, hire top management, recruit knowledgeable board members. Moreover, general partners are active at the strategic level as generators of strategic initiatives. They may also be involved on the personal level as friends, mentors and confidants of venture CEOs (Sapienza et al. 1994).

As Tyebjee and Bruno 1984 say "....once the deal has been consummated, the role of the venture capitalist expands from investor to collaborator..."[11]. Once the investment has reached its exit stage, the venture partnership may exit either by taking the company public or via a private sale. In the latter case, stocks may be acquired either by an outside investor or bought back by the company.

A public offering is usually the favourite mean to exit, as it generally allows the highest return on the venture partnership's investment[12]. It is also welcomed by the firm's management, as it preserves its independence. Moreover, it provides the company with a cheaper access to capital by creating a liquid secondary market for its securities. However, a public offering seldom means a complete liquidation of the venture partnership's interest in the firm. In the US, the partnership is usually restricted from selling its shares either by law or by the offering underwriter.

A private sale can be very attractive if the new investor is willing to pay cash the securities and if her entering the firm's capital ends the partnership involvement in the company. On the other hand, this exit route may be unsuitable to the company's management, as the company might lose its independence. In order to avoid that, the partnership's exit may take place through either a put of stock back to the firm or through a mandatory redemption. The former mechanism is used when the company has been funded through a common stock purchase and it is based on a valuation algorithm agreed in advance. The latter is used in case of preferred shares.

---

[11] Cit. in Tyebee and Bruno 1984. Recent empirical evidence of venture capitalists' monitoring activity in the US is provided in Lerner 1995; Hellman and Puri 2000; Kaplan and Strömberg 2000. Lerner 1995 finds that venture capitalists are more likely to join or be added to the boards of private companies in periods when the chief executive officer of the company changes. He also documents that the number of general partners serving as company directors increases when the CEO lacks entrepreneurial experience. Hellman and Puri 2000, studying a sample of 173 high technology start-up firms find that venture capital is associated with a significant reduction in the time to bring a product to market. Kaplan and Strömberg 2000, find that venture capitalists play a large role in shaping and recruiting the senior management team either by replacing a founding member or by hiring experienced executives.

[12] While many mechanisms exist to liquidate a fund, the most attractive option seems to be on average an IPO. A study conducted by Venture Economics and cited in Jeng and Wells 2000, finds that one dollar invested in a firm that eventually goes public yields a 195% average return for a 4.2 year holding period. The same amount invested in an acquired firm only provides an average return of 40% over a 3.7 year average holding period.

Partnerships choose how and when to exit in order to obtain the maximum value for their stake of the firm. Several studies document their valuable role in public offerings (Megginson and Weiss 1991; Barry et al. 1990). On average, companies backed by venture capitalists public offerings' underpricing is lower than other companies'. This probably means the market rewards the company with a certification premium, as "...reputable partnerships do not bring lemons to market..." (Fenn 1995). Moreover, as usually the partnership that originally funded the company does not completely liquidates its stake at this stage, a lower underpricing reflects the value of the venture capitalist's ongoing management activities.

4.4 The Venture Capital Industry Development in Europe

Even though in the United Kingdom the first venture capital firm was started in 1945, the European venture capital market size was not comparable with the US one until the 1980s[13]. However, the European industry took the lead in 1988 in terms of the investments made and almost had the same pool of available funds in 1989 (Ooghe et al. 1991). At that time though, more than half of the fund raising and investment activity was concentrated in the UK, while the Spanish, the Italian and especially the German industries lacked a significant size[14].

During the 1990s the market extent increased in every European country, thanks to several factors. The creation of parallel secondary stock markets provided an exit mechanism that stimulated new funds raising[15]. Moreover, many governments began to consider venture capital as beneficial to economic growth and started funding sponsorship programs.

---

[13] 3i – Investing in industry – backed by the most important British banks and by the Bank of England, was created in 1945 in order to fund high growth businesses through equity and debt financing.

[14] However, as mentioned in the introduction, Germany venture capital market size may not be correctly estimated by EVCA due to the strict definition of venture capital used when collecting data.

[15] The Unlisted Securities Market was created in the United Kingdom in the 1985, the Parallelmarkt in The Netherlands in 1982, the Second Marché in France in 1983 and in Belgium in 1985, the Geregelter Markt in Germany in 1988. In 1998 the Euro.NM circuit started working.

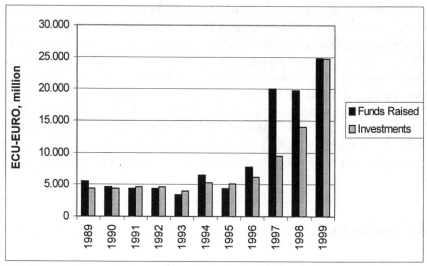

Fig. 4.2. Evolution of new funds raised and investments in the EU, 1989- 1999 (Source: Christofidis and Debande 2001)

As Black and Gilson 1998 documented, there is a positive correlation between stock markets activity and the venture capital industry size. The growth in new funds raised during the 1990s matched the increase in the number of newly listed companies in parallel markets both in Europe and the US (see Tables 4.1 - 4.6).

Stock markets attractiveness for young, fast-growing companies is crucial as it makes floating an effective divestment mechanism for venture capitalists. As returns achieved on investments exited through IPOs are higher than those attained on investments exited through alternative routes, IPOs availability reduces the perceived risk by investors, thus fostering their commitment to venture capital and therefore its cost.

IPOs also create an incentive for the portfolio companies management. As mentioned, if the venture capitalist sells its stake through a flotation, the entrepreneur gets cash only if she sells some of her shares in the offering, but she obtains an increased value and liquidity for unsold shares. Moreover, the contract between the venture capital fund or partnership and the portfolio company generally guarantee that the extra control rights that were initially given to the fund - including board membership and veto power over business decision – are released on an initial public offering whether or not the fund sells any shares at all in the IPOs. The venture capitalist thus retain only the control rights related to his proprietary shares. As she gradually relinquish his stake in the company, control is transferred to the entrepreneur. If regaining control is important to the entrepreneur, as this latter obtains it contingent on the firm's success, she will exert more effort.

On the other hand, when a trade sale is the only exit way, unless the company is able to buy back its shares, the control rights will shift from the venture capitalist to the acquirer of the venture capital fund's stake.

Table 4.1. Venture capital activity in Europe and the US: new funds raised

Year	1985/19 89 Avg.	1996	1997	1998	1999	2000	(data in billions of dollars)	
							Last five years avg.	Last five yrs. Avg. growth
Europe°	3,41	8,51	21,15	18,78	22,08	40,63	19,09	59,74%
France	0,84	3,23	0,83	2,96	3,06	5,64	2,76	67,12%
Germany	0,11	0,43	3,29	2,14	4,33	5,57	2,64	189,16%
Italy	0,23	0,81	1,12	1,01	1,51	2,62	1,22	38,00%
Spain	0,07	0,07	0,51	0,66	0,59	1,79	0,61	220,47%
UK	2,04	4,21	15,12	10,17	9,25	9,63	8,41	55,33%
US	3,05	11,78	17,1	29,42	59,9	92,92	35,69	68,98%

° European countries considered: Austria, Belgium, Denmark, Finland, France, Germany, Greece, Iceland, Ireland, Italy, The Netherlands, Norway, Portugal, Spain, Sweden, Switzerland, United Kingdom.

Source: Ooghe et al. 1991; EVCA Yearbook and NVCA Yearbook, various issues.

Table 4.2. Venture Capital Activity in Europe and the US: New Funds Raised/GDP

Year	1985/1989 Avg.	1996	1997	1998	1999	2000	1996/00 Avg.
France	0,07%	0,21%	0,06%	0,20%	0,24%	0,44%	0,23%
Germany	0,01%	0,02%	0,16%	0,10%	0,22%	0,30%	0,16%
Italy	0,03%	0,07%	0,10%	0,09%	0,14%	0,24%	0,13%
Spain	0,02%	0,01%	0,09%	0,19%	0,11%	0,32%	0,14%
UK	0,03%	0,36%	1,27%	0,73%	0,64%	0,68%	0,73%
US	0,07%	0,15%	0,21%	0,34%	0,64%	0,94%	0,46%

Source: Ooghe et al. 1991; EVCA Yearbook and NVCA Yearbook.

Table 4.3. Venture capital activity in Europe and the US: total available funds for investment (data expressed in billions of dollars)

Year	1985/1989 Avg.	1996	1997	1998	1999	2000	Last five years avg.	Last five yrs. Avg. Growth
Europe°	12,87	10,5	24,04	34,39	29,34	44,60	28,58	98,39%
France	2,08	1,42	1,49	2,99	10,88	8,26	5,01	88,76%
Germany	0,55	0,40	2,44	0,69	4,27	7,00	2,96	291,22%
Italy	0,40	0,97	1,49	1,58	1,65	2,68	1,67	45,61%
Spain	0,34	0,29	0,29	0,94	0,90	2,25	0,94	91,16%
UK	7,07	5,04	14,86	18,91	11,33	14,8	12,54	115,02%
US	25,59	48,9	65,1	90,9	142,9	209,8	111,52	55,76%

° European countries considered: Austria, Belgium, Denmark, Finland, France, Germany, Greece, Iceland, Ireland, Italy, The Netherlands, Norway, Portugal, Spain, Sweden, Switzerland, United Kingdom.

Source: Ooghe et al. 1991; EVCA Yearbook and NVCA Yearbook.

Table 4.4. Venture Capital Activity in Europe and the US: Total Funds available/GDP

Year	1984/1989 Avg.	1996	1997	1998	1999	2000	1996/00 Avg.
France	0,22%	0,09%	0,11%	0,21%	0,84%	0,64%	0,38%
Germany	0,06%	0,02%	0,12%	0,03%	0,23%	0,37%	0,15%
Italy	0,10%	0,08%	0,13%	0,15%	0,16%	0,25%	0,15%
Spain	0,13%	0,05%	0,05%	0,27%	0,17%	0,40%	0,19%
UK	1,22%	0,43%	1,24%	1,35%	0,97%	1,04%	1,01%
US	0,59%	0,63%	0,79%	1,04%	1,55%	2,12%	1,23%

Source: Ooghe et al. 1991; EVCA Yearbook and NVCA Yearbook.

Table 4.5. Number of domestic companies newly listed – parallel markets

Year	1996	1997	1998	1999	2000
Europe	240	316	317	380	560
France	49	62	109	62	67
Germany	13	24	51	121	118
Italy	1	0	1	6	33
Spain	1	2	3	1	3
UK	131	100	69	97	265

Source: FIBV, Annual Statistics.

Table 4.6. Total capital raised by newly listed companies over GDP – parallel markets

Year	1996	1997	1998	1999	2000
France	0,06%	0,06%	0,10%	0,08%	0,13%
Germany	na	na	na	0,38%	0,62%
Italy	0,00%	na	0,00%	na	0,34%
Spain	0,00%	0,00%	na	na	na
UK	0,04%	0,05%	0,03%	0,05%	0,19%

Source: FIBV, Annual Statistics.

Government programs for assisting venture capital over the same period had a positive effect on the industry activity too.

Even though no empirical analysis on this topic have been conducted for the European countries, an appropriate legal environment is deemed to have increased the level of private equity investment in Portugal, Ireland, the UK[16], while the lack of any special legal structure for private equity firms and no tax and other incentives aimed at the industry are considered one of the main causes of the feeble signs of growth of the small Austrian market (Jeng and Wells 2000; Christofidis and Debande 2001).

European governments intervening in the market varied widely. In some countries the government only provided a favorable legal and tax infrastructure while in other cases direct public funding was supplied.

Where this happened, besides the creation of public sponsored venture capital funds, government support measures encompassed upside leverage schemes, provisions for the funds' operating costs and downside protection schemes. All these measures were usually targeted to small funds, as they tend to invest more money in early stage businesses and in high-tech sectors while having less access to large institutional investors' money.

---

[16] In the UK, the Inland Revenue started the Venture Capital Trust scheme in 1995 in order to encourage individuals to invest in small, higher-risk, non listed trading companies. Venture Capital Trusts (VCTs) are similar to investment trusts. After their initial fund raising VCTs shares are quoted on the London Stock Exchange. Individuals who invest in VCTs are entitled to some important tax advantages on shares acquired up to a maximum of £100,000 per tax year. In order for a company to qualify as a VCT, at least 70% of its investment by value throughout its most recent accounting period must have been newly issued shares or securities in companies not listed on the official list of any stock exchange. The gross asset value of the company funded by the VCT must not exceed £15 million.

Leverage schemes usually allow the sources of a venture capitalist's fund to borrow an equivalent amount of soft money. This type of scheme enables the fund to gear up the scale and the returns from investing to the benefit of the private equity investors. Support programs for operating costs coverage are tapped by small funds to make more of the money raised available for investing. 4 to 5% of finance raised are required to cover the operating costs of the venture capital management activity for specialized, early stage technology funds. Thus, without any external support, a substantial proportion of the total of a small fund could be "lost" in payments to its management rather than being allocated to the portfolio firms. This would discourage potential backers. The down-side protection schemes are mechanisms, either under the form of an instituted public supported insurance scheme or under the form of public sharing in the cost of investment, so that the loss burden on the venture capital fund in case the project fails decreases.

However, the general activity growth in the 1990s did not cancel out the deep disparities among European countries. Venture capital activity is unevenly spread across Europe. Moreover, after a period of crisis lasting until the first half of the 1990s, the US market started an impressive recovery. As a result, the European market now lags behind the US.

In cumulative terms, the UK market represents half of the private equity funds raised in the EU. France and Germany follow. Even measuring investment activity in percentage of gross domestic product, the UK ranks first in Europe both in the 1984/1989 period and in the 1996/2000 one. However France and Germany's average growth of new funds raised in the last five years is higher (especially in the case of Germany).

The comparative size of the private equity markets can be appraised also comparing the annual amount of private equity investment (see Table 4.7). The gap between the amount invested in venture capital per inhabitant in the US and in the EU has been increasing since the mid 1990s.

Besides size, there are other remarkable disparities across countries, starting from the sources of funding (see Table 4.9).

As for Europe, both bank subsidiaries and pension funds are the major sources of cash (contributing for 29% and 25% of the total amount of money raised over the 1996/2000 period), while in the US banks subsidiaries' cash infusion in venture capital funds reaches the 14% of the total funds collected only. In the US, pension funds are the main contributors to venture capital backing (46% of the total amount raised).

Within Europe however, while in the UK most of the cash raised over the 1996/2000 period came from pension funds, banks were the main sources in Germany, France, Italy and Spain. Apparently, pension funds tend to be less active in the latter two. This is probably due both to the limited amount of money pension funds raise and to the limits the law imposes to their investing activity. In Germany and France though, there is a notable increase in pension funds contribution to venture capital financing over the most recent years (see Table 4.9). In Germany, even insurance companies have increased their participation in the market.

Table 4.7. Investments in year (Source: Ooghe et al. 1991; EVCA Yearbook and NVCA Yearbook)

	Avg. 1984/ 89	1996	1997	1998	1999	2000	Avg. 1996/00	Avg. 1996/00 growth
Europe°	2,59	8,62	10,96	16,20	23,55	32,30	18,33	39,37%
France	0,44	1,08	1,43	2,00	2,64	4,90	2,41	47,49%
Germany	0,10	0,91	1,51	2,20	2,96	4,40	2,40	48,76%
Italy	0,16	0,66	0,68	1,05	1,67	2,74	1,36	45,25%
Spain	0,07	0,25	0,30	0,41	0,68	1,04	0,53	44,51%
UK	1,48	3,78	5,70	8,92	10,79	12,18	8,27	35,32%
US	3,29	11,21	17,21	21,98	59,37	103,49	42,65	81,41%

° European countries considered: Austria, Belgium, Denmark, Finland, France, Germany, Greece, Iceland, Ireland, Italy, The Netherlands, Norway, Portugal, Spain, Sweden, Switzerland, United Kingdom.

Table 4.8. Investments in year/GDP

	Avg. 1984/89	1996	1997	1998	1999	2000	Avg. 1996/00
France	0,03%	0,07%	0,10%	0,14%	0,21%	0,38%	0,18%
Germany	0,01%	0,04%	0,07%	0,10%	0,16%	0,23%	0,12%
Italy	0,02%	0,05%	0,06%	0,10%	0,16%	0,25%	0,13%
Spain	0,03%	0,04%	0,05%	0,12%	0,13%	0,19%	0,11%
UK	0,22%	0,32%	0,48%	0,64%	0,92%	0,86%	0,64%
US	0,07%	0,14%	0,21%	0,25%	0,64%	1,05%	0,46%

Source: Ooghe et al. 1991; EVCA Yearbook and NVCA Yearbook.

Table 4.9. Sources of funds by investor type

Avg. 1984/89	Cor- porate Inves- tors	Private Indivi- duals	Gov. Agen- cies	Bank subsidia- ries.	Pen- sion Fund s	Insu- rance Comp.	Aca- demic Institut.	Other
Europe°	8%	5%	10%	28%	17%	12%	0%	18%****
France	11%	4%	2%	37%	3%	13%	0%	31%****
Germa-ny*	32%	3%	5%	41%	4%	7%	0%	8%****
Italy	8%	5%	15%	36%	1%	3%	0%	31%****
Spain	14%	3%	22%	40%	13%	1%	0%	7%****
UK	5%	4%	2%	19%	20%	11%	0%	39%****
US	13%	11%	Na	12%***	40%	12%***	Na	9%**

Avg. 1996/00	Corp orate Inves tors	Private Indi- viduals	Gov. Agen- cies	Bank subs.	Pen- sion Funds	Insur- ance Comp.	Aca- demi c In- sti- tut.	Other
Europe°	10%	7%	4%	29%	25%	14%	1%	10,6%****
France	11%	4%	2%	32%	10%	10%	0%	31,7%****
Germa-ny*	8%	8%	9%	43%	14%	13%	0%	5,72%****
Italy	14%	14%	1%	43%	7%	4%	0%	17,05%****
Spain	9%	6%	7%	42%	9%	5%	0%	21,83%****
UK	10%	5%	4%	19%	36%	14%	1%	12,38%****
US	14%	11%	Na	14%***	46%	14%***	Na	15%**

° European countries considered: Austria, Belgium, Denmark, Finland, France, Germany, Greece, Iceland, Ireland, Italy, The Netherlands, Norway, Portugal, Spain, Sweden, Switzerland, United Kingdom.
*for 1984/1989 data onWest Germany only
** Endowments and Foundations.
***Insurance and financial institutions.
****Fund of funds, capital markets and not available data.

Source: Ooghe et al., 1991, EVCA Yearbook and NVCA Yearbook.

The distribution of investments by funding stage is documented in Table 4.10.

Compared to the US, European venture capitalists as a whole tend to favor later stage financing, with seed and start-up investments collecting only a tiny fraction of funds committed. Investors seems to be attracted by the less risky later stages as they offer apparently easier and faster opportunities to achieve returns. Among the European countries, Germany has the highest portion of early stage investments, while UK and Italy have the lowest. Notably in the UK early stage investing has dropped dramatically over the last 10 years[17].

Table 4.10. Investment Characteristics: Stages in % by amount invested

1984/1989	Early (Seed + Start-up)	Expansion	Repl. Capital	Buyout
Europe°	14%	44%	5%	36%
France	12%	56%	11%	17%
Germany*	20%	44%	1%	34%
Italy	11%	45%	34%	10%
Spain	36%	41%	11%	10%
UK	11%	37%	1%	50%
US	30%	46%	4%	21%
1996/2000	Early (Seed + Start-up)	Expansion	Repl. Capital	Buyout
Europe°	11%	34%	6%	48%
France	15%	37%	14%	34%
Germany	38%	50%	3%	19%
Italy	12%	32%	14%	41%
Spain	12%	59%	5%	24%
UK	4%	24%	5%	67%
US	25%	43%	21%	10%

Source: Ooghe et al. 1991; EVCA Yearbook and NVCA Yearbook.

Analyzing investment patterns by sector, even though the absolute amount of money channelled towards high technology sectors has increased dramatically in the last years, in Europe consumer-related products and services account for the main share of total investments and investments in non high-tech sectors are 71% of the total.

---

[17] In this respect, Christofidis and Debande 2001, underscore a correlation existing between the funds raised by pension funds and the allocation of investment to later stages, reflecting the greater risk aversion of pension funds.

While in the US the investing share in high-tech sectors has been 80% over the 1996-2000 period, in Europe the combined technology based sectors represent around 29%. However, there are wide discrepancies among European countries. The share of high tech investing in Germany is almost 37% of the total (though decreasing from 59% during the 1984/1989 years) whereas Italy ranks lowest with 14%.

Among high-technology sectors, Internet/digital content has attracted the bulk of venture capital investment. Biotechnology has attracted investors mainly in Germany (where the percentage of high-tech investing is the highest among European countries)[18].

---

[18] New technology based firms have specific financing needs linked to the type of products under development. For instance, in the biotechnology businesses, product inception to market is traditionally a time consuming process, which covers several years. In such a case very little cash is available to the financial backers of the firm before divesting. As a result, equity finance is the almost only usable source of money. On the other hand, in the software business, the product has to be brought to the market in a very short timeframe. Income generation is rapid and allows the funding with a mixture of debt and equity. Moreover, ventures operating in the software business need a shorter lapse of time to get to the divesting stage. This may make this industry more attractive than other high-tech sectors for risk averse investors. Therefore, comparing European venture capitalists' investment activity by sector with Americans', the lower investing in biotechnology in Europe confirms the higher risk aversion of European venture capitalists.

Table 4.11. Investment Characteristics: Sector distribution in % by amount invested

Average Investment Characteristics over the period 1984/1989

	Communications	Computer related	Other Electronics Related	Bio-tech.	Medical/Health Related	Total high-tech invest.	Industrial/Energy	Consumer related	Other Products*	Total non high-tech invest.
Europe	3%	9%	6%	4%	3%	25%	2%	23%	50%	75%
France	3%	11%	11%	3%	5%	33%	0%	18%	50%	68%
Germany	11%	30%	6%	11%	1%	59%	0%	7%	34%	41%
Italy	2%	5%	4%	1%	1%	13%	2%	23%	33%	58%
Spain	3%	11%	2%	0%	5%	21%	4%	14%	61%	79%
UK	3%	7%	4%	2%	3%	19%	1%	30%	50%	81%
US	11%	25%	10%	6%	12%	64%	1%	11%	24%	36%

Average Investment Characteristics over the period 1996/2000

	Communications	Computer related	Other Electronics Related	Bio-tech.	Medical/Health Related	Total high-tech invest.	Industrial/Energy	Consumer related	Other Products*	Total non high-tech invest.
Europe	9%	9%	4%	3%	5%	29%	27%	19%	25%	71%
France	12%	9%	5%	2%	5%	33%	23%	19%	26%	67%
Germany	8%	14%	4%	8%	3%	37%	25%	10%	29%	64%
Italy	9%	2%	1%	1%	1%	14%	33%	17%	36%	86%
Spain	9%	8%	2%	1%	3%	22%	25%	21%	32%	78%
UK	7%	7%	3%	2%	7%	26%	25%	25%	25%	74%
US	42%	18%	5%	6%	9%	80%	3%	6%	11%	20%

° European countries considered: Austria, Belgium, Denmark, Finland, France, Germany, Greece, Iceland, Ireland, Italy, The Netherlands, Norway, Portugal, Spain, Sweden, Switzerland, United Kingdom.
Other products comprises: industrial products and services, chemical and materials, industrial automation, other manufacturing, transportation, financial services, other services, agriculture, construction, mining, utilities and conglomerates.

Source: Ooghe et al. 1991; EVCA Yearbook and NVCA Yearbook.

Divestment routes take also different forms in the European countries and the US. While in the US the ratio divestments by IPOs over divestments by trade sale is 0,84, in Europe the number of divestments by trade sales over the 1996/2000 period has been fivefold the number of public offerings (of which IPOs are a fraction only). However, besides the UK case, public offerings in Europe are becoming a more significant exit route (see Table 4.15)[19].

The recourse to exit mechanisms differs among European countries. In the UK, trade sales remained the most important exit route until 1998. Trade sales then dropped in 1999 to 22% from 51%, while divestment by other means (by repayment of preferential shares and principal loans) became the preferred exit route in 1999 and 2000. Until 1999, the UK had the highest level of exit through public offerings. The 2000 bear IPO market discouraged exit by public offering, which in 2000 accounted for 7% of divestments only.

In France, even if trade sales are the most frequent divestment way, their amount fell considerably between 1998 and 2000, while in spite of a bear IPOs market, the amount divested through public offerings kept on being almost stable over the same years. In Germany, the 1998-2000 period witnessed a strong growth of divestment by means of public offerings, mostly due to the increasing number of IPOs. This reflects the dynamism of the Neuter Market, established in Frankfurt in March 1997 to enable young fast-growing companies to float their shares more easily. In Italy trade sales remain the most popular exit route, representing more than 50% of the amount cashed by venture capitalists. Public offerings and IPOs are becoming a more important exit route as the Italian stock exchange has gone through significant reforms leading to the creation of the "Nuovo Mercato" in June 1999, to sustain the flotation of new technology based firms.

---

[19] The percentage of divestments by public offering in the UK was 35% over the 1991/1995 period. It dramatically dropped to 16% on average in the following years (1997/2000).

Table 4.12. Divestment Characteristics: amount per type (data expressed in billions of dollars)

1997/00 avg.	Di-vestment by trade sale	Divestment by public offering		Sale of quoted equity	Di-vestment by write-off	Di-vestment by Other means	Total Di-vestment
		Total	Di-vestment by flotation				
Europe	3,28	1,33	0,47	0,61	0,59	2,53	7,72
France	1,00	0,31	0,04	0,20	0,08	0,29	1,68
Germany	0,28	0,10	0,05	0,02	0,13	0,34	0,86
Italy	0,20	0,05	0,02	0,02	0,01	0,13	0,39
Spain	0,09	0,03	0,02	0,01	0,01	0,07	0,19
UK	1,26	0,56	0,32	0,28	0,22	1,50	3,54

° European countries considered: Austria, Belgium, Denmark, Finland, France, Germany, Greece, Iceland, Ireland, Italy, The Netherlands, Norway, Portugal, Spain, Sweden, Switzerland, United Kingdom.

Source: EVCA Yearbook, various issues.

Table 4.13. Divestment Characteristics: percentage per type

1997/00 avg.	Di-vestment by trade sale	Divestment by public offering		Sale of quoted equity	Di-vestment by write-off	Di-vestment by Other means
		Total	Di-vestment by flotation			
Europe	42,46%	17,18%	6,13%	7,85%	7,65%	32,70%
France	59,57%	18,18%	2,42%	12,05%	4,77%	17,47%
Germany	32,52%	11,40%	5,85%	2,44%	15,04%	39,48%
Italy	51,39%	11,94%	5,19%	5,11%	3,47%	33,21%
Spain	44,90%	15,72%	10,58%	5,78%	4,73%	34,69%
UK	35,56%	15,74%	8,95%	7,78%	6,34%	42,36%

° European countries considered: Austria, Belgium, Denmark, Finland, France, Germany, Greece, Iceland, Ireland, Italy, The Netherlands, Norway, Portugal, Spain, Sweden, Switzerland, United Kingdom.

Source: EVCA Yearbook, various issues.

Table 4.14.

Divestment Characteristics: Number of Venture Backed IPOs

	1996	1997	1998	1999	2000
Europe°*	Na	na	239	149	249
US	279	137	78	258	231

Divestments Characteristics: Number of Venture Backed M&As

	1996	1997	1998	1999	2000
Europe°*	na	Na	965	1241	1308
US	115	159	197	225	275

° European countries considered: Austria, Belgium, Denmark, Finland, France, Germany, Greece, Iceland, Ireland, Italy, The Netherlands, Norway, Portugal, Spain, Sweden, Switzerland, United Kingdom.
* European Data refers to private equity divestments by flotation and trade sale.

Source: NVCA Yearbook and EVCA Yearbook, various issues.

Table 4.15. Percentage of divestments that are public offerings

	Average 1991/1995 (1)	1997	1998	1999	2000	Avg. 1997/00
Europe°	Na	15%	19%	21%	14%	18%
France	11%	19%	17%	16%	16%	18%
Germa-ny	Na	3%	15%	18%	12%	11%
Italy	4%	11%	9%	10%	17%	12%
Spain	1%	11%	28%	16%	12%	16%
UK	35%	15%	19%	26%	7%	16%

° European countries considered: Austria, Belgium, Denmark, Finland, France, Germany, Greece, Iceland, Ireland, Italy, The Netherlands, Norway, Portugal, Spain, Sweden, Switzerland, United Kingdom.

(1) Source: Jeng and Wells 2000.
(2) Source: EVCA Yearbook, various issues.

Table 4.16. Percentage of divestments that are trade sales

	Average 1991/1995	1997	1998	1999	2000	Avg. 1997/00
Europe°	na	49%	54%	37%	33%	42%
France	47%	56%	72%	58%	39%	60%
Germany	na	35%	25%	24%	40%	33%
Italy	61%	66%	58%	37%	56%	51%
Spain	42%	47%	36%	47%	47%	45%
UK	30%	50%	51%	22%	24%	36%

° European countries considered: Austria, Belgium, Denmark, Finland, France, Germany, Greece, Iceland, Ireland, Italy, The Netherlands, Norway, Portugal, Spain, Sweden, Switzerland, United Kingdom.

Source: Ooghe et al. 1991; EVCA Yearbook and NVCA Yearbook.

Table 4.17. Percentage of divestments that are write offs

	1997	1998	1999	2000	Avg. 1997/00
Europe°	12%	6%	7%	8%	8%
France	12%	2%	3%	6%	5%
Germany	16%	2%	19%	18%	15%
Italy	4%	7%	2%	2%	3%
Spain	5%	6%	4%	4%	5%
UK	11%	5%	5%	5%	6%

° European countries considered: Austria, Belgium, Denmark, Finland, France, Germany, Greece, Iceland, Ireland, Italy, The Netherlands, Norway, Portugal, Spain, Sweden, Switzerland, United Kingdom.

Source: EVCA Yearbook, various issues.

US and European venture capital funds' performance also differ. First of all, European venture capital funds return show a better result for later stage investments than for early stage ones (see Table 20). This may explain the decreasing percentage of seed and start-up investments over the total amount devoted to venture capital activity during the 1990s (see Table 4.10).

Moreover, comparing the European venture capital funds' return with other public asset classes, their performance looks poor for early stage funds. Even if results get better for later stage funds, they are probably still not good enough to compensate investors for the risk taken. In fact, lower risk asset classes' return is comparable.

On the other hand, in the US early stage investments return is higher than less risky later stage ones. The long term returns to US venture capital funds demonstrated a better performance than the European ones as well, probably thanks to the higher recourse to IPOs in exiting.

Table 4.18. Cumulative net IRRs of European mature private equity funds

	Cumulative IRR (%) from inception to 31st December 1996			Cumulative IRR (%) from inception to 31st December 1999		
	Pooled	Median	To p quarter	Pooled	Median	To p quarter
Fund Type						
Early stage	5,7	4,5	27,2	10,8	9,2	42,9
Development*	7,3	5,4	18,7	11,9	7,2	29,8
Buyout	17,6	15,5	41,9	19,6	12,6	43,9
Generalist	19,4	1,3	22,9	12,4	6	22,6
All private equity	18,6	6,6	29,1	14,5	8,8	33,9

Source: Christofidis and Debande 2001.

Table 4.19. Mature private equity and comparators, net investment horizon returns to 31st December 1999

	1 Year	3 Years	5 Years	10 Years
Early stage				
European Venture Capital	17,6	19,9	19,6	15,0
US Venture Capital	247,9	75,6	63,2	31,5
MSCI Equity	28,8	29,8	24,9	16,9
HBSC Small Cap	81,4	33,2	24,6	15,4
JP Morgan bond	-1,9	7,1	10,2	7,9
Development				
European Venture Capital	56,0	35,1	27,0	15,8
US Venture Capital	70,2	33,8	36,4	26,5
MSCI Equity	29,2	30,4	25,3	15,6
HBSC Small Cap	83,6	35,3	25,8	13,6
JP Morgan bond	-2,1	6,9	10,1	7,5
Generalist				
European Venture Capital	65,0	31,4	25,2	17,5
US Venture Capital	122,0	46,8	39,8	21,9
MSCI Equity	29,6	29,8	24,8	16,7
HBSC Small Cap	85,5	36,6	26,4	15,7
JP Morgan bond	-2,0	5,9	9,3	7,2
All equity				
European Venture Capital	54,3	29,5	24,9	16,3
US Venture Capital	146,2	53,8	46,4	25,2
MSCI Equity	29,0	28,9	24,7	17,0
HBSC Small Cap	83,8	34,5	26,1	16,4
JP Morgan bond	-2,6	5,4	8,4	6,9

Source: Christofidis and Debande 2001.

4.5 Conclusions

Venture capital markets in the US and in Europe differ substantially both in terms of size and substance. Various combined considerations can spell out the gap between the two: the efficiency of exit mechanisms, the regulatory environment, cultural differences in entrepreneurship, the maturity of the market.

The high efficiency of IPOs as an exit mechanism compared to the available alternatives partly explains the bigger size of US venture capital industry. Venture capitalists are willing to finance young businesses, knowing that an active IPOs market will allow them to cash out if the start-up firm succeeds. While parallel secondary markets activity in Europe is taking off and therefore exiting investments through flotation is becoming a feasible option, the importance of other factors, where the European lag is considerable, must not be understated.

The development of the venture capital industry is affected by regulatory changes in terms of capital gains taxation, investors legal obligation, labor regulation and the degree of intellectual property rights protection.

Reduction of capital gains taxation has a significant effect on fund raising, especially from small savers, as the British experience with Venture Capital Trusts has shown. Moreover, it can induce more corporate employees to become entrepreneurs, as most of the reward from starting a business comes from the appreciation on the equity of the company. However, only the UK and the Irish legislation provides for specific tax breaks for venture capital investments.

Investors' legal obligations affect venture capital fund raising as well. For example, the modification by the US Department of Labor of the Employment Retirement Income Security Act's "prudent man rule" in 1978, by allowing pension funds to invest in venture capital, made venture capitalists able to cut the costs of raising money. Nowadays, pension funds are the main source of money for venture capital in the US. On the other hand, in Europe, only Finland, Ireland, the Netherlands and the UK have no legal restrictions on investment in equities by pension funds.

A more flexible labor market encourage US employees to work for young, unstable companies and makes it easier to start a firm. Where strong layoff protections exist, as in Germany and Italy, the reduced mobility of workers impose costs on start-up businesses and discourage their formation.

Table 4.19. Average job tenure, 1999

	Level (years) all sectors
France	11,2
Germany	10,3
Italy	12,1
Spain	10,1
UK	8,3
US	6,7

Source: OECD.

Property rights protection a is a key issue in the development of a venture capital market oriented to high-technology sectors or early stage investment. By either allowing venture capitalists to patent results of R&D activity or have copyright protection, an effective intellectual property rights legislation provides a collateral in case of default thus reducing the risk backers perceive and therefore the cost of funds for venture capitalists. The supply of venture capital is also affected by risk and return patterns. Compared to other asset classes venture capital is considered to be at the more risky end of the investment spectrum. Therefore a higher return is required by investors, especially for early stage investments, which are perceived as riskier. While past performance for US funds justify the risk incurred by investors, in Europe venture capital funds results over the last ten years have been poor.

Finally, cultural differences in entrepreneurship are often quoted to explain the difference in the dynamism of venture capital market in the US and in Europe. European managers are considered to be less entrepreneurial and less willing to risk failure than their US counterparts (Bank of England 2000; Schefczyk and Gerpott 2000). However, the importance of this factor may be overstated. It is probably the family-oriented organization of small and medium European firms that do not encourage venture capitalists activity, as entrepreneurs are not willing to give up their firms' control, even if that would allow to acquire a higher financial stability and better managerial skills. As flotation becomes more frequent though, the additional value brought by venture capitalists clashes less with entrepreneurs' claims.

More worrying than the supposed lack of entrepreneurial spirit in Europe, is the lack of training in high-technology field of European venture capitalists, documented in some recent studies, which hinders investments in high-technology sectors (Van Osnabrugge and Robinson 1999).

Part 3

Venture Capital in Italy: Regulatory and Legal Issues

1 Regulations and Supervision: The Role of Central Bank

Anna Giuiusa

1.1 Introduction

The venture capital and private equity market in Italy, i.e. the offer of risk capital and strategic consultancy for development of business- has expanded considerably in the last years, even if the dimensions are still modest compared to the main countries in western Europe (France and Germany), without considering the Anglo-Saxon experience. The amount of the total private equity investments in Italy was around 9.048 billions of Lire[1] on June 30th, 2000.

The Italian private equity and venture capital investors belong to a market sector which is quite fragmented. The intermediaries supervised by the Bank of Italy are the commercial banks, the merchant banks (ex article 107 of Legislative Decree 385/93, hereafter referred to as Italian Banking Law) and the asset management companies (SGR) that manage closed-end funds. Investors of other categories than the mentioned ones are: subsidiaries of foreign banks, international funds with some investments in the domestic territory in order to get a diversified portfolio, insurance companies, initiatives by Italians constituted abroad, companies that invest "captive"[2] resources and private investors who invest their own available funds (so-called business angels).

1.2 The Risk Capital Market: The Importance of the Banks

A number of analyses of the financial structure of companies in the principal industrial countries[3] highlight some key characteristics of the Italian business system:

---

[1] Source: "The Italian association of venture capital investors".
[2] This kind of intermediaries collects financial resources from the companies in their group and invest in similar activities with the scope of developing the group through the growth of the companies appertaining to it.
[3] Bank of Italy 1988; Barca and Magnani 1989; Signorini 1993; Gersandi 1994

- the important role of the commercial bank system and the scarce percentage of the financial instruments listed on the regulated markets (bonds and commercial papers);

- a high degree of leverage with a large component of short term debt;

- the low level of share capital traded on the regulatory markets.

The explanations are numerous and can be summarised in the insufficient development of a risk capital market and institutional investors, and in the fiscal advantage of indebtedness. The ownership structure of the Italian companies, which is characterised by the high presence of family members who are reluctant to open up to the market with fear of losing control, also explains a lot, as well as the significant public presence in many companies.

The last years development of venture capital and private equity activities in Italy is a part of the assumption that the financial resources based on self- financing and debts are exhausted. The reduction of the gap between bank and market financing circuits requires a restructuring of the relationship between banks and companies which depends on variables connected to the development of the markets and a gradual change in the credit intermediary's role.

Regarding to the market variables there are three profiles that influence can be detected: the offer, the demand and their meeting point.

The essential elements that influence the offer are: the gradual reduction of public shareholdings in companies and the opening up to a wider shareholder base, as well as the introduction of fiscal incentives for quotations on the regulated markets.

The essential elements that influence the demand are: the gradual development of institutional investors and of a structured system for safeguarding minority shareholders.

The regulation and administration of the markets in a way to make it as fluid and secure as possible for trading and regulatory operations constitute the elements that influence the meeting point of the offer and the demand.

Regarding the credit institutions business, the relationship between the banks and the industrial companies has to be analysed from two aspects: the first one is related to the ownership links between the banking system and the industrial companies and the second one is related to the consequences these links can have on the industrial sector's credit facility system .

Under the first aspect the choice of Italian legislator is to separate the banks from the industrial companies; in order to realize this, a limit of shareholdings by non-financial companies in bank's capital is imposed. In particular article 19 in the Italian Banking Law provides for the prohibition of parties, carrying out business activities in sectors other than banking (even through controlled companies), to acquire shares when, added to those already held, would result in a holding that exceeds 15% of the voting capital of a bank or, in any case, result in the control of the bank.

Behind these provisions lies the need to guarantee the independence of capital allocation for the banking system in their selection of profitable initiatives avoiding the pressure that can be exerted by customers owning a significant part of the shareholdings.

It is also provided that the Supervisory authority (Bank of Italy) can request information about the composition of the shareholdings of banks both in the genetic phase and during the life of the credit institution. Bank of Italy has also to be informed whenever shareholdings of significant size are transferred; in some cases an authorization is required.

In fact, the shareholdings, held directly or indirectly, inferior to 15% but superior to 5%, or in any case- without considering these limits- resulting in the control of the bank, have to be authorized by the Bank of Italy (article 19, paragraph 1, 2, 3 Italian Banking Law). A notification requirement also come into force when reaching various levels of shareholdings (article 20 paragraph 1, Italian Banking Law.). The Supervisory authority base its valuation on the criterion of "sound an prudent management": shareholders with possible conflicts of interests in valuating risks correctly could influence negatively, not only the efficiency of the capital allocation, but also the bank's economical and financial balance.

The defence of the stability is also important in the rules related to banks and banking group's shareholdings (articles 53 and 67 of the Italian Banking Law). The current regulations[4] enacted by the Italian central bank, in accordance with the Credit Committee's resolutions, are in line with the principles of the Decree of the Minister of Finance of June 22, 1993, which preceded the Italian Banking Law, but are in accordance with the article 22 of the legislative Decree 481/92. This regulation, in contrast to the past, allows the credit institutions to acquire shareholdings in industrial companies, respecting the established limits. The different kind of investments can be separated in financial companies, industrial companies and "functional" companies.

According to the European Community law - article 51 of the 2000/12/ECD, that substitutes article 12 of the 89/646 ECD - credit institutions are prohibited to hold participations in non financial companies equal to an amount greater than 15% of the total value of the institution's own funds. The total amount of shareholdings in industrial companies cannot be

---

[4] The Supervisory Regulations for banks, Title IV, Chapter 9.

greater than 60% of these funds. In accordance with the European Community norms, italian regulations do not provide for an authorization of the Supervisory authority, but have a fixed a grid with maximum limits that may be held. The reason for having these limits lies in the need to avoid an excessive level of locked- up capital as well as to avoid the risk linked to the fluctuations in value of the shareholdings held and to the relayed repayment on shares compared to debt.

The general quantitative limit of undertaking shareholdings is equal to the so-called "free capital" arising from the difference between supervisory capital[5] and the total of assets and participations held by the company.

Following this, there is a limit related to single shareholdings ("concentration limit"), and a total limit, related to the total "non-financial" shareholdings ("overall limit"). Both limits are related to the supervisory capital of the bank and are divided into three levels, corresponding to three different categories of banks or banking groups: "ordinary banks", "qualified banks" and "specialised banks". In particular, the qualification of "qualified bank" can be obtained, with prior authorisation from the Bank of Italy, by intermediaries with a supervisory capital not inferior to Euro 1 billion and which fulfil the requirements related to the prudential rules on capital adequacy[6]. Institutions can obtain the qualification of "specialised bank" if they, in addition to the above mentioned characteristics, also have a liability structure characterized by a significant level of medium and long term deposits. The "ordinary banks" are identified by those that remain. The granting of the authorization by the Bank of Italy depends on an evaluation concerning: 1) the experience matured in the field of financial assistance to industrial companies as well as the results achieved 2) the adequacy of the organisation structure in customer selection 3) the bank's technical situation with reference to the concentration of risks, financial equilibrium and market risks.

---

[5] The supervisory capital is composed of the primary capital and the supplementary capital, after the deduction of participations that do not take part of the consolidated group. The capital, the reserves, the provisions for general banking risks and other funds and reserves form the primary capital which is included in the supervisory capital without any limits. The revaluation reserves, the subordinated liabilities and the credit risk reserves make up the aggregated supplementary capital which cannot exceed the amount of the primary capital.

[6] The major risks in the banking sector are credit and market risks, which contain the exchange risks, the risks connected to regulations, interest rate risks and concentration risks. For every risk there has been settled minimum capital requirements: the sum of all these requirements make up the total capital requirements for a bank (Supervisory regulations for banks, Title IV, Chapter 4, Section I, Paragraph 1)

The "concentration limit" is 3% of the supervisory capital for the "ordinary banks", 6% for the "qualified banks" and 15% for the "specialised banks". The "overall limit" is equal to 15% of the supervisory capital for the "ordinary banks", 50% for the "qualified banks" and 60% for the "specialised banks". These limits are less stringent taking into account that this kind of banks present an adequate financial structure and have matured a significant experience in the field of financial assistance to companies, and are therefore considered capable of managing the risks connected to a greater concentration of their capital in the industrial sector.

Finally, the conversion of receivables from companies in crisis into risk capital is recognised within the same context as restructuring projects subject to Supervision.

A further quantitative limit is the so-called "separation limit", equal to 15% of the share capital held in one company. This threshold − equal for all banks, independent of their size − may be overcome when the value of each holding and the sum of excess is within the limit of 1% of the supervisory capital of the bank or banking group (2% for the "qualified" and "specialised" banks). The "separation limit", unlike those ones previously mentioned, does not have its origins in the European Community regulations but was introduced directly by secondary regulations enacted by the Credit Committee and by the Bank of Italy.

The regulations are completed by:

- rules which limits the amount of medium/long term financing to companies (30% of the funds raised); the surpassing of this limit may be authorised for banks with particular financial and organisational structures;

- regulation which limits loans to a single large borrower or group of connected customers;

- rules that establish the support (under the form of guarantees) by the banks for the placement of debt securities on behalf of non-financial companies.

1.3 The Role of Non-Banking Intermediaries: the Merchant Banking Companies

Article 106 of the Italian Banking Law identifies those intermediaries who, performing one of the activities included therein (investments in shareholdings, provision of financing, provision of payment services or exchange intermediation), in compliance with certain conditions[7] are required to enrol in the relevant register held by the Ufficio Italiano dei Cambi (so-called "General Register"). Within this register, the successive article 107 identifies a subset of intermediaries who, in relation to the method of carrying out the activities, the size of the operations as well as the relationship between equity and debt, must enrol in a "Special Register", managed by the Bank of Italy. These intermediaries are subjected to a prudential rules' system.

Among the members of the Special Register, are found companies which, in their activity, carry out acquisitions of shareholdings on behalf of the public (merchant banking companies).

Article 6 of the Minister of Finance Decree of July 6, 1994 (issued based on article 106, paragraph 4 of the Italian Banking Law) defines the activity of acquisition of shareholdings as: the acquisition, holding and management of share capital rights in companies which create a situation of dependence between the two companies in order to the development of the acquired one. This activity, which should be considered performed on behalf of the public, must:

• be performed on behalf of third parties with professional manners;

• have a prepared way- out and the scope of transferring the shareholdings;

• have the purpose of company restructuring, production development or satisfy the financial requirements of the acquired company even through the raising of risk capital.

The enrolment in the Special Register is governed by the Ministry of Finance Decree (The Ministry of Finance Decree 13.5.96) that establishes the limit of size at which the company is obliged to register (certified volumes of financial activity equal to or superior of 100 billion of Lire or equity equal to or superior of 50 billion of Lire with reference to approved financial statements and valid for at least six months after the approval).

The intermediaries enrolled in the Special Register are supervised by the Bank of Italy, according to an amplified regulation which is connected to the regulation for the intermediaries enrolled only in the Register held by

---

[7] The conditions are:
– Legal form of società per azioni, accomandita per azioni, a responsabilità limitata o cooperativa;
– share capital non less than five times the lowest limit for a limited company;
– members and corporate officers satisfying experience and integrity requirements established by articles 108 and 109.

the UIC (so-called General Register), and is characterised by a higher level of strictness so as to reach the targets stated in article 5 of the Italian Banking Law[8].

The supervisory controls' structure on these intermediaries is quite simple. Other than integrity requirements for the shareholders and experience and integrity requirements for the corporate managers (that already must be held within the enrolment in the General Register), are requested only the notification of the shareholdings to the Bank of Italy. There are no limitations about the nature of the controlling shareholders of the intermediary, nor limits to the undertaking of shareholding by the intermediary. It is obviously understood that when the intermediary is part of a banking group the regulation, noted in the previous paragraphs, will be applicable also for the merchant banking companies, having as its reference parameter the consolidated supervisory capital.

Article 107 in the Italian Banking Law establishes that the financial intermediaries enrolled in the Special Register shall give important information to the Bank of Italy such as periodic results, financial statements and reports from extraordinary board meetings of special importance[9]. The supervisory authority also has the right to carry out on-site inspections.

According to the resolution of Credit Committee, the Bank of Italy has issued technical instructions for the financial statements and introduced regulations concerning the level of risks held by these intermediaries by establishing a method to calculate the supervisory capital, similar to the one used by the banks. The Bank of Italy has also established norms on risk concentration[10], on derivatives[11] and on foreign exchange risk[12]. Mini-

---

[8] The targets of article 5 of Italian Banking Law are: the sound and prudent management of the persons subject to supervision, the overall stability, efficiency and competitiveness of the financial system and the compliance with provisions concerning credit.

[9] Changes in regulations or statutes, mergers, disunions, liquidations.

[10] The scope of the regulations in risk concentration is to limit the financial intermediaries risk of instability deriving from single borrowers of amounts relevant in proportion to the supervisory capital. In order to keep a satisfying degree of diversification of the credit risk there is a global limit of the total amount of the so called "large risks" (financial positions that exceed 10% of the intermediary's supervisory capital) and also an individual limit for every single position. In particular, the financial intermediaries are requested to keep:

1. the total amount of large risks within the limit of eight times the supervisory capital (global limit);
2. every single risk position towards single borrowers within the limit of 25 percent of the supervisory capital (individual limit).

mum organisational requirements were also introduced for the intermediaries in the Special Register, in May 2000.

The merchant banking companies can be authorised (article 18 paragraph 3 of the Legislative Decree 58/98) to provide investments services such as dealing for own account on derivative instruments and placement of financial instruments. In this case the regulations related to the prudential rules on market risks established for the SIM (security- trading companies) will be applicable, other than the norms of investment services.

The Bank of Italy has the right to take the following extraordinary measures towards the intermediaries enrolled in the Special Register :

- prohibit the intermediaries to undertake new operations, due to violation of laws or regulations issued in accordance with the Italian Banking Law;

- erase the intermediary from the Special Register in the event of non compliance with the provision of article 106, paragraph 2 of the Italian Banking Law, where one of the conditions referred to in article 106, paragraph 3 is no longer met or in the event of serious violations of law or norms issued in accordance with the Italian Banking Law.

These limits can be exceeded (and instead be replaced by an individual limit of 40 percent of the supervisory capital) by the intermediaries that:
1. appertain to a banking group enrolled in the Special Register provide by art. 64 of Italian Banking Law)
2. appertain to EU banking groups on the conditions that:
    - they communicate to the Bank of Italy, enclosing all documents necessary to demonstrate their belonging to an EU banking group, to be subjected to the consolidated supervision provided by decree 92/30/CEE;
    - the parent company acts as guarantor for the controlled Italian company (Supervisory regulations for intermediaries appertaining to the Special register 1st part, Chapter V section IV).

[11] Hedge operations with derivatives are carried through by financial intermediaries to protect the ordinary operations from interest rate risks, exchange rate risks or market price fluctuations. The total amount of operations in derivatives, with or without underlying security (with the exception for instruments on exchange rates), without the scope of hedging, cannot exceed twice the supervisory capital (Supervisory regulations for intermediaries appertaining to the Special register 1st part, Chapter V section II).

[12] To limit the exchange rate risk, the intermediaries are obliged to keep their "open net position in currencies" under a limit of twice the supervisory capital (for an explanation of how to calculate the "open net position in currencies" see Supervisory regulations for intermediaries appertaining to the Special register 1st part, Chapter V section III).

At December 31, 2000 there were 18 merchant banking companies enrolled in the Special Register; of which 11 were part of banking groups, 3 were part of private financial groups, 3 belonged to public bodies, and 1 to the co-operative system. From an analysis of the companies' financial statements resulted that the composition of the financial structure was strongly unbalanced towards the banking sector, which reflects the ownership structure of the companies: 66% of the reserves were held by the banking groups which also were financing the merchant banking companies up to 89%.; corporate bonds are very rarely used.

The assets were characterized by considerable financial support through direct financing as well as syndicated loans, supplied by the merchant banking companies to the companies in which they have holdings (the loans paid at the end of 2000 amounted to Lire 2,965 billion). Many merchant banking companies also supply a multitude of services to the market, not always related to the activity of shareholding undertakings, that end up being one of their principal income sources.

1.4 The Role of the Non-Banking Intermediaries: Stock Market Closed-End Funds

The stock market closed-end funds were introduced to our financial system with the law of August 14, 1993 n. 344 in order to favour investments in companies non listed on regulatory markets. In these kind of funds, the stability of the resources necessary when making medium/long term investments– typical for equity activities– is assured by the fact that the right to reimbursement of the quota is recognised at pre-determined expiry dates.

The discipline of asset management – and thus also of stock market closed-end funds – has been simplified by the legislative decree 58/98 and the implementation of the related norms. Other than the "de-specialisation" of the asset management companies that now simultaneously can manage open and closed funds, and take on mandates to manage individuals, following is recalled with particular reference to closed funds:

1. the possibility to hold majority shareholdings in the share capital of non-listed companies, previously excluded by the Law 344/93;

2. the faculty to reserve the participations in the fund to only "qualified investors"[13], benefiting in this way of a significant space of self-regulation in the investment activity with exception to the general limits imposed by the Supervision with regard to the funds offered to the public;

3. greater flexibility consented in the procedure of participation in funds with reference to, for example, the possibility to contemplate payment on call for reserved closed- end funds, the acceleration of the procedures connected to a possible increase or re-dimensioning of the fund's initial capital, the elimination of the general obligation of quotation for non-reserved closed-end funds (this obligation remains only in case the relative quotas are inferior to a pre- determined minimum amount).

The other principal characteristics of the closed-end funds provided for by the regulations are:

- the duration of the initiative cannot exceed thirty years, except for a maximum extension of three years in order to complete the disposal of the investments, on conditions that this faculty is provided for in the regulations and approved by the Bank of Italy;

- the minimum amount of each subscription that for the funds which are prevalently invested in non-quoted financial instruments, may not be inferior to Euro 50,000;

- a frequent (at least twice a year) valuation of the unit and publication of the same;

- the quotation, that is obligatory when the unit's amount is inferior to Euro 25,000.

The current supervisory mechanism for closed-end funds is essentially centred on:

- the asset management company's possession of the regulation requirements for intermediaries provide collective portfolio management services (so called SGR)[14]. The SGR, once authorised by the Bank of Italy

---

[13] Art. 37 paragraph 1 lett.B9 Legislative Decree 58/98 and art. 15 Ministry of Finance Decree 24.5.99 n.228.
[14] The authorization requirements consist in:
- the SGR's possession of a capital that exceeds the established minimum limit (one million of Euro), the corporate officer's possession of the experience and integrity requirements, and the member's possession of integrity requirements;

and registered in the relevant register, must subject the fund's internal rules to the of the Supervisory authority and follow the prudential rules in keeping the supervisory capital equal to the equity coverage of "other risks" (established as 25% of fixed operating costs resulting from the last financial statement) and the equity requisites for closed-end funds (obligatory investment quota in each closed-end fund managed, equal to 2% of the equity of the fund; this investment however is not obligatory for the reserved funds for qualified investors);

the limits of risk diversification in the funds: in particular, the investments in a single non- quoted company must be maintained within 20% of the total investments of the fund, while those in shares of more than one issuer belonging to the same group may not be greater than 30% of the activity. The regulations for closed-end funds reserved for "qualified investors" however, as already mentioned, provide for different limits;

- the notification requirements when the quotas in the fund are offered to the public or when the fund is listed on a regulatory market. The preparation of an information memorandum is not required when the offering is turned to a number of interested parties not greater than two hundred and a minimum unitary investment not inferior to Euro 250,000 is requested;

- the prohibition of providing loans different from forward operations on financial instruments, of taking on loans greater than 10% of the total net value of the fund, of selling uncovered financial instruments, of investing in financial instruments issued by the SGR and non-quoted financial instruments issued by companies of the group to which the SGR belongs, of acquiring – with reference to each operation in financial instruments made by companies of the group to which the SGR belongs – financial instruments superior to 60% of the total amount committed for placement by each company;

- the same limits for diversification and holding of voting rights provided for open funds on the part of the portfolio of the fund invested in quoted financial instruments and in derivatives.

At December 31, 2000 there were 14 operating closed-end funds, of which 11 were issued by the banking sector (Annual Reports 2000). The funds' regulations have similar characteristics such as duration of the fund

---

- the shareholders' capacity to guarantee sound and prudent management of the SGR and a group structure that not adventures the supervisory authority's effectiveness.

(one decade), minimum threshold of subscription (Lire 100 billion) and investment policy. This latter, in most of the cases, is directed towards development operations in medium size companies operating in industrial, commercial and service sectors; in general investments in newly created companies are excluded. Some funds are specialised in one market sector or geographic region.

The composition of the sources of the closed-end funds (equal to Lire 1,454 billion) by typology of subscriber, shows that the majority comes from private investors (approximately 55%) and banks (34%) while the weight of foundations, insurance companies and pension funds is negligible. 46% of the resources were invested in non-quoted companies with a large concentration on the North-Western parts of the country and operating in traditional production sectors. The strategy applied by the asset management companies was that of acquiring minority interests. It is recalled that Italian funds, up to the issuing of the new regulations implemented by the Legislative Decree 58/98 could not hold interests in non-quoted companies superior to 30% of the capital or could not , in any case, assume the control of a company.

1.5 The Activity of Private Equity in Banking Groups: a Vision at the Consolidated Level

The actors in the private equity sector are all related in some way to banks, intermediaries with origins in banks or in any case related to poly-functional banking groups[15].

The norms and regulations outlined are set, on the one hand, in a manner to avoid the possibility of "regulatory arbitration" – through homogenisation of the prudential rules, established for all intermediaries independent of their legal form or organisational model – on the other hand, with a disposition of a wide array of instruments that allow them to structure their relationship with the companies in ways that are different from the traditional system of providing credit facilities.

The activity of private equity, carried out within the "universal bank" or within a banking group by specialised intermediaries, develop a broader and more enduring relationship between intermediary and company, where

---

[15] Organizational structures alternative to multidivisional model; common in all productive sectors and in companies of all different sizes. Only in the banking sector there are regulations for this kind of organic structures. The poly- functional banking group is defined as the together of companies - controlled by the same parent company – operating in different areas provided by EC Directive.

the intermediary is transformed from an undifferentiated credit provider into an organiser of the whole financial structure of the company, developing financial partnerships in different shapes, adequate to the specific operations and to development projects and commercial strategies of the companies. The strategic and financial advice (planning of the investment and development initiatives) can also be concentrated in assisting with the issuing of bonds on the market, in financing extraordinary operations (M&A, LBO) or in quotation of the company on the Stock Exchange.

This approach, that is based on the company's choice of an intermediary as reference and of a greater disclosure of the company to this intermediary, results in a greater consciousness which makes it easier in many ways for the intermediary to manage the various risks undertaken.

Firstly, the practice of several credit lines - characterised by a co-partnership by several banks in the financing of the company - lowers the amount of credit given by each bank and thus the costs of selection and monitoring sustained. This encourages "assurance" behaviour in the risk management between the banks providing the credit and the maintaining of significant information asymmetry that reduces the possibility of control and discipline by the bank towards the company.

The "reference" bank could impose a management more aware of risks by controlling the productive use of the funds and assuring greater possibility to recover the amounts paid in case of crisis through a timely evaluation of its reversibility and a following preparation of restructuring measures. The bank could exercise a real "corporate governance" function in the financed company, through the control of the strategies and the development prospects by requesting, in protection of its rights and reputation, a sound and prudent management.

The banks suffer from low capitalisation of the financed companies and are exposed to the trends in the economical cycles: a more balanced structure between risk capital and indebtedness of the companies would allow the banking system to share with the market the risk connected to the financing, placing themselves on a more favourable point of the risk-return curve.

Apart from improving the conditions for risk management, the approach of the bank as a reference for the company, also evidences positive aspects concerning the profitability from the relationship.

The undertaking of the role as bank of reference consolidates the relationship with the client favouring the evolution towards a model of financial assistance that includes a wide range of services such as introduction of the company on the financial markets through placing of debt or capital securities. The "dis-intermediation" of the banks is compensated by advantages of diversification of the income sources: the offer of high value added services such as consultancy to the companies may allow the banks to stabilise cash flows and to free the structure of their income statement from the dependence on interest margins.

The detailed knowledge of the company's financial requirements also favours adoption of cross selling policies; the possibility of a unitary evaluation of the relationship consent to make a pricing based on the overall correlation between risk and return.

The banks' or banking groups' use of operating leverages placed at disposal by the legislator is rather limited. In fact, from an analysis of the banks' financial statements related to year 2000 it is noted that, almost eight years after the introduction of the regulation regarding shareholdings in industrial companies, the national banking system does not appear to have taken on particular commitments with the industrial world.

In fact, with reference to the "separation" limit approximately one third of the industrial shareholdings exceed the limit of 15% of the share capital, although they remain below the "ceiling" of the separation threshold (1% or 2% of the participating bank's supervisory capital). Regarding the "overall" and "concentration" limits, the thresholds for the three categories of banks have been largely respected.

The free capital that makes up the bank's margin for further acquisition of interests in other companies was, in June 2000, equal to Lire 59,047 (the consolidated supervisory capital at the same date amounted to Lire 188,589). The shareholdings in non-financial companies amounted to Lire 10,074 billion prevalently concentrated to the banks in Central-Northern Italy (Lire 9,956 billion) which have much more assets than the banks in Southern Italy: the supervisory capital allocated by the banks in the Central-North was equal to Lire 177,788 with a free-capital of 95% of the total consolidated free capital (Lire 54.987 billion) (Bank of Italy 2000).

Within the context of th current growth process and of reorganisation and strategic repositioning of the banks and banking groups, the private equity and venture capital sector is under development: these activities are carried out by autonomous business units within "universal bank" or by specialised operators constituted by separate legal entities within the multifunctional banking groups (merchant banking companies and closed-end funds).

In both cases the basic idea is to focalise on the single components of the bank or banking group in activities that requires specific expertise, creating "centres of excellence" with productive and managerial competence that respond to critical factors of success in the different business areas.

These activities cover specialised connotations both under the profile of investment considerations in the various alternatives of financing of start-up, expansion financing, turn-around financing and choice of target companies – which in some cases present high growth prospects but are characterised by high risk - and under the profile of managing the relationship with the company centred on the attainment of a return on capital invested through the sale of a part of the share capital, with a private agreement (trade sale) or through quotation on the stock market (IPO).

Thus it appears evident that the evaluation parameters of the relationship between risk and return of these activities are different, and at times antithetical, to those characterised by a stable presence in the share capital of a company providing assistance at 360 degrees, which is the approach of the "reference bank".

It is also necessary to take into account the characteristics of the financial sources of the assets invested by the SGR that manages closed-end funds: in fact, these are not own funds but raised by third parties. Therefore the SGR must, in any case, be guaranteed independence in making allocation choices, even if the logic of the productive processes within the banking group is characterised by realisation of synergies through an integrated management of the customer relationships. The banking group's synergies can be utilized, in the context of the production process of the management services, in the phase of acquisition of investment proposals (deal flow) in which the channel of the banking group's territorial network and knowledge of the entrepreneurial structure in the geographic area certainly have strategic importance.

The contribution of the commercial banks could also be extended to a first screening of the proposals received from the market, but often they do not have the experience and professional skills necessary to perform this role. After the screening follows analysis of the investment project and its possibility to succeed, evaluation of aspects relative to the financial equilibrium of the target company, of the market sector it belongs to and its growth prospects. These activities are carried out by the management team who prepare an investment proposal for the board of directors of the fund. It is in this phase of decision-making that it is necessary to guarantee total autonomy to the managerial and control functions of the fund. It is necessary to exclude the cumulating responsibilities of the parties involved in the deliberation phases within the banking group (board of directors, management bodies) in order to eliminate possible conflicts of interest that could appear in the hypothesis of intervention between a customer committed to the banking group through the funds raised by other institutional investors.

Also the monitoring phase is characterised by a peculiar approach: forms of controls are realised through the mechanisms of corporate governance in the companies in which shareholdings are held. In fact, the intermediary normally acquires the role of "insider" in the company through a presence in the principal decision-making bodies. Occasionally provision is made for the preparation of a periodic report by the company that allows the intermediary to control the trend of the company's principal economic- and financial indicators and the state of progress of the business plan.

1.6 Conclusions

The Supervisory authority's approach towards the questions of the relationship between banks and companies – in which the problems related to the activities of private equity and venture capital enter – is much more complex than those related to the separation regulations. The allocation of the public's financial resources has to be consistent with a balanced development of the productive structure. Once the parameters and the rules by the Supervisory authority have been defined, it is up to the intermediaries to make the moves with their own strategy, exploiting the operating possibilities available.

The role of the Supervisory authority will be realised in the ascertainment of the existence of the conditions that assure an aware management of the risks both under the profile of the respect of the norms that impose capital adequacy against the risks undertaken and under the profile of the appropriateness of the organisational and procedural measures taken in place by the banks and financial intermediaries in consideration of their current activity.

With reference to the above mentioned, some principal keys have been fixed[16] that, while the autonomous responsibility of the intermediaries in relation to the choice of internal control remain, are centred on:

- the necessary separation between operational and controlling functions and the elimination of conflicts of interest in the assignment of authority;

- the provision of organisational solutions that are capable of identifying, measuring and monitoring adequately all the risks undertaken or that will be undertaken in the different operating segments (monitoring, measuring and control system of the risks adequate for the complexity and size of the activity performed);

---

[16] The supervisory regulations for banks, Title IV, Chapter 11, Section II Paragraph .1.

- the structuring of reliable information systems and appropriate reporting procedures to the different management levels to which the different control functions are assigned.

By analysing practices adopted by the most efficient intermediaries (the "best practices") the Supervisory authority identifies benchmarks which the intermediaries can use in their evaluation of managerial solutions and the instruments adopted in governing risks. This favours the development of organisational structure, control procedures and information systems in a way to guarantee a sound and prudent management.

Table 1.1. Summary overview for the holdings in non-financial companies

	HOLDING LIMITS		
	"Concentration" limit	"Overall" limit	"Separation" limit
"Ordinary" Banks	3% of the supervisory capital	15% of the supervisory capital	15% of the supervisory capital (2)
"Qualified" Banks	6% of the supervisory capital	50% of the supervisory capital	15% of the capital of the invested company (3)
"Specialised" Banks	15% of the supervisory capital	60% of the supervisory capital (4)	15% of the capital of the invested company (3)

(1)   At least 50% of the total ceiling must be utilised for the acquisition of shareholdings in companies quoted on the regulatory markets.

(2)   The limit may be surpassed if the value of the shareholding is contained with 1% of the supervisory capital of the participant and the sum of the excesses of the 15% limit is within 1% of the supervisory capital.

(3)   The limit may be surpassed if the value of the shareholding is contained with 2% of the of the participant and the sum of the excesses of the 15% limit is within 2% of the supervisory capital .

(4)   This limits refer only to the "qualified" investments.

Table 1.2. Merchant banks registered in the Special List in accordance with Article 107 of the Legislative Decree 385/93 at December 31, 2000

Name	Group
Arca Merchant Spa	Banking consortium
Euromobiliare Corporate Finance Spa	Credem Group
Finanziaria Senese di Sviluppo Spa	Public Body
Finanziaria Banca Agricola Mantovana Spa	MPS Group
Finec Merchant Spa	Cooperative
Fin-Eco Merchant Spa	Bipop-Carire Group
Intek Spa	Non-banking consortium
MB Finstrutture Spa	Mediobanca Group
Mittel Generale Investimenti Spa	Mittel
NHS – Nuova Holding Sanpaolo Spa	Sanpaolo IMI Group
Nordest Merchant Spa	Banca Pop. Vicenza Group
Sade Finanziaria Spa	Mediobanca Group
Sanpaolo Imi Private Equity Spa	Sanpaolo IMI Group
SFIRS Spa	Enti pubblici territoriali
Sofipa Spa	Bancaroma Group
SO.PA.F Spa	Vender family
UBS Capital spa	UBS
Veneto Sviluppo Spa	Enti pubblici territoriali

Table 1.3. Operative closed-end funds at December 31, 2000

Name of management company	Name of fund	Group
Arca Impresa gestioni SGR	Arca Impresa Arca Impresa 2000	Banking consortium
Eptafund Sgr	Eptasviluppo	Banking consortium
S+R Investimenti e Gestioni Sgr	Obiettivo impresa Rolo Impresa	Unicredito Group
Fidia Spa	Prudentia	Banking consortium
BNL Gestioni Sgr	BNL Investire impresa	BNL Group
BPC Investimenti Spa	Maestrale	BPC Spa
Gestnord Fondi Spa	Sella Banking Investment Technology Investment Fund	Banca Sella Group
Sofipa Sgr	Mezzogiorno impresa	Bancaroma Group
Ducato Gestioni Sgr	Ducato Venture	MPS Group
Interbanca Gestioni Investimenti Spa	Interbanca Investimenti Interbanca Investimenti Due	Banca Antonveneta Group
Sviluppo Imprese Centro Italia Spa	Centroinvest	Banking consortium

Table 1.3. (cont.)

Name of management company	Name of fund	Group
Kairos Partner Sgr	Kairos Partner Private Equity Fund	Private
Mediolanum State Street Sgr	Fondamenta MSS	Mediolanum Group State Street Group

2 The Constitution of a Venture Capital Company

Vincenzo Capizzi

2.1 Introduction

The objective of the present work is to perform a detailed exploration of the different models in which it is possible to carry out institutional venture capital activity in Italy, i.e. the investment by qualified financial intermediaries in the risk capital of non-financial companies. This investment, in line with the terminology used in economic literature and professional practice, must also show significant growth prospects and be in the initial stages of thei company's development process[1].

Vice versa, the term private equity is used to describe the investment activity performed in the target company's life cycle after the initial stage; e.g. shareholdings acquired with the main objective of facilitating the resolution of a problem of generational change or, buy-out/in operations (so-called leveraged acquisitions), which refer to different companies in relation to size, maturity, control and financial fundamentals[2].

In particular, the subject examined in this paragraphconcerns the constitution process of the intermediary wishing to operate in the venture capital business in the Italian financial market; moreover, the analysis will also include the operating models with which the venture capitalists raise capital for their core activity (fundraising), select projects/companies deserving a financing (investment process), and finally, proceed to exiting from the shareholdings acquired (way-out).

However, the main element of the contribution is represented by the centrality of the closed-end investment funds in the national venture capital industry. As a matter of fact, this intermediary is particularly suited for providing a balanced stream of financial resources to non-quoted companies, characterized with high growth and profit potential, but that do not have the necessary financial resources to support them. The international

---

[1] Sahlman 1990; Gompers 1995; Abbot and Hay 1995; Gompers and Lerner 1999; Gervasoni and Sattin 2000.

[2] In some recent definitions both venture capital and private equity, apart from the company's life cycle phase, refer to merchant banking, i.e. the division institutionally appointed to the direct undertaking of shareholdings in non-financial companies. (Caselli and Gatti 2000).

experience underlines both the importance of this category of intermediary, and the benefits in terms of allocation efficiency that can derive from this financial system. The Italy closed-end funds, although rather recent, after an initial period of stagnation have started to develop and realize their productive activity on a professional basis, following the "fundraising of capital-investment-monitoring-exiting-distribution to subscribers" scheme.

In fact, since the first regulation establishing the Italian closed-end funds (Legislative Decree 344/93) has been revised and homogenized with the disciplines in force in other countries, where the venture capital sector is more developed, the risk capital market and its actors have increased significantly.

For this reason, the Italian experience is very interesting as useful lessons may be obtained on designing an intermediary system and financial instruments that will allow an efficient flow of capital from the parties with positive financial balances to those in deficit of financial resources who, by their nature, do not have an immediate and direct access to the stock markets.

The rest of chapter is structured as follows: in paragraph 2 a close examination of the venture capital and private equity market in Italy is performed, in order to highlight the trend of investments made in the last years and their division by typology, size of the target company and sector of activity. Paragraph 3 focuses on the main operators in the national risk capital market, clarifying the actual and prospective role of the Italian closed-end funds. These intermediaries are the subject of specific examination in the second part of the work, from paragraph 4 on, where the institutional, regulatory, organisational, and managerial specificities of the Italian closed-end funds are examined, with special reference to the cases in which for some time the closed-end funds perform their activities. In particular, we will stress the most critical aspects of the legislation measures that in 1993 established them in Italy and show how the successive measures enacted by the national legislator modified and improved the present situation. Finally, in paragraph 5, we analyse a series of observations emerging from the Italian experience of closed-end funds and supply some points for consideration by the policymakers.

2.2 The Venture Capital Market in Italy

From the beginning of the '90 and, in particular, in the last five years, the venture capital and private equity market has seen, in Italy, an uninterrupted growth, in terms of the total value of investments made, the actual number of operations and, finally, the number of companies financed (see Fig. 2.1).

In fact, at the end of 2000, the total value of risk capital investments in non-financial companies reached 5,750 billion Lire, showing a 67%increase compared with the approximate 3,444 billion in the previous year and a 450% increase compared with 1997 (a little more than Lire 1,164 billion). The number of companies financed was 490 in 2000, largely superior to the 309 ones in 1999 and 175 in 1997.

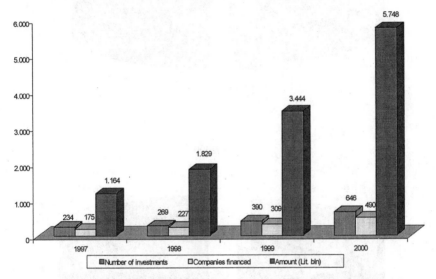

Fig. 2.1. The Italian venture capital and private equity market: growth trend in the 1997-2000 period

It is worth underlying again that the venture capital term used hare refers to risk capital investment in companies that are in the initial phase of their life cycle (so-called early-stage investments); this means either the investor's intervention in the experimentation phase, where the technical validity of the product/service has still to be demonstrated (seed financing), or in the commencement phase of the production activity, where the commercial validity of the product/service is not yet known (start-up financing). However, the investments made in order to accelerate the companies' development (expansion financing), favor a restructuring of the shareholding base (replacement capital financing), and permit the company's acquisition by a new entrepreneurial group (leveraged/management buy-out financing)[3] belong to the private equity category.

a)1994

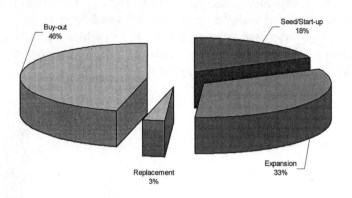

b)2000

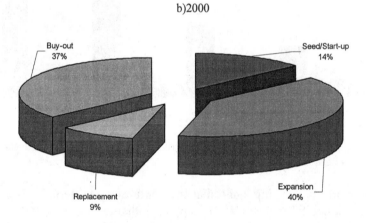

Fig. 2.2. Distribution of the amount of investments in venture capital and private equity by typology

---

[3] The definition of venture capital in economic literature is not unequivocal; for example, some authors also include expansion financing investments (Jeng and Wells 2000). Moreover, the term venture capital is frequently used in professional practice as a synonymous for private equity, therefore including all the typologies of investments listed above. The terminological solution adopted in the present contribution is coherent with the definitions chosen by the main official sources of data and information (NVCA; EVCA; AIFI).

If we then adopt a disaggregated vision of the venture capital and private equity market (Fig. 2.2) it is possible to understand how, in the 1994-2000 period, the segment with the largest growth rate relates to the seed/start-up investments (+ 32.7%), followed by the buy-out operations (+ 25.0%), while the risk capital investments in expansion and replacement capital financing show a reduction. However, in terms of the number (and not the value, in billion Lire) of investments, Table 2.1 shows how, in 2000, over half of the total investments made related to "pure venture capital" (seed/start-up): comparing the result with the three previous years, the growth in this segment of the market appears truly considerable.

Table 2.1. Percentage distribution of the investments in venture capital and private equity in 2000

	Seed/Start-up		Expansion		Replacement		Buy-out		Total
Number	1997	2000	1997	2000	1997	2000	1997	2000	1997
	40%	53%	35%	36%	14%	3%	11%	8%	2000
									100%
									100%
Value	12%	18%	27%	33%	29%	3%	32%	46%	100%
									100%

Source: AIFI 2001

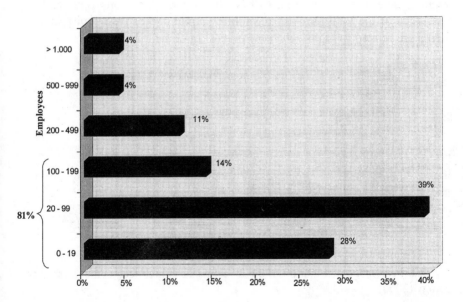

Fig. 2.3. Percentage distribution of the number of venture capital and private equity investments by size of the target company

In addition, these investments, apart from the typology adopted, are particularly suitable for developing small and medium sized companies, that can count on alternative financial (and often managerial and industrial) resources compared to bank financing. Data from the Italian Association of Institutional Investors in risk capital (A.I.F.I.) shows how, in 2000, most venture capital and private equity investments (over 81%) addressed companies with less than 200 employees; 28% of the investments financed companies belonging to the minimum sized category, i.e. less than 20 employees (Fig. 2.3). Although lacking precise details on the typology of investments, it is likely that all - or most – pure venture capital investments (seed/start-up) related to smaller sized companies. It is the case of the Italian industry, world-wide known for the preponderance of small and medium sized companies[4].

Finally, analysing the destination of the investment flows realised by operators in the domestic equity capital market, the typologies of economic activity preferred by the venture capitalist are characterised by capital intensive production processes, not much standardised and with high technological content; e.g. the internet-related businesses, telecommunications, information science and technology, automation, electronics, biotechnology and medical equipment industries. In particular, empirical evidence relating to the national venture capital industry shows how seed/start-up financing investments represent 62% of the value (in billions Lire) and, in fact, 82% of the total number of venture capital and private equity investments are related to high-tech companies (Fig. 2.4).

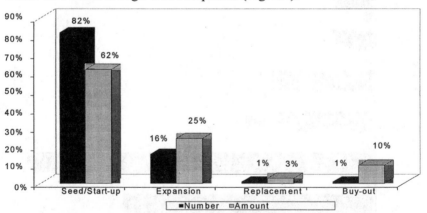

Fig. 2.4. Percentage distribution of the investments in high-tech companies by typology (2000)

---

[4] The statistics elaborated by AIFI not always permit the breakdown by typology of investment (seed, start-up, expansion, replacement, buy-out).

2.3 The Main Actors in the National Venture Capital Industry

A review of the main actors in this market is useful in order to have a precise picture of the operating – current and prospective – possibilities for the different typologies of professional risk capital investors in non-financial companies.

Even today, the general denomination of institutional investors in risk capital describes operators, with different profiles and organisations, belonging to the following categories:

1. Banks. Since a few years, following the Consolidation Act related to the banking and credit legislation (Legislative Decree 385/93), the national banks, respecting pre-determined quantitative limits, are allowed to acquire risk capital in non-financial companies[5]. However, a lot of banks were already operating in the venture capital and private equity market – through their own merchant banks – before the application of this regulatory measure, and internally possessed the appropriate expertise and professional skills for operating in the risk capital market. As the venture capital activity implies the possibility of becoming a company's shareholder, and not just a creditor, requires a clear separation within the bank between the units in charge of the lending and investment activities, whose production processes and economic-financial profiles of the services provided are very different.

2. Merchant banks and holding companies. They are typically national banking operators, that, in line with the Anglo-Saxon model of merchant banks, professionally perform risk capital interventions, as well as intermediary and consultancy activities in the corporate finance, asset management and risk management segments. Mediobanca is the Italian main intermediary for tradition and volumes traded (within this category).

---

[5] In relation to this, the Consolidation Act identifies a total limit (relationship between the total of non-financial shareholdings held and the equity for Bank of Italy purposes), a concentration limit (relationship between the size of the individual non-financial shareholdings and the equity for Bank of Italy purposes) and a separation limit (relationship between the size of individual shareholdings and the net equity of the company invested in); these limits, however, assume different values in function of belonging to a data bank of one of three typologies specifically identified in the regulation: ordinary banks, qualified banks and specialised banks. For more detailed information, see Forestieri and Mottura 2000.

3. Industrial operators and holdings of industrial groups, i.e. holding companies held – singularly or in joint ventures – by industrial operators with the objective of investing in young companies, generally with a high growth potential. The larger company succeeds in financing innovative technology and defending its original business from the threat of substitute products, without internalising risky investments, therefore not compatible with the managerial policies agreed with its financiars and market. In Italy, the presence of corporate venture capitalists is of little significance, i.e. large industrial groups directly operating in the risk capital market.

4. Public operators, that supply financial risk capital to companies promoting and developing specific economic activities or geographic areas, especially on the employment side. Therefore, although operating in the venture capital and private equity market, their objectives are partially different from the other operators', and significantly influence the selection of companies to invest in, the investment duration (often a long-term one) and the strategies for investment monitoring.

5. Regional financial companies. Constituted towards the end of the '70s as agencies for development and regional programming instruments, they have gradually undertaken an important role in financing the growth of small and medium sized companies. Their main areas of activity relates to financial services, industries and territory interventions. With reference to the venture capital market, the regional financial companies grant medium and long term financing, actively provide subsidized financing and acquire minority shareholdings. In general, they do not have the necessary expertise in order to intervene actively in the management of the participated company and operate in the most important sectors for the development of a particular geographical area.

6. Operators originated from cooperative companies. The cooperation too, through special financial companies, are active in the risk capital market, and intervene in order to stimulate the promotion and creation of new entrepreneurial cooperative enterprises, without assuming any control. As in the case of public operators, the objectives pursued with the risk capital intervention are partially different from the traditional value maximization seeked by the other intermediaries. As a matter of fact, the initiatives financed relate prevalently to innovative technological programmes, occupational increases and development of particular geographic areas.

7. Private operators. This category includes the so-called business angels, i.e. private and "informal" investors (generally entrepreneurs, ex con-

sultants or ex managers) with financial means and a good network of contacts and managerial skills, in order to invest in new entrepreneurial initiatives. Moreover, important operators are the incubators of companies, who supply resources to young companies, helping them to survive and grow when they are most vulnerable, i.e. during the seed and start-up phases. Apart from supplying financial and managerial resources, as the venture capitalists, the incubators rpovide office, machinery and equipment services, and also large amounts of space for carrying out the production activity.

8. Italian closed-end funds: financial intermediaries who raise capital mostly from institutional investors or private individuals with considerable personal wealth. They aim at investing, in the medium-long term period, in low liquidity and high risk companies, e.g. non-quoted ones. We will examine the institutional and operating characteristics of these intermediaries in the following paragraphs, with particular reference to the Italian case.

9. Foreign banks, investment banks and closed-end funds. Finally, the foreign intermediaries have been operating in Italy for many years: Anglo-Saxon/US investment banks; subsidiaries of foreign banks through related companies; foreign closed-end funds, managed singularly or in collaboration with Italian partners. These operators, due to their greater experience in the risk capital business and the different regulatory environments in which they carry out their institutional activities, are the most active players in the market, in terms of numbers of operations concluded and value of capital paid (Figs. 2.5 and 2.6).

With reference to the Italian venture capital market, several operators are excluded, as they in fact play a significant role in other countries: e.g. pension funds, which have not yet completely taken off in Italy due to the structural deficiencies which have historically troubled the national social security system. However, pension fund are likely to become more important in the coming years within the national financial system, therefore contributing to the strengthening of the venture capital and private equity market.

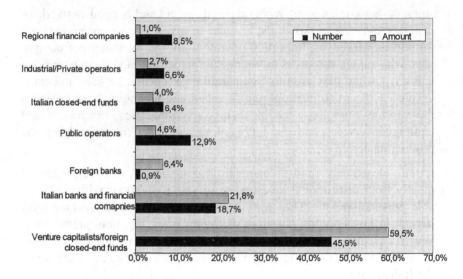

Fig. 2.5. Venture capital and private equity market by investor category (2000)

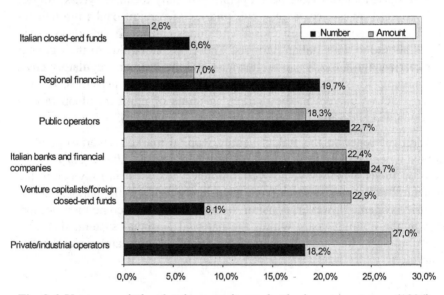

Fig. 2.6. Venture capital and private equity market by investor category (1996)

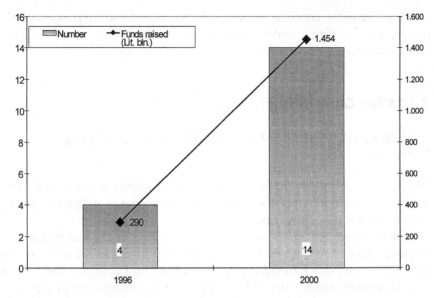

Fig. 2.7. Italian closed-end funds during the 1996 – 2000 period

In the Italian market, several tendencies have contributed to bring the national venture capital industry closer to the more developed indirect risk capital markets, where the financing opportunities are greater for the small and medium sized companies with high growth potential. In particular, they refer to the following structural changes:

– the weight of public operators, regional financiers and companies origi-nated from cooperatives has decreased significantly, as proven by the fact that these operators' professional and operating characteristics and expertise is not the engine of a developed and efficient risk capital mar-ket;

– the Italian closed-end funds have increased significantly – by number and amount of funds raised – in particular since the regulations on the subject were modified (Fig 2.7). As a matter of fact, in the perod from Sept. 1999 to Sept. 2000, 5 new closed-end funds were established, while from Sept. 1995 to Sept. 1999 nine new nes were added;

– the strong growth in foreign closed-end funds, also managed by venture capitalists: this trend suggests greater competitiveness compared to na-tional closed-end funds, that in part seem to have diminished from the previous year.

As in the other contries, also in Italy the closed-end investment funds represent a significant option for performing the venture capital activity, and it is destined to assume higher importance over the next years.

2.4 Italian Closed-End Funds

2.4.1 Mutual Investment Funds and Closed-End Funds: A Summary

The closed-end investment funds represent worldwide - and now also in Italy - one of the main operators capable of performing professional venture capital and private equity activity.

Mutual investment funds, together with the investment companies with variable (marked to market) net worth (in Italy so-called SICAV) are the alternative solutions for offering the management of collective savings in Italy (Forestieri and Mottura 2000). This service is disciplined by the Consolidation Act related to financial intermediaries (Legislative Decree of February 24, 1998, n. 58). They are autonomous equities divided into quotas, belonging to a number of participants and managed by a management savings company (hereafter Sgr), authorised by the supervision authority to manage collective savings. Legally and substatially, the term "mutual" indicates that the equity raised through the transfer of different participations is an undivided joint estate of assets, and each investor is co-owner of the quotas paid in. Other peculiar elements to this financial intermediary are:

1. the fund equity is distinct from the Sgr's (or management company's), and from the participants to the fund itself;

2. all the fund's quotas have equal values and rights, and are represented by registered or bearer certificates;

3. the subscribers to the fund's quotas cannot influence its management activities;

4. the fund's quotas – also called "liability to market" as they are calculated at market value and not book value – are not liabilities of the Sgr.

The mutual investment funds are different from the target market for their method of operating and the typology of assets subject to the collective investment (Fig. 2.8).

First, it is possible to distinguish between the funds offered to the public (retail), destined to all the potential savers, and the reserved funds, focused on the "qualified investors", i.e. professional investors that, for expertise and volumes, institutionally perform intermediary activity and individual savings management and are particularly suited for subscribing fund's quotas[6] for participating in non-quoted financial instruments – e.g. the venture capital funds. These operators generally are[7]:

- investment companies, banks, exchange agents, Sgr, SICAV, pension funds, insurance companies, financial companies belonging to banking groups and parties registered in the lists as mentioned in art. 106, 107 and 113 of the Banking Consolidation Act (Legislative Decree 385/93);

- foreign parties authorised by the regulations of their own country to carry out the same activities as in the previous point;

- banking foundations;

- individuals (or bodies) with the specific skills and experience in financial instrument operations, as expressly declared in writing by the individual or his legal representative.

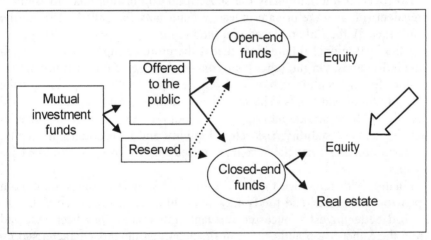

Fig. 2.8. Typology of mutual investment funds

With reference to the operational method, we must distinguish between "open-end" and "closed-end" funds. A fund is defined as "open-end" if the participant-subscriber has, at any time, the faculty to invest in (thus acquir-

---

[6] Reference is made to the chapter by Geranio, in this volume.
[7] This list is held in the Bank of Italy Regulations of September 20, 1999 (Section I – Measures of a general character) as stated by art. 6, 34, 36, 38 and 50 of the Legislative Decree of February 24, 1998, n. 58 (Finance Consolidation Act).

ing newly issued quotas at current prices) or exit from the fund (exercising the redemption of the quota and timely obtaining the monetary equivalent at current prices). In a "closed-end" fund, instead, the subscribers' right to the reimbursement of the quota is only recognised at a pre-determined expiry date, as stated in the information prospectus. Thus, closed-end funds do not provide the participants the possibility to subscribe new quotas and "exit" from the fund at any moment. This is why it is particularly suited to allocating financial resources to small and medium sized companies: mainly non-quoted; with high growth potential; in need of "patient" capital, stably available to management and that does not oblige from the beginning – when the absorption of cash required for the investments is high – periodic payments of pre-determined cash flows to repay the financing.

Therefore, the typical subscriber of a closed-end fund is an investor – either individual or institutional – capable of assuming a rather high level of risk, in expectation of a similarly high return to be realised in the medium-long term. Hence, the fund manager has stable resources and invests them in low liquidity assets.

However, if internationally the subscription by institutional investors is prevalent (on average private investors contribute for only 10-15% to the funds raised), the Italian closed-end fund regulation (law n. 344/93) penalises the institutional fundraising, that is therefore strongly focused on private individuals. On the other hand, the funds raised from institutional investors for the venture capital and private equity activities privileges the foreign closed-end funds. The main consequence is that the capital raised prevalently from private individuals does not possess the same characteristics, in terms of stability, risk/return/liquidity and investment philosophy, of corresponding financial resources raised largely from institutional investors.

Finally, with reference to the type of asset subject to investment, both open-end and closed-end funds may invest in liquid assets, even if the latter had better invest – once the relevant authorisation has been received from the Supervision authority – in financial instruments characterised by reduced liquidity and a greater risk profile, e.g. by risk capital in non-quoted companies. The closed-end funds, in addition, are the only vehicles permitted by the law, within the management of collective saving services, to invest in property and real estate activities, which show interesting development potential, as at the end of 2000, the 6 closed-end real estate[8] funds operating in Italy managed resources above Lire 3,100 billion.

---

[8] Ibidem.

2.4.2 The Structure of Italian Closed-End Funds

Different solutions permit to carry out institutional venture capital activity through the creation of a "vehicle" in order to raise financial resources from investors and channel them, under the form of participations, towards companies presenting, in virtue of a product or innovative technology, a high growth potential. In particular, the markets where the venture capital is more developed historically and by size, present the following legal structures (EVCA 1999; Gervasoni 2000):

1. partnership: the shareholders are responsible for the management or agree to delegate the collective management of the funds raised to external professionals (management company). In any case, the total sharefolders' and fund's reserves are equal, and the fund itself does not have an autonomous legal form;

2. limited partnership: a scheme similar to the previous one, that provides the presence of two clearly defined categories of shareholders. In fact, the limited partners, i.e. subscribers to the fund's quotas, have a limited responsibility in the fund's management and investment decisions; the general partners, apart from transferring capital, are responsible for organising the fundraising, managing the funds raised and proceeding, at the expiry date, to the reimbursement of the quotas to the subscribers.

3. corporation: a company whose shareholders are the investors. The main disadvantage with respect of the revious organisational froms is due to the fact that the corporation is subjected to taxation on the capital gains realised (and not distributed), whereas the partnership and limited partnership are characterised by full "fiscal transparency". They are therefore typically constituted when the law does not permit certain parties to operate through partnership or limited partnership[9];

4. closed-end fund[10]: an autonomous legal body independent from both the subscribers and the company that manages the resources. The subscribers, however, do not have the possibility to intervene in the management or investment choice, which exclusively concerns the management company, assisted on specific aspects if necessary by one or more advi-

---

[9] In the United States, for example, the banks cannot maintain holdings in limited partnerships: the only possibility they have to create a closed-end fund is through a corporate entity.

[10] This term refers to the legal structures operating in the United States; similar to closed-end fund are, in Great Britain, the venture capital trusts; in France, the fonds communs de placement à risque; in Germany the unternehmensbeteiligungsgesellschaft; in Spain, the fondos de capital-riesgo.

sory companies. This is the legal solution envisaged by the Italian regulations to carry on in a professional manner the activity of managing collective savings with reference to non-quoted companies[11].

However, it is worth underlining how, setting aside the particular legal structure chosen, there are common characteristics that distinguish a venture capital fund compared to the other financial intermediaries presented in paragraph 3:

1. limited life: the fund has a pre-established expiry date, at the end of which the exiting and redemption of the quotas subscribed are returned to the investors. This characteristic, apart from minimizing the risks of differences between venture capitalists and investors on the timing and methods of redistributing the funds invested, is a powerful incentive for optimising the efficiency of the management company's investment policies, that, in case of worse results than expected, will seriously compromise its ability to raise funds in the future;

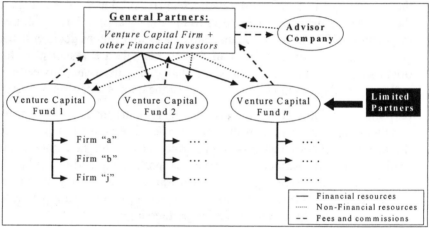

Fig. 2.9. The typical structure of a limited partnership

---

[11] The Finance Consolidation Act (Legislative Decree 58/98) also provides, in the management of collective savings, the investment in the variable capital of a company (sicav) solution, whose share capital represents the assets that must be invested and managed. The same normative corpus, however, assimilates the assets of the sicav to the open-end mutual investment funds, thus imposing the same limits to the concentration of risks for the latter, the most important of which prevents them from making investments for more than 20% in non-quoted financial instruments from the same issuer. This prevents to consider the sicav as a professional operator in the venture capital market.

2. flexibility. The management company can launch several funds simultaneously, each one characterised with a distinctive duration, capital and investment philosophy; therefore, it is possible to satisfy a variety of investor categories, each with a specific risk/return/liquidity profile, widening the depth of the risk capital market. This flexibility, in addition, allows the manager to delegate to other parties (advisor companies) some of their institutional activity (fundraising, identification of the target companies, investment selection and/or monitoring, analysis of the exit opportunities). He is therefore always able to supply the clientele with highly specialised and sophisticated products, without necessarily possessing internally a wide and specialised expertise;

3. remuneration mechanisms. The parties appointed to the fund management receive, first of all, a fixed management fee, generally varying between 2% and 3% of the total capital raised; in addition, the management company also participates in the final result of the fund, through the carried interest mechanism, that allows it to receive a certain percentage (usually 20%) of the total capital gains realised in the exit phase. Hence, the venture capitalists are more responsible in the investment selection and management activities, as these affect an important part of their own remuneration[12].

We will now examine the major differences between the structure of the limited partnership, widespread in the Anglo-Saxon world due to its advantageous fiscal asymmetry, and the closed-end fund, that represents the only option allowed by the Italian legislator (Figs. 2.9 and 2.10).

First, the venture capital funds constituted through limited partnerships provide the existence of two distinct investment categories – general partners and limited partners – with involvement and responsibility levels differentiated in the management of the capital raised. This separation does not occur, but it is typical in the closed-end funds, between the reserves of the venture capital fund and of the fund managers – the general partners: it allows the management team to act autonomously in the selection of the best investment opportunities, accelerating the decision-making process relative to the preparation and conclusion of the investments.

---

[12] For further details on the typical incentive and remuneration mechanisms in the venture capital industry, see Liaw 1999.

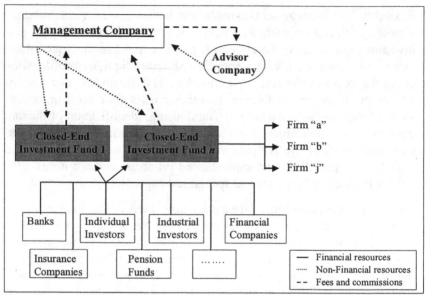

Fig. 2.10. The typical structure of a closed-end fund

In order to avoid the emergence of opportunistic behaviour by the general partners, they are explicitly prohibited to perform trading operations on their own behalf (self-dealing), which, for example, could allow them to receive benefits not available to the limited partners.

Second, in contrast to the closed-end funds, the subscribers can exit their investment before the end of the fund's life, i.e. the limited partners can ask at any moment for the reimbursement of the quota subscribed. It is thus possible that, against the limited partners' request of exiting, there might arise liquidity risks within the limited partnership, which could jeopardize the stability of the financial resources given to the companies financed[13].

Finally, a further element of differentiation relates to the legal nature of the limited partnership: as it is not a company with share capital, it is not eligible, in the countries whose legal regulations provide for such a company structure, to be admitted for quotation on the official stock markets.

The quotas of a closed-fund, on the other hand, can be traded on a regulatory market, and in addition, as stated in the Italian regulations, in case of quotation it is possible to subdivide the quotas, with the aim of permitting

---

[13] In practice, however, the Anglo-Saxon experience shows how the limited partners' request for an anticipated exit of the investment is extremely infrequent, as in the case of scarce satisfaction for the general partners' management activities, thay will not participate in the latter's future initiates; see Jeng and Wells 2000.

greater marketability of the certificates representing them and increasing the liquidity profile.

In the countries where it is possible to constitute a limited partnership, the largest part of its success is due to the favourable fiscal regimes, differing from the ones applicable to the other intermediaries operating in the venture capital market, e.g. the closed-end fund[14].

2.4.3 The Evolution of the Applicable Legislation

This paragraph focuses on the most critical aspects characterising the legislative measure (law 344/93) that, for the first time, established in Italy this typology of financial intermediary and that has significantly affected the operation and diffusion on a vast scale[15]. Moreover, we will analyse how the following measures enacted by the national legislator were able to intervene in the problems highlighted. In the final paragraph, we will make a final evaluation and stress the questions not yet solved.

The constant reference point will be the regulatory solutions adopted in those countries where the closed-end funds have operated most and have been able to fully perform their potentiality.

In particular, the greatest weakness relating to the characteristics of the legislation of the Italian closed-end funds – only recently modified – can be summarised in the following points:

– legislation related to the fund raising;

– limits to the holding of voting rights;

– limits to the fund duration;

– fiscal regime applicable.

Legislation in Relation to the Fund Raising

The legislation for closed-end funds, which did not distinguish between retail and reserved funds, establishes that the fund's reserves must be raised through one single emission of quotas, equal in unitary values. These quotas, as later on specified by the Treasury Ministerial Decree n.228 of May 24, 1999 (hereafter D.M. 228/99), must be subscribed within 18 months

---

[14] In the United States, in fact, if at the beginning of the eighties less than 40% of the venture capital funds were structured as limited partnerships, this percentage was over 80% in the middle of the nineties; see. Gompers and Lerner 1999.

[15] For a systematic discussion on the discipline currently in force, see Gervasoni 2000.

from the Consob's approval of the fund's information prospectus. The payments of the quotas subscribed must be made in one unique step, within the terms provided in the fund's regulations.

The problem emerging from such a legislative provision arises from the fact that the management company may have to manage considerable amounts of liquidity for a non-negligible period before the fund's resources can be actually invested. However, the investment in non-quoted companies, which often have an extremely limited track record, presupposes a screening and analysis activity that is time consuming. All this could result in the significant reduction of annual percentage returns of the investment guaranteed to the subscriber after the reimbursement of the quota, with consequential detriment to the manager's reputation.

To solve this problem, the D.M. 228/99, after having allowed the possibility of establishing funds, beside the retail ones, whose participation are reserved exclusively to qualified investors – as listed in paragraph 4.1. – provides the draw down mechanism, i.e. the payments of the quotas can be made in different occasions, following the subscriber's commitment to pay on the management company's request and based on the effective investments decided each time.

In addition, law 344/93 also establishes that the management company must invest its own reserves in quotas of funds managed, within a minimum of 5% and a maximum of 10% of the value of each fund. This limit was, first, lowered to 2% by the Regulations of the Bank of Italy of July 1, 1998 and, afterwards, eliminated in relation to reserved funds by the D.M. 228/99. In fact, in the case of retail funds the subscription of at least 2% of the reserves of each fund aims at transmitting a strong signal to the average investor, in order underline that the management company, before carrying out solicitation activity of savings to the public, firmly believes in the success of the initiative. In the case of reserved funds, however, it is legitimate to presume other and more sophisticated parameters for the professional investor to take into consideration in evaluating the profile of the investment's expected profitability.

Limits to Holding Voting Rights

The Treasury Ministerial Decree n. 228 of May 24, 1999 and the successive Regulation of the Bank of Italy of September 20, 1999, implemented by article 6 of the Finance Consolidation Act (Legislative Decree 58/98), radically changed the operating possibilities of closed-end funds, removing those pre-existing restraints to investments, that limited the effectiveness and diffusion of the national venture capital industry.

In particular, law 344/93 states special limits on the risk concentration and holding of voting rights, so that the fund's reserves could not be invested:

1. for more than 80% in non-quoted shares (with a minimum limit, however, of 40%);

2. for more than 20% in Italian or foreign government securities;

3. for more than 20% in shares, or quotas with voting rights, in quoted companies;

4. in liquid assets sold to another fund of the same group;

5. in shares or quotas of companies if, as a whole, 51% of the capital is owned by less than three companies managing mutual funds (open and closed), belonging to different groups;

6. in shares, or quotas with voting rights, issued by the same company for a value superior than 5% of all the shares issued by the company, if quoted;

7. in shares, or quotas with voting rights, issued by the same company for a value superior than 30% of all the shares issued by the company, if not quoted.

The Regulation of the Bank of Italy of 1999 substantially revised and modified the above-mentioned limits, benefiting closed-end retail funds, and allowed the reserved funds a broad faculty to deviate from the aforesaid norms of risk hedging and subdivision, thus widening the investment possibilities.

The current regulations provide only for a minimum investment limit in financial instruments of retail funds in non-quoted companies – 10% of the fund's reserves. However, there are not any maximum limits, and as a consequence there is nothing to prevent these funds to invest the entire capital raised in a non-quoted company. In compliance with the diversification principle, the reserves of the fund cannot be invested:

1. for more than 20% in non-quoted financial instruments by the same issuer;

2. for more than 15% in quoted financial instruments by the same issuer;

3. for more than 30% in financial instruments issued by several companies belonging to the same group (limit reduced to 20% when the group belongs to the Sgr);

In relation to the reserved funds, the regulation of the Bank of Italy of September 20, 1999 establishes that "in the fund regulations different norms from the general ones can be prudentially fixed", thus leaving significant faculty in separating from each of the above concentration limits.

However, the aspect of greatest significance and originality of the new regulations related to the holding of voting rights is the fact that closed-end funds – both offered to the public and reserved – can acquire majority shareholdings in the risk capital of non-quoted companies, ususally object of the venture capital and private equity activity. With reference to retail funds, there is only the limit, and for quoted companies only, to hold a maximum of 10% of the quotas with voting rights from the same issuer. The reserved funds for professional investors can, instead, also depart from this regulation. Law n. 344/93 excluded in all cases the possibility to hold controlling shareholdings, even if not meeting the limits above.

The removal of the prohibition for Italian closed-end funds to hold controlling interests in   non-quoted companies represents a real point of discontinuity from the past, as it eliminates the most critical element encounted by the intermediaries in performing their institutional activity in the risk capital market, especially considering that the prohibition was obviously not applicable to operators subject to other regulations.

First, greater weight is attributed to the closed-end fund within the corporate governance of the companies financed, as the financial intermediary "counts" more on the board of directors and no longer finds himself in a position of inferiority compared to the majority shareholder. A similar condition should benefit the company financed, as the transfer of industrial experience and skills is easier. Moreover, the assistance supplied to the entrepreneur is more effective in relation to formulating and implementing the company's competitive strategies and adaptating the structural layout to the changes required by the choice of particularly rapid growth paths.

Second, a more equilibrated division of the voting rights, where the entrepreneur does not maintain the majority of his own company even after the entry of the closed-end fund in the risk capital, allows to perform remuneration and incentive mechanisms in order to induce the fund to avoid opportunistic behaviour and to concentrate on the company's growth (Sahlman 1990; Gompers 1995). The entry of the venture capitalist in a company is preceded by rather elaborate contractual negotiations, in which, typically, a series of appropriate criteria and mechanisms are fixed to allow the entrepreneur to increase, over time, the quota of risk capital held, in order to reach particular growth objectives; therefore, forms of equity-based remuneration, share buy-back programmes and the issue of convertible shares are some of the possible mechanisms permitting the entrepreneur and the majority shareholder to align their objectives.

Third, as shown by empirical experience, a greater level of involvement in the investment outcome gives the venture capitalist greater incentives and better operative possibilities in monitoring the entrepreneur's behaviour; this is a very important aspect in a market where moral hazard represents the principal factor of imperfection in trading (Jeng and Wells 2000). Moreover, the typical target of the closed-end funds' activity, i.e. small and medium sized non-quoted companies with high ownership concentration, is characterised by a high level of "information opacity"; it is therefore important to benefit from privileged observation points, in order to gain full access to the information sources of the business.

Fourth, the possession of a majority interest is often an essential requisite in managing the exit (way-out) coherently with the timing and profitability objectives of the closed-end fund. If the financial investor does not have full discretionary powers in exiting the investment through trade sales operations or mergers with other companies, and the recourse to the stock market is not practicable, the only alternative available is the participation's sale to the majority shareholder (or to the company's management), which are a possible detriment to the investor's economic returns. In fact, at the end of the programmed investment period of the closed-end fund, the company might not yet be ready for quotation on an official regulatory market, which however, in the Italian case, does not present the efficiency and liquidity typical of the Anglo-Saxon stock markets. On the other hand, the shareholding's sale on the private market of ownership reallocation makes the requisite of the company's control particularly stringent, in order to manage the private sale (and thus the choice of the buyer, the determination of the operation's timing, sales price and method of payment). Moreover, it is desirable to avoid the application of a "minority discount" to the share value (Capizzi 2000).

Finally, the possibilities to assume majority shareholdings in the risk capital of non-financial companies, increasing the investment's control and effectiveness, ends up by increasing the broadness and flexibility of the options available to the management company, and thus favours the development of the national venture capital industry, in line with the experience of other financial systems.

Limits to the Fund's Duration

Law n. 344/93 states that the fund's duration could not be less tham five or more than ten years, except for a possible "grace period" of three years at the end of the fund's duration, expressly authorised by the Bank of Italy in order to allow the investment exit.

The D.M. 228/99 extended to all closed-end funds – retail and reserved – the maximum duration limit from ten to thirty years (plus the three grace years that mighy be requested to the Bank of Italy to allow the exit from the investment).

Such a modification to the regulations concerning the fund's duration further increases the flexibility of the investment choices, as in the absence of stringent expiry dates, the exit process of the investment can be postponed to a moment when greater value has been acquired.

Therefore, a longer duration is more coherent with the nature of the venture capital activity. Differing from the private equity investments, addressed towards companies that have already undergone the commencement and development phases and are often ready for quotation on the stock exchange, the path that the venture capitalist must take together with the company in the start-up phase is long and difficult. In fact, in some cases, ten years cannot be considered as a sufficient time horizon to permit the company to overcome the initial phases of its development (early-stage), characterised by the structural incapacity to product cash flows which are positive or, however, superior to the cost of financial resources obtained from debt securities: e.g. the high capital intensive businesses, biotechnologies, telecommunications, some segments of the pharmaceutical industry.

Other than reducing the possibility of behaviour oriented towards the short term (short-termism), aimed at searching those investment opportunities capable to produce high capital gains in a "reasonable" time period – and, therefore, not necessarily compatible with the seed/start-up financing – lengthening the maximum duration period of the closed-end funds is more coherent with the investment time horizons of institutional investors such as pension funds, whose weight in the Italian financial system is bound to increase in the coming years.

Fiscal Regime Applicable to Closed-End Funds

The fiscal regime of closed-end funds, as disciplined by law 344/93, art. 11, valid until the entry in force on July 1, 1998 of art. 8, paragraph 3 of the Legislative Decree 461/97, stated the tax exemption on the income of corporates, individuals and on the local income tax. A single tax was established equal to 0.25% of the fund's net value, with a reduction to 0.10% where the fund had invested no less than 50% of its own reserves in shares or equivalent securities issued by non-quoted Italian companies considered as "small" in compliance with law n. 317/91, i.e.:

- industrial companies with no more than 200 employees and a capital invested not higher than Lire 20 billion;

- commercial and service companies with no more than 75 employees and a capital invested not higher than Lire 7.5 billion.

The gains of the fund, in addition, did not contribute to the formation of the assessable income of the receivers with the exclusion of those exercising commercial activities; however, in this case, if the fund's quotas were held for a minimum period of three years, a tax credit was established equal to 25% of the gains. This regime was only applicable to those parties who held a quota of the fund inferior to 2%, and then increased to 10% with a special Treasury Ministerial Decree for each of the categories of institutional investors[16].

The ordinary taxation regime for corporate entities did not provide, however, any allowments for the capital gains, which entered in the company's normal income. This regulation significantly penalised the national venture capitalists, at least with reference to the regimes in place simultaneously in other European countries (such as France, Belgium, Holland, Denmark). The dividends distributed by the fund entered within the company's normal income and were not subject to ILOR. In case they were originated from a foreign resident company, with a holding greater than 20% (10% if quoted), the dividends contributed to assessable income only for 40% of their value. There is clearly, therefore, a fiscal asymmetry in favour of management companies resident outside the national boundaries, that benefit from a much better treatment than the Italian closed-end funds.

The new fiscal regime currently in place, that acknowledges the regulations of the Legislative Decree of November 21, 1997, n. 461 and of the Legislative Decree of December 23, 1999, n. 505, stands within a more general revision of the taxation system of mutual investment funds.

In particular, in relation to the fiscal regime applicable the fund's gains, as in the previous regime, the funds are not subject to tax on income of corporates, individuals and to the local income tax (now substituted by the regional tax on production); the withholdings on the income of capital, in addition, are only a lump sum.

The Legislative Decree n.505 of December 23, 1999 stated that on the management result matured each year, and relevant to the qualified participations held by the collective investment organisms (disciplined by ar-

---

[16] Article 3 of the Treasury ministerial decree of February 9, 1994 identifies as "institutional investors" the following categories: a) obligatory pension schemes; b) insurance companies; c) banks; d) holding companies; e) management companies of closed-end funds; f) equipollent foreign organisations.

ticle 8 of the n. 461 of November 21, 1997), a substitute tax is applicable, i.e. 27% rather than 12.5%. This measure, however, considers qualified all participations in capital or reserves with a voting right in the company or body higher than 10% for the participations traded on regulatory markets, or 50% for the other ones.

Compared to the previous regime, the tax credit applied on the gains from participations in collective investment organisms, undertaken in the same year by commercial companies and related to the management result subject to the substitute tax of 27%, is 36.98% of the total amount. The fiscal treatment of the closed-end funds has been equalized to that of the open-end mutual investment funds. The measures contained in the legislative decree are applicable to the collective investment organisms with at least 500 participants, starting from January 1, 2000.

In relation to the fiscal regime applicable to the Italian participants in the fund, the gains distributed are not part of the assessable income, with the exception of the participations in the fund undertaken in the same year by commercial companies. The gains deriving from these shareholdings affect the income only in the year in which they are received. The gains received are subject to a tax credit equal to 15% of the amount.

The corporate bodies cannot, by express regulatory exclusion, opt for the substitute tax regime for the capital gains (that establishes a minimum 12.5% and a maximum 27% rate), in case these derive from participations in investment funds.

The fiscal regime applicable to the foreign participants of the fund, depends on two categories:

− resident in countries in the white list[17];

− resident in other countries.

The first list benefit from a few advantages: where non residents have received gains paid by collective management organisms subject to substitute tax, they have the right to the reimbursement of the tax paid by the fund (equal to 15% of the gains paid). In the Italian collective investment organisms whose quotas or shares are subscribed exclusively by non-resident parties, the organisms are exempt from paying the substitute tax on the management result.

---

[17] The white list includes all the countries with an agreement guaranteeing the exchange of information, e.g. USA pension funds, Dutch holding companies, Luxembourg soparfi and Madeira companies; otherwise − e.g. "fiscal heavens" − they are included in the black list: Swiss and Luxembourg holdings, English pension funds, Anglo-Saxon limited partnerships.

The residents of countries on the so-called black list are applied, instead, a 12.5% taxation.

Table 2.2. Principal legislative measures related to closed-end funds

Legislation August 14, 1993, n. 344
Legislative Decree November 21, 1997, n. 461
Legislative Decree February 24, 1998, n. 58 (Finance Consolidation Act)
Regulation of the Bank of Italy July 1, 1998
Treasury Ministry Decree May 24, 1999, n. 228
Regulation of the Bank of Italy September 20, 1999
Legislative Decree December 23, 1999, n. 505
Bank of Italy regulation December 24, 1999

2.5 Final Considerations: Open Problems and Policy Indications

We have attempted to critically analyse the main factors that hindered the diffusion of closed-end funds in the Italian market, i.e. the category of intermediaries most suitable, by nature, to perform a professional venture capital activity.

The mechanism used at a regulatory level in attempting to solve most of these problems consisted in the identification of two distinct categories of closed-end funds: offered to the public (retail) and reserved. The latter includes the funds destined to institutional investors, and a series of exemptions are applicable to the limits that conditioned the Italian closed-end fund, especially if compared to the other countries, both European and non.

Table 2.2 shows chronologically the principal legislative measures that have from time to time modified the discipline of closed-end funds, starting from the institutive law (law n. 344/93).

In particular, the main lessons that can be taken from the Italian experience of closed-end funds can be summarised as follows:

1. It is necessary to provide closed-end funds with a certain amount of flexibility in relation to the methodology of raising funds from the investors, in order to avoid problems related to managing considerable

amounts of excess financial resources for a non-negligible period of time;

2. In order for these intermediaries to contribute strongly to the development of a broad and efficient venture capital market, they must be also provided with the possibility to acquire majority interests in the risk capital of the companies financed. This is often the only way to be able to manage with maximum discretionary powers the exit phase (way out) and monitor effectively the behaviour of the entrepreneur financed;

3. The fund's duration must be compatible with the financial requirements and the growth path of the companies financed, that often need significant periods of time to be able to overcome the seed and start-up phases, and thus begin to free financial resources for their investors. Lengthening of the fund's duration from 10 to 30 years guarantees a more stable flow of resources to the company financed and, in addition, makes possible a future participation in this market by the pension fund operators, that have a medium-long term investment time horizon;

4. The returns received by the investors should be considered from the fiscal point of view as those deriving from other stock market investments, and in particular should not be penalised compared to the results of the activity carried out by similar types of financial intermediaries with reference to the other countries.

Finally, there remain certain elements of weakness characterising the discipline and operation of Italian closed-end funds not explicitly analysed in this paper, as they are not the subject of legislative modifications planned in relation to this category of operators.

Reference is made, first, to the problem of admission and trading on the official regulatory markets of the closed-end funds' participation quotas. The current discipline does not place a real and proper quotation obligation on Italian closed-end funds – as envisaged in law 344/93 – , therefore transferring the choice of requesting the quotation to the discretion of the Sgr, that must decide when the fund is created.

The experience of other countries indicates that most operating closed-end funds are regularly quoted on official stock markets (EVCA various years); the suggestion that can be derived is to institutionalize and quicken the quotation procedures of the closed-end funds' quotas, in order to achieve significant benefits in terms of liquidity and accessibility of the investment for the potential subscribers[18].

A further aspect not examined in this paper relates to the potential problem of a single bank or financial institution conditioning the management company in performing its institutional activity. This is the case of the "captive" funds, different from the "independent" ones as their sources of supply are exclusively or prevalently parent companies or other related companies, and thus do not turn directly to the capital markets for raising their financial resources. The reduced operating autonomy of the captive funds - widespread in Italy – could significantly affect the investment policies of the closed-end fund's management company and reduce the weight of the non-financial resources transferred from the fund to the companies financed (mainly business expertise and managerial resources).

The international experience underlines how:

- Mostly independent funds perform a more evident role of active investor towards the companies financed (A.I.F.I. various years);

- In the Anglo-Saxon markets the independent operators are prevalent compared to the captive ones.

It should be therefore considered if it is not apropprioate to adopt some measure that stimulates the constitution of independent management companies, capable of operating autonomously with the rules and judgement of the market and of realising fundraising /commitment/management investment policies fully coherent with the expectations of the subscribers and receivers of the financial resources intermediated.

The first Italian closed-end funds constituted according to the new discipline have only started recently. In a few years, when some of these funds end their fundraising-investment-exiting cycle, it will be possible to verify the real potential of this typology of intermediaries within the Italian economic and financial context and, consequently, to know if also in Italy the closed-end funds can really become the main operator in the venture capital industry.

---

[18] The regulations for the placing and quotation of the closed-end funds' quotas are stated in the issuer's regulations, enacted by Consob – in implementation of the Legislative Decree 58/98 – with resolution n. 11971 of May 14, 1999.

3 Legal Issues for Italian Venture Capital Investment Schemes

Enzo Schiavello and Jonathan de Lance Holmes[1]

3.1 Introduction

This chapter will address the main legal issues in structuring, establishing and placing interests in Italian venture capital investment schemes with domestic and international investors. Some initial remarks will help clarify the main focus of this work and its underlying options.

The reference to "Italian" investment schemes (or funds) encompasses all investment schemes whose primary investment focus is on Italian business enterprises, whether or not this is stated in the fund documents. This notion includes investment schemes established under Italian law as well as other schemes established under the laws of different jurisdictions and serving the same market purposes. The aim is to provide a comprehensive view of the legal market for these investment schemes rather than just a particular segment thereof.

Our main focus will be concentrated on investment schemes intended for, and placed with, qualified investors as opposed to retail investors. As a function of the legal scheme adopted and the circumstances of each individual fund, this feature may derive from a requirement that investors fall within certain categories to join the investment scheme or undertake a robust commitment (normally not less than Euro 1 million) or a combination of both.

This chapter is divided in two parts. Paragraphs from 3.2 to 3.5 will examine the key drivers of the fund industry and the material terms normally found in fund documentation in line with standards developed in the international market. Paragraphs from 3.6 to 3.9 will more specifically deal with Italian investment schemes. A preliminary review of the international context will help readers better understand: (a) the current state of the domestic market and particularly the options available to fund promoters seeking to effect substantial international fundraising; (b) certain recent

---

[1] Although the chapter is based on a strict cooperation between the two authors, paragraphes from 3.2 to 3.5.9 is attributed to Jonathan de Lance Holmes and paragraphes from 3.6 to 3.9 to Enzo Schiavello.

changes to the legal rules governing Italian closed-ended funds; and (c) the foreseeable development of the Italian legal system and market for these investment schemes.

Finally, although venture capital and private equity are two industries with fairly distinctive features in terms of financed business stages and related types of investment transactions, the differences between the two industries are not so significant from the perspective of fund structuring. Thus, the following description largely applies to venture capital and private equity investment schemes without distinction. Legal features that are peculiar to venture capital investment schemes will be noted.

3.2 Background: Development of the International Private Equity Market

The modern private equity market is a relatively new phenomenon, dating back essentially to the 1960's in the USA. In its beginning, the market concentrated on early stage investments in start-up and developing companies ("venture capital"), involving relatively small overall amounts of money raised and invested. As the market has developed to cover transactions ranging from the very small to the top end (multi-billion dollar deals) of the international M&A market it has also segmented into market sectors.

Some funds still concentrate on early stage or "venture capital" investments in start-up or rapidly growing companies. These investments generally take the form of minority equity stakes, designed to grow in value and be repaid as companies grow and obtain later rounds of financing. Overlapping with this are "development capital" funds targeting these later rounds of equity financing for companies which are no longer start-up but which require additional capital to finance growth. Again, these investments are normally in the form of minority equity or preference share capital. Buy-out funds concentrate on the outright purchase (together with an equity stake for a management team) of mature and often very large companies on the public or private M&A markets, with a view to increasing equity value through restructuring, financial engineering or more efficient management from an incentivised team, and realising this value through initial public offers, securitisations and/or trade sales. Additionally, there are specialist funds established to provide particular types of finance, such as mezzanine finance.

The private equity fund market has grown up alongside, and to support the activities of the private equity managers. Some private equity managers operate within, and invest the capital of, financial institutions such as investment banks and insurance companies, and therefore do not need to raise external equity finance. However, a substantial portion of the market is represented by managers who must raise money from external investors. These are sometimes part of a larger financial group, but are generally independent (often employee-owned) entities.

The market for private equity fund raising developed initially in the United States. It is a market dominated by large institutions, who can afford to commit substantial amounts of money over an extended timescale to essentially illiquid investments. The principal industry players have therefore traditionally been corporate and public sector (including US State) pension funds, insurance companies and charitable foundations. It has not generally been a market directed at retail or corporate investors.

Although the European market is many times smaller than its American counterpart, it is the most developed private equity market outside the United States. Private equity houses operating in Europe have generally modelled their fund structures and terms on US lines. This is partly a result of the managers coming from a US background (either European branches of US private equity houses or having worked in the US industry) and partly because the investor market for such funds remains dominated by US institutions. The U.K. is the most developed European private equity market, and its legal system and tax regime have also enabled the basic US fund model (described below) to be used. As more continental European private equity managers have established themselves and European institutions become a greater share of the investor market, other fund structures and terms have been developed (for example, in relation to Italy). However, the US/U.K. model remains the starting point and the benchmark for the expectations of most international fund managers and investors regarding fund terms and structures.

3.3 Key Drivers of Fund Industry

The private equity fund industry is driven by the commercial and tax requirements of fund managers and their investor base. A fundamental of private equity investing is that the underlying investments are by their nature illiquid and will take a substantial time to realise. The lack of a liquid market makes it difficult to value such investments until they are disposed of. Set against this is the potential for a higher overall rate of return on these investments than that available in the public securities markets. It is a market in which the skills of individual managers are seen as key to generating returns to investors. The investment process itself is time consuming, and it can take several years to invest a fund.

This combination of factors has produced the following fundamental drivers for the private equity fund industry.

3.3.1 Performance - Internal Rate of Return

Investors look to achieve a higher return on private equity investments than available on the public markets and are willing to incentivise management provided that these returns are achieved. Management, in turn, wish to benefit personally from the commitment of their skills and time over what can be a long period from the first investment to the last divestment of a fund. This has produced a two tier structure of remuneration: an annual fee (often structured as a profit share) of a percentage of investors' commitments to the fund, plus a performance fee or "carried interest". The latter is where managers expect to make most of their returns, and it is generally 20% of profits (either of the fund as a whole, or on a deal by deal basis).

Investors are willing to pay these levels of fees only if they have received a minimum return themselves, which has led to the concept of a "preferred return". No carried interest or performance fee is paid unless investors have received back their invested capital plus a notional rate of interest (generally 8 or 10% per annum, compounding) on that invested capital.

This rate of return is not calculated on investors' commitments (much of which will not be invested for some time) but on invested capital for the period from draw down to return, expressed as an internal rate of return ("IRR"). As well as being the benchmark for the payment of carried interest, IRR figures are the most commonly used marketing tool for private equity houses wishing to advertise their returns (and for investors to show the rewards that justify the illiquidity and increased risk of private equity investment).

3.3.2 Draw Downs

As noted above, it generally takes several years to invest a private equity fund. In order not to depress IRRs, and to allow investors to effectively manage their cash, private equity funds draw down committed capital from investors over a period of time. This is generally around five years, with potential for later drawings for follow-on investments, fees and expenses.

3.3.3 Return of Capital

Investors in private equity expect the return of their invested capital as and when investments are disposed of. Managers also generally wish to achieve this, in order to maximise IRRs. Private equity fund structures will therefore generally allow for the full return of capital (not just profits) to investors on a flexible basis.

3.3.4 Tax Transparency

As for all investment fund structures, investors expect private equity funds to be tax efficient that is, that an investment through the fund will incur no more tax than a direct investment in the underlying investments. Ideally therefore, a fund structure will involve the fund itself not paying tax whilst allowing the investor to benefit from whatever double tax treaties or other tax exemptions it would be entitled to on a direct investment.

In the case of the private equity fund industry, as noted above a substantial proportion of the investor base is either fully or partially tax exempt (e.g., state and corporate pension funds, charities and foundations). These entities expect to benefit from at least very favourable rates of tax under double tax treaties on investments outside their home jurisdictions and not to pay tax on investments in their home countries at all. Given this investor base, and these requirements, it is not surprising that private equity funds investing in the US, the U.K. and many other jurisdictions are normally structured as tax transparent entities.

3.4 Structures Most Frequently Used in International Fund Raising: Limited Partnerships

By far the most current fund structure in the international private equity market is the limited partnership. These entities are generally established under the laws of a US state (in particular, Delaware) under English law or under the (English-based) laws of offshore fund centres such as Jersey, Guernsey and the Cayman Islands. Limited partnerships set up in any of these jurisdictions are broadly similar. This is not to say that there are not important differences for some purposes, such as the separate legal personality of a Delaware or Scottish partnership in contrast to the lack of separate legal personality for partnerships in the other jurisdictions.

3.4.1 Advantages

Limited partnerships have the following advantages for investors and managers:

- Tax Transparency: Limited partnerships are tax transparent in many (but not all) jurisdictions, crucially including the US and the U.K. To the extent that they are not transparent in other jurisdictions, it may be possible to structure for tax efficiency through investment vehicles below the level of the limited partnership.

- Flexible Capital Structure: Limited partnerships are not subject to rules which apply to companies on the maintenance of capital, restrictions on distributions, and the like. It is therefore easier to achieve flexibility both on the timing of drawdowns of capital (or cancellation of undrawn capital) and on the distribution of capital proceeds to investors, than would be possible in a corporate structure.

- Commercial Flexibility: There is generally very little prescriptive legislation regarding the terms of a limited partnership agreement. Complex profit sharing, voting and other arrangements can therefore be written into a limited partnership agreement in a way which can be difficult to replicate in a corporate structure.

- Confidentiality: The limited partnership agreement does not need to be filed for public inspection, so its terms can remain confidential. It is worth noting however that the identity of limited partners and their invested capital is often available on a public register.

- Limited Liability: As the name suggests, unlike most other tax transparent structures limited partnerships afford limited liability to investors (providing that they comply with restrictions on taking part in management - see below).

- Accounting: Although investors, as a commercial matter, expect to see audited private equity fund accounts, there is generally no legal requirement as to the accounting standards to be followed, and accounts are not publicly available.

3.4.2 Disadvantages

- Restriction on taking part in management: Limited partners lose their limited liability if they take part in the management of a limited partnership. This can be an important restriction on allowing investors to participate in decisions (although many jurisdictions have some form of "safe harbour" in their limited partnership law to allow for matters such as vetos on conflicts of interest).

– Unlimited Liability of the General Partner: The general partner of a limited partnership has unlimited liability for its debts and obligations. However, it is generally possible to insulate management from the worst effects of this (through using limited liability companies as general partners, and by investing through limited liability structures below the level of the limited partnership).

– Uncertainties Regarding Limited Liability: It is unclear whether certain jurisdictions will recognise the limited liability of limited partnerships, at least in the case of those without separate legal personality (such as the U.K., Channel Islands and Cayman Islands).

– Complexity: Since the entire terms of a limited partnership have to be written out in full in its limited partnership agreement (often together with side letters and ancillary documents) these are generally highly complex documents. Some of this complexity could be avoided by the under-pinning of corporate law principles; however, this complexity is a by-product of the increased flexibility of a limited partnership structure.

3.5 Material Fund Terms

The following is a summary of the commercial terms of a "typical" private equity fund. These terms will be substantially the same whether a fund invests in venture capital, development capital or buyouts, and regardless of the size of the fund or the target size of deals. Fund terms vary and are generally negotiated in detail between the managers and the larger investors.

3.5.1 Commitments by Investors

Each investor commits at the outset to subscribe up to a maximum "total commitment". This is the figure on which the annual management fee or general partners profit share is based. The commitment to subscribe will generally be made by executing and returning a subscription agreement to the general partner.

3.5.2 Multiple Closings

Investors are generally admitted to a fund in a series of closings. This is partly because of the time it takes to market and negotiate a private equity fund (substantial investors usually insist on negotiating the terms of fund documentation and on obtaining protections and commitments in the form of side letters and it may take several weeks to months for each of these negotiations to be completed). The multiple closing approach is taken in order to allow the fund managers to close a fund and commence investing once a minimum critical level of capital has been committed and without having to wait for the final investment negotiations to be completed. In addition, some investors prefer only to invest (or to devote time to the negotiation of an investment) once they see that a fund has achieved this critical mass and has closed. Still other investors wish to wait further and see the results of the negotiation process and the final terms of a fund before making a commitment.

The multiple closing structure raises a number of issues. First, any changes to the terms of the fund between first and final closing must be made available equally to all investors. Second, a fund may have made investments between the first and the final closing, which will have been paid for with capital drawn down from the investors who came in at the first closing. To achieve equality going forward it is necessary to rebalance investors' share in these investments. At the same time investors who have funded those investments will expect some financial compensation for having done so. Third, there is a point in time beyond which investors who have funded these investments, and seen them rise in value, will be unwilling to allow later investors to participate in the uplift in value.

For these reasons, the market norm is for there to be a limit (generally 9 or 12 months) for the time elapsed between first and final closing and for all amendments to fund documents to be made equally available to all investors. Investors who come in at subsequent closings will reimburse (with interest) earlier investors and the general partner to put all investors in the same parties they would have been in had they all invested in the fund at first closing.

3.5.3 Draw Downs

Out of the Investor's total commitment, only a small proportion will be drawn down at the outset, to cover establishment expenses and the first instalment of the annual management fee or the general partners profit share.

The remainder can be called as and when required by the general partner for the purpose of making investments, or to pay fees and expenses.

Commitments are drawable in full at any time during the "commitment period". This is a period of (generally) about 5 years, during which the fund is expected to become fully invested. At the end of this period the general partner is prohibited from drawing down capital to make new investments and investor's remaining commitments are cancelled. other than a small amount (generally 10% of total commitments) for "follow-on" investment in existing portfolio companies, or to pay fees and expenses going forward.

There may be circumstances in which the right to drawdown can be cancelled before the end of the commitment period. These will generally cover events such as the departure of one or more "key men" whose expertise is considered essential for the management of the fund. It could also include a change of control of the fund manager.

3.5.4 Default by Investors

Because of the reliance on drawdowns to fund investments (rather than capital being paid up in full at the outset) a private equity fund is highly dependent upon the credit of its investors. A default by an investor in meeting a drawdown could have serious consequences if the fund has committed to purchase investments and will require other investors to bear an increased share of each investment. For this reason, fund documents generally subject defaulting investors to severe consequences. These vary from fund to fund, but a typical example would be for a defaulting investor to become subject to a high rate of interest during a "cure period" in which a late drawdown would become possible. If the default is not remedied within this period the investor's interest in the partnership would be frozen and the investor would cease to be a partner. The investor's only right thereafter would be to obtain back its invested capital to the date of default (with either no profit element or a reduced share of profit) at the end of the life of the fund, subordinated to payments to other investors.

To balance these consequences there are generally some "excuse" provisions to cater for situations where investors are legally prohibited from meeting drawdowns to invest in particular types of investments. It is usual for these to include to be a "change of law" excuse provision, and for investors subject to particular legal regimes (such as the US ERISA pensions legislation) to be given specific exemptions. Operating these "excuse" provisions can lead to complications; those investors' undrawn commitments must either be cancelled or their pro rata participation in other investments must be increased so as to ensure equal treatment.

3.5.5 Management/Advisory Structure

A limited partnership consists of at least one general partner and at least one limited partner. Only the general partner(s) can take part in the management of the partnership's business, or have the authority to represent and commit the partnership in dealing with third parties. The general partner has unlimited liability for the debts and obligations of the partnership. There may also (depending on tax and regulatory issues) be a manager and/or one or more investment advisers or sub-advisers in any private equity fund structure. The role of the respective parties will be as follows:

General Partner

The limited partnership agreement will specify the powers of the general partner and any restrictions on the exercise of these powers. The general partner will be given (subject to the appointment of a manager if there is to be one - see below) responsibility for all investment and divestment decisions, and power generally to manage the business of the partnership. These powers will be limited in particular respects - for example, to ensure that the investment policies and any limits on borrowing are followed. The partnership agreement will also contain provisions regulating the liability of the general partner, its staff and delegates and indemnifying them against liabilities incurred in the performance of these roles, regulating conflicts of interest and dealing with the removal or withdrawal of the general partner.

This last point is obviously very commercially sensitive. The market norm is for limited partners (the investors) to be able to vote to remove the general partner for 'cause'. This is generally defined to include insolvency, criminal conduct, material breach of fund documents and/or acts of wilful misconduct, fraud or gross negligence in relation to the fund. It is also becoming more common (but is still not the market norm) for a specified majority of investors to be able to vote to remove the general partner regardless of fault (this provision is known as 'no fault divorce'). Following removal of the general partner the limited partners are generally able either to vote to appoint a replacement general partner or to terminate the partnership and distribute its assets. Market practice varies greatly as to whether and to what extent the management group will retain its carried interest following termination for cause, or whether this can be allocated to incentivise a replacement general partner to manage the fund.

Manager

In many funds responsibility for management is vested in a separate management company, either appointed by the partnership itself or by the general partner under delegated authority. Reasons for this include tax planning, regulatory requirements and a desire to insulate the manager (which may be a company with substantial assets employing the management group's employees) from the unlimited liability which comes with general partner status.

Advisor and Sub-Advisors

In structures involving offshore (for example, Jersey, Guernsey and Cayman Islands) based managers or general partners, it is common for investment advice (and execution of deals) to be provided by an onshore (e.g., London based) advisor. It is also common for sub-advisers to be appointed in particular jurisdictions.

Investors - the Advisory Board

As noted above, limited partners are prohibited by law from taking part in the management of the business of a limited partnership. It is however common for there to be an advisory board or committee consisting of representatives of certain investors. Conflicts of interest involving the management group are often referred to the advisory board, which may also have a role in approving valuations of investments. The advisory board's powers are generally however quite limited, and the general partner often has control over the board's composition.

3.5.6 Investment Policies

The fund's investment policies will be set out in its placing memorandum and there will often (but not always) be an obligation in the partnership agreement for the general partner to adhere to these policies (sometimes with an ability for these to be varied with the consent of the advisory board). As well as general policies (geographical area, deal size and stage of investment) there will generally be the following specific restrictions on the investment powers of the general partner:

– Diversification: no more than a specified percentage of total commitments (commonly 25%) to be invested in a single investment or linked investments.

– Bridge financing: where the general partner intends to onsell part of an investment within a specified period after its acquisition (e.g., by syndicating or introducing co-investors) the above limit is often raised (e.g., to 35% of total commitments).

– Hedging: not permitted except for efficient portfolio management, as opposed to as an investment driver in its own right.

– Borrowings: these are generally only permitted at the level of the fund (as opposed to leverage in the investment structures through which the fund acquires its portfolio companies) for limited periods and purposes, such as to 'bridge' a drawdown of capital from investors.

– Publicly traded securities: since investors are not willing to pay private equity levels of fees for the management of publicly traded securities, there are often strict limits circumstances in which these may be held by the fund (e.g., the accumulation of a stake in a public company as a prelude to a takeover bid).

– Fund documents may also regulate the terms on, and circumstances in which co-investment opportunities may be offered to investors.

3.5.7 Distributions

The partnership agreement will prescribe the way in which income and gains are distributed among the partners. This is how the carried interest payable to management will be paid since (for tax reasons) this is not normally paid as a fee but is structured as a share in partnership profits. There will generally be a vehicle (itself possibly a limited partnership) which participates in the partnership as a limited partner with special rights to receive the carried interest. Individual managers will participate in carried interest received by this vehicle (through often complicated arrangements structured with tax and regulatory requirements in mind, as well as designed to incentivise managers to remain with the private equity house rather than to leave and join competitors). Managers generally expect to receive carried interest not as employment income, but to be taxed on it as capital gains on an investment in the fund (and to pay tax at lower rates and/or to be able to accumulate these gains offshore free of tax).

A typical distributions clause will provide that (after payment of fees and expenses or a general partner's profit share designed to replicate such fees) cash received by the partnership will be applied first in repayment to investors of their invested capital plus a preferred return, and thereafter will be split between the investors and the carried interest vehicle in the ratio 80:20. In negotiating these clauses there are obvious tensions between the interests of the managers and those of investors, and there are many variations on such provisions. Managers wish to receive carried interest as early as possible in the life of the fund, as and when investments are disposed of at a profit. From an investor's point of view there are risks associated with this 'deal by deal' approach to carried interest, mainly that profits on such deals will be offset by losses on later deals, meaning that overall the management team may have received more than 20% of the overall profits on the fund. From their point of view a 'whole fund' carried interest scheme is preferable, under which no carried interest is payable until investors have received an amount equal to their total commitments plus a preferred return over the expected life of the fund. Often a compromise involves accrual of carried interest on a deal by deal basis but with a percentage of this being held back or paid into an escrow account until the point when a 'whole of fund' threshold has been passed.

3.5.8 Fees and Expenses

Apart from carried interest, the management group will expect to receive an annual management fee (or equivalent profit share) and reimbursement of expenses incurred in a connection with the management of the fund. The fund (and therefore the investors) will be expected to bear the expenses of establishment (up to a cap) but this will generally not include the fees and expenses of placing agents. The reimbursement of expenses will also not include employment costs and general overheads of the management group, which are expected to be borne out of the management fee.

At the end of the commitment period the management fee will generally be reduced from a percentage of total commitments to a percentage of the (reducing) amount of capital still invested in the fund.

Managers of private equity funds are often in a position to benefit from their role by receiving fees and other benefits such as break fees on transactions involving the fund, or directors, corporate finance and advisory fees charged to the fund's portfolio companies. This creates an obvious potential conflict of interest with the fund and investors. To mitigate this, fund terms generally provide the management fees are reduced by all or a substantial percentage of these other fees and benefits.

3.5.9 Transfer of Interests

Because of the fund's dependence on future drawdowns from investors, their credit status is important. The fund's interests could also be harmed by admission of limited partners in circumstances leading to the need to register under securities and other laws (in particular the US Securities Act and Investment Companies Act) or causing the management group to be in breach of other legislation (e.g., the US ERISA pensions legislation). For these reasons (and also to preserve confidentiality and to prevent information about the management group and the fund being available to competitors) limited partnership interests are not transferable except with the consent of the general partner. This is often at the absolute discretion of the general partner, but sometimes the general partner will agree not to unreasonably withhold its consent to a transfer. It is also common for general partners to agree in particular cases to consent to transfers, for example to replacement trustees of an investor which is a pension scheme.

3.6 Legal Structuring of Italian Funds: General Overview

The venture capital fund industry is much newer in Italy than in other European countries and, especially, the United States. The first known venture capital investment schemes established to invest primarily on the Italian market is traced back to the second half of the nineties. These early investment schemes are almost invariably established in corporate form and are often located in tax-efficient European jurisdictions. This reflects a general trend which has also characterized the private equity fund industry in Italy from the outset. Indeed, if one does not consider certain recent developments, the prevailing legal structures underlying these investment schemes revolve around a corporate entity functioning both as a recipient of the investors' equity funds and as an investment vehicle. These corporate structures are normally governed by contractual documents including a series of Subscription Agreements signed by the individual investors and a Shareholders' Agreement entered into by all of the investors, the investment manager, and the investment vehicle. Only in a few known cases have private equity investment schemes with a primary focus on Italy been established in the form of a limited partnership. As well as limited partnership structures, Italian closed-ended funds have not enjoyed great popularity at least until law no. 344 of 14 August 1993 ("Law 344/93", which first set forth provisions governing these funds) was repealed by legislative decree no. 58 of 24 February 1998 ("Legislative Decree 58/98") and implementing regulations issued in 1999 (see paragraph 3.7) set forth rules on funds reserved to qualified investors.

The main factors typically taken into account by those involved in the Italian fund industry when planning the legal structure of a new investment scheme are:

- flexibility;

- investor base;

- tax regime; and

- set-up fees.

Flexibility has been a major drawback of Italian closed-ended funds for quite some time. For example, until the regulation of Italian closed-ended funds substantially changed in 1999, the return of the invested capital upon divestment could not be distributed to the investors until 5 years had elapsed from completion of the drawdown of the investor commitments[2]. Funds could not acquire a controlling interest in, or otherwise more than 30% of the capital of, any investee company. In addition, management companies were subject to very burdensome capital requirements. Because the legal regulation of closed-ended funds provided by Law 344/93 was not designed to meet the needs of investors and managers, these funds were de facto ignored by the market. When sponsors and managers target a significant number of international institutional investors (pension funds, funds-of-funds, etc.) in the planning phase of a new investment scheme, they normally consider the limited partnership structure for the reasons described in paragraph 3.4. Flexibility as well as the international investors' familiarity with limited partnerships is a factor playing in favor of this structure which are weighed against other factors. These normally include a comparison with corporate structures. The perceived disadvantages of corporate structures are:

- complexity of certain procedures;

- tax regime.

Some of the objectives described in paragraph 3.3, such as designing efficient mechanisms to govern financial flows in and out of the investment scheme or dealing with an investor default, are more difficult (although not impossible) to achieve with a corporate structure. In fact, corporate structures have generally been able to substantially meet all standards discussed in paragraph 3.4 even though more burdensome procedures than those entailed in a limited partnership may be involved. Also, corporate structures do not afford investors the benefit of tax transparency. However, the tax planning associated with the use of a corporate structure is often able to mitigate the downside of international institutional investors being unable to treat the selected vehicle as transparent for tax purposes. Because cultural reasons and generally higher set-up costs often play against structur-

---

[2] Only capital gains could be distributed without any time restriction during the funds' life.

ing the investment scheme as a limited partnership (a contractual scheme developed in Anglo-Saxon jurisdictions), sponsors and managers are normally reluctant to accept this structure unless there is clear evidence that it will significantly enhance their fundraising chances. This is unlikely to occur when the planned fundraising includes a vast majority of Italian investors who are more accustomed to investing through a corporate vehicle. The structuring of an investment scheme is largely driven by tax considerations. However, it cannot be affirmed that a given structure can deliver the best tax benefits to investors under any circumstances. Rather, the tax planning of an investment scheme depends on many factors to be analyzed on a case-by-case basis including the jurisdiction in which most investments are expected to be made and the prevailing investor base. For example, a non-domestic tax transparent entity is not the most efficient vehicle for making investments in Italian companies. Italian tax authorities treat these entities (including limited partnerships) as being subject to income taxes (IRPEG and IRAP) on capital gains and income earned from an Italian source. Since Italian taxation applies to limited partnerships (as opposed to their partners):

- the investors are unable to benefit from Tax Treaties;

- the benefit of Tax Treaties is not available to limited partnerships either, because they are not subject to tax in their own jurisdiction.

Thus, full Italian taxation applies to capital gains and income deriving from an Italian investment. To avoid this, limited partnerships normally structure their Italian investments by interposing a corporate vehicle located in a tax-efficient jurisdiction. However, this may result in the procedures and timing of distributions (upon total or partial divestments) as well as the tax characterization of the distribution proceeds for investors, being less efficient than investments made directly by the limited partnership. Also, capital gains and income distributed by limited partnerships to Italian tax residents are characterized as dividends for Italian tax purposes. However, since limited partnerships - as opposed to companies - are the distributing entities, certain tax benefits associated with dividends received by an Italian company from a non-domestic affiliate are not available. When a corporate vehicle is used in the structuring of an investment scheme, the tax planning typically seeks to achieve the following targets:

1. Treaty protection of investments made in other jurisdictions (i.e. no taxation of capital gains and limited withholding tax on dividend distributions in the jurisdiction in which the investment is made);

2. no taxation of capital gains and dividends in the jurisdiction of which the investment vehicle is a tax resident; and

3. no withholding tax on the distributions of divestment proceeds to investors.

Targets (1) and (2) are rather easy to achieve by using corporate vehicles located in several EU jurisdictions. Target (3), on the other hand, is definitely harder to pursue with a company than with a limited partnership unless multi-tier corporate structures are put in place and/or certain conditions are met by the investors in structuring their own investments in the scheme. In some cases, corporate EU investors have been able to repatriate divestment proceeds in the form of dividends under the so-called Parent-Subsidiary Directive[3]. Another important factor that is often considered when the legal structure of a new investment scheme is planned is the forecast size of the set-up costs. The structuring and establishment of a new investment scheme requires a substantial amount of legal work. When the scheme is structured as a limited partnership, the legal fees involved are generally higher than those entailed by a corporate structure since the relevant work is performed in a different legal market. This aspect, of course, may play in favour of a corporate structure.

As a matter of fact, venture capital and private equity investment schemes have predominantly been established in corporate form, often in jurisdictions other than Italy, until the end of the nineties. This trend began to change around 2000, i.e. shortly after the law and regulations governing Italian closed-ended funds were re-shaped by Legislative Decree 58/98 and the implementing rules[4]. The modified legal scene has had a strong impact on the decisions taken by market players when planning new investment schemes. The lack of flexibility affecting Italian closed-ended funds in the "old" legal environment is no longer a stumbling block today. The current tax regime (see paragraph 3.7.4) applicable to Italian closed-ended funds is viewed as a sensible trade-off between the lower taxation levels sometimes available under different legal schemes and the benefit of tax certainty provided by a scheme entirely located in one jurisdiction. Similar considerations apply to regulatory issues (see paragraph 3.8). One could argue

---

[3] EC Directive 90/435. As applied to the investment schemes considered here, this Directive may enable corporate EU investors to earn capital gains and other investment income with virtually no taxation in the jurisdiction of the distributing corporate vehicle and in their own jurisdiction.

[4] An analysis of the current regulation of Italian closed-ended funds will be provided in paragraph 3.7 below.

that financial burdens can be higher[5] and the set-up procedures more time-consuming[6] than those involved with different legal schemes. However, significant steps towards reducing these burdens were recently taken by the Bank of Italy, also as a result of the efforts exercised by A.I.F.I.[7], and further improvements can be reasonably expected in the future.

3.7 Italian Closed-Ended Funds

3.7.1 Background

Closed-ended funds were introduced in Italy by Law 344/93. Implementing rules were issued by the Bank of Italy on 14 March 1994 and by the Ministry of Treasury on 30 April 1994. The basic characteristics of these funds have remained unaltered through the subsequent legal changes. These are the following:

- Each fund is an unincorporated pool of assets managed by a licensed management company. The fund's assets are legally separated from the assets of the managing company and from those of the investors. The assets are beneficially owned pro-indiviso by the investors subject to the rules governing their management and appropriation.

- Management companies are supervised by the Bank of Italy.

- Funds are governed by contractual documents approved by their managing companies subject to the authorization of the Bank of Italy.

- The fund's cash and financial assets are held under the responsibility of a custodian bank to safeguard investors' rights.

Closed-ended funds and their management companies were faced with severe operational restrictions under Law 344/93 including the following:

1. Each management company was required to have a share capital of at least Itl. 5 billion (more than Euro 2.5 million) and to invest own funds in at least 5% of the amount of any managed fund.

---

[5] In relation to the capital requirements that licensed management companies (so-called SGRs) must meet.

[6] Due to the need to obtain an autorization from the Bank of Italy.

[7] The Italian association of the firms involved in the investment industry (Associazione Italiana degli Investitori Istituzionali nel Capitale di Rischio).

2. A fund could not survive unless investor commitments totaling at least 60% of its targeted size were obtained. Subject to this condition being met fund downsizing was subject to authorization by the Ministry of Treasury, and investors were free to withdraw their commitments.

3. The funds' regulation was essentially designed for retail investors. For example, managers were required to seek quotation on the Stock Exchange of the certificates representing investors' interests in the fund within 36 months of the end of the subscription period. While this prescription did not apply to funds placed exclusively with institutional investors, this class of investors was narrowly defined[8].

4. Multiple closings were not allowed.

5. Funds were subject to several penalizing investment and related restrictions including the following:

   ~ the acquisition of a controlling interest in, or otherwise more than 30% of the capital of, an investee company was not permitted;

   ~ investments in non-quoted shares or other equity securities could not be lower than 40% or greater than 80% of the fund amount; and

   ~ funds were not permitted to assume or grant loans.

   In the event of funds exclusively placed with institutional investors, these and other restrictions could be derogated from, subject however to authorization from the Bank of Italy.

6. The return of the invested capital upon divestment could not be distributed to the investors until 5 years had elapsed from completion of the drawdown of the investor commitments[9].

7. The carried interest was capped at 20% of the cumulative profits exceeding a hurdle rate to be determined based on objective criteria (subject to review by the Bank of Italy) and could only be paid to the managing company upon liquidation of the fund.

Since private equity and venture capital fund industries have developed and thrived in other jurisdictions as essentially unregulated businesses, these restrictions made Italian closed-ended funds as governed by Law

---

[8] For instance, large corporates and high net worth individuals were not included.
[9] See footnote 4.

344/93 very unattractive for market players. As stated above, the market basically set these funds aside.

Law 344/93 was repealed by Legislative Decree 58/98, but the bulk of the former law provisions continued to apply until rules implementing the new legislation dealing with management companies and closed-ended funds were issued by the Ministry of Treasury and the Bank of Italy. Unlike Law 344/93, Legislative Decree 58/98 restricted itself to essentially setting out the main principles of the reformed legal system and delegated wide regulatory powers to the Ministry of Treasury and the Bank of Italy[10]. Among others:

1. the Bank of Italy was empowered to establish:

   a.   general criteria and prohibitions concerning the carrying out of the investment activity, also covering the relationship between management companies and other companies of their groups;

   b.   prudential rules on risk limitation and diversification;

   c.   capital requirements to be met by management companies;

   d.   general criteria to be followed in the drawing up of fund documents and their contents[11];

2. the Ministry of Treasury was empowered to establish general criteria to be followed with respect to:

   a.   permitted investment objects;

   b.   categories of investors to whom funds may be offered for subscription;

   c.   frequency of issue and redemption of investors' interests in a fund, minimum commitments, relevant procedures and, more generally, manners in which investors can participate in a fund;

   d.   maximum duration of funds;

   e.   cases in which the prudential rules on risk limitation and diversification set forth by the Bank of Italy can be derogated

---

[10] Regulatory powers were also delegated to CONSOB, but these are of no relevance here.

[11] These are intended to supplement the minimum contents set out by section 39.2 of Legislative Decree 58/98.

from also taking into account the nature and professional experience of the investors; and

f.   cases in which the management companies must seek quotation on a regulated market of certificates representing investors' interests in a fund.

The Bank of Italy issued Regulations dealing, among others, with points (1) (c) and (1) (d) above on 1 July 1998[12]. Thereafter, Decree no. 228 was issued by the Ministry of Treasury on 24 May 1999, dealing with the matters delegated to same by Legislative Decree 58/98, and Regulations dealing, among others, with points (1) (a) and (1) (b) above were issued by the Bank of Italy on 20 September 1999.

3.7.2 A Comparison Between the Different Regulatory Systems

The regulatory system established under Legislative Decree 58/98 marks a clear election for a more liberal approach to the regulation of closed-ended funds reserved to qualified investors than that inspiring Law 344/93. The main changes are as follows:

1. Management companies - now called "Società di Gestione del Risparmio" or "SGRs" - are regulated as a new type of licensed financial intermediary which may, among others, manage open and closed-ended funds as well as pension funds. Their capital may not be lower than Euro 1 million.

2. The minimum size of a fund is set out in the fund documents. If investor commitments totaling at least the fund's minimum size as specified in the fund documents are obtained by the promoting SGR, the investors may not withdraw their commitments.

---

[12] The contents and general criteria to be followed in the drawing up of fund documents - as set forth by the Bank of Italy Regulations of 1 July 1998 - are to some extent influenced by certain provisions of Law 344/93 which ceased to have effect only when new rules were issued by the Ministry of Treasury and the Bank of Italy in 1999 under Legislative Decree 58/98. Hence, the rules dealing with fund documents set forth by the Bank of Italy Regulations of 1 July 1998 and the rules subsequently issued by the Ministry of Treasury and the Bank of Italy do not appear to be perfectly co-ordinated. These latter rules should prevail in the event of any inconsistencies, as indirectly confirmed by a release of the Bank of Italy dated 14 September 2001 which is commented on in paragraph 3.7.3.

3. SGRs are no longer required to seek quotation on a regulated market of certificates representing investors' interests in a fund also in relation to retail funds.

4. Rules applicable to funds exclusively placed with qualified investors are definitely more flexible. Qualified investors are defined more liberally than under Law 344/93 and include, among others, all natural and legal persons declaring in writing to have specific competence and experience in securities transactions.

5. Funds exclusively placed with qualified investors may freely determine policies and restrictions to be abided by in their investments and borrowings[13].

6. By contrast, all funds - including those reserved to qualified investors - must abide by the general criteria and prohibitions affecting their investment activity set forth by the Bank of Italy (see the Bank of Italy's regulatory powers under paragraph 3.7.1). Prohibitions include:

   ~ lending funds other than in the form of forward transactions on securities;

   ~ short sales of securities;

   ~ acquiring securities issued by the promoting/managing SGR or unquoted securities issued by companies of the SGR group; and

   ~ acquiring or selling assets to or from any shareholder, director, general manager or statutory auditor of the promoting/managing SGR or any other company of the SGR group.

7. Funds may distribute the return of invested capital to investors upon divestment without time restrictions[14].

Open points remained as follows:

1. Multiple Closings. These were not allowed[15].

2. Replacement of the SGR. This requires a modification to the fund documents. Any such modification is subject to the Bank of Italy's ap-

---

[13] Indeed, these funds may derogate from the prudential rules on risk limitation and diversification established by the Bank of Italy - see the Ministry of Treasury's regulatory powers under paragraph 3.7.1.

[14] These distributions are effected through a pro-quota reduction in the investors' interests in the fund whereas capital gains are payable to investors as distribution of fund income.

[15] See Section 14.2 of the Ministry of Treasury Decree no. 228 of 24 May 1999.

proval which, as a general rule, is released at the request of the SGR[16]. Since investors may wish, under certain circumstances (see paragraph 3.5.5), to be able to control this process and force the SGR to accept their decisions, questions can be raised as to whether investors may be granted these powers under the current regulatory framework.

3. General Prohibitions. Being able to lend funds to an SPV or an investee company is sometimes important to appropriately structure an investment transaction. Also, the existing prohibition on purchasing or selling assets is a way to deal with conflicts of interests. However, conflicts are handled through more flexible mechanisms in the international arena (normally by submitting transactions potentially involving conflict issues to the approval of the Advisory Board, i.e. a body at which the investors are represented).

4. Costs. It is unclear whether broken deal expenses can be borne, at least partially, by a fund as opposed to the managing SGR[17].

5. Carried Interest. Several questions remain unclear under the Bank of Italy Regulations of 1 July 1998 including the following:

   ˜ maximum amount (i.e. whether a 20% cap applies);

   ˜ computation base (i.e. whether this can include the preferential return - so-called hurdle rate - granted to investors);

   ˜ time of payments (i.e. whether carried interest payments can be effected prior to the liquidation phase);

   ˜ whether a "deal by deal carry" (see paragraph 3.5.7) is permitted;

   ˜ whether carried interest payments can be effected directly to parties other than the managing SGR (e.g. individual managers, advisors, etc.).

6. Transferability of Investors' Interests. The interests of investors in limited partnership are normally not freely transferable, as the general partner wishes to retain control over factors that are material to all investors and the investment scheme itself. Some of these factors include reviewing the creditworthiness of a proposed transferee or forbidding a sale when the proposed transferee is a benefit plan (or invests benefit plan monies) and the fund is not operating as a VCOC under ERISA rules, etc. Investors' interests in an Italian closed-ended fund are in principle

---

[16] See the Bank of Italy Regulations of 1 July 1998, point 12.
[17] See the Bank of Italy Regulations of 1 July 1998, points 9.1 and 9.2.

freely transferable[18], the only exception being that if a fund is reserved to certain qualified investors, transfers can only occur in favour of qualified parties. Whether additional restrictions or other control mechanisms can be introduced remains an open question.

3.7.3 Recent Developments

On 23 July 2001, the Bank of Italy issued Regulations designed to foster the creation of venture capital funds by universities, research centres established as legal persons, local authorities, university and banking foundations, and consortia formed by universities, local authorities and chambers of commerce. These venture capital funds may be established and managed by SGRs having a minimum capital of Euro 100,000 as opposed to the Euro 1 million capital required of SGRs under ordinary rules. This benefit is available when the following conditions are met:

1. the majority of the SGR's capital is held by parties falling within the categories specified above;

2. funds are reserved to qualified investors[19];

3. funds exclusively invest in start-ups carrying out research work with a view to subsequent industrial exploitation in high-tech business sectors; and

4. the total size of all such funds established and managed by the SGR does not exceed Euro 25 million.

A noted above, some of the points addressed in the last part of paragraph 3.7.2 probably arise as a consequence of the fact that the Bank of Italy Regulations of 1 July 1998 were issued when several provisions of Law 344/93 were still in force[20] pursuant to the transitional rules contained in Legislative Decree 58/98. This is indirectly confirmed by a release issued by the Bank of Italy on 14 September 2001. Said release was issued in response to a request made by A.I.F.I.[21] by submitting to the Bank of Italy a proposed standard form for fund documents reserved to qualified investors. The release acknowledges that this standard form conforms to the

---

[18] See section 36.8 of Legislative Decree 58/98.

[19] Individuals declaring to have specific competence and experience in securities transactions are required to undertake a commitment amounting to not less than Euro 250,000.

[20] See footnote 12.

[21] See footnote 7.

existing legislative and regulatory framework. Many clauses suggested by the form proposed by A.I.F.I. depart from the provisions of the Bank of Italy Regulations of 1 July 1998. For instance, it is envisaged that broken deal expenses can be shared between a fund and its managing SGR. Also, most points raised under paragraph 3.3 above with respect to carried interest payments do not appear to be an issue when viewed in the light of this form. The above release implicitly recognizes that certain provisions contained in chapter 4 of the Bank of Italy Regulations of 1 July 1998 are not applicable to closed-ended funds reserved to qualified investors since they are inconsistent with the principles of greater flexibility introduced by the 1999 regulatory measures for these funds. By the same token, one could argue that the above principles should also apply to other issues considered above. For instance, the current legal and regulatory environment does not seem to prevent fund documents from providing that investors may elect to replace the managing SGR with another SGR (upon the occurrence of certain circumstances) and accordingly file a request with the Bank of Italy to approve the relevant modifications to the fund documents.

Another step ahead in the relaxation of the rules applying to closed-ended funds reserved to qualified investors was marked by law no. 410 of 23 November 2001[22]. This law:

1. lifted the previously existing ban on multiple closings; and

2. delegated the Ministry of Treasury to establish the precautions that should apply in relation to any purchase, sale and contribution of assets to or from shareholders of the promoting/managing SGR or companies of the SGR group including the possible appointment of independent experts to determine the fair value of the assets being transferred.

Law no. 410 of 23 November 2001, deals essentially with real estate funds, although the relaxation of the rules on multiple closings applies to all closed-ended funds. Whether the new regulatory powers of the Ministry of Treasury under point 2 above will be exercised with regard to all closed-ended funds or real estate funds only remains an open question.

3.7.4 Conclusions

With the legal and regulatory changes discussed in the above paragraphs, Italian closed-ended funds will predictably become the norm in the domestic market. The only doubt remaining is whether the current regulatory rules establish a suitable environment for these funds to undertake substan-

---

[22] Converting into law decree-law no. 351 of 25 September 2001.

tial international fundraising. As seen above, closed-ended funds reserved to qualified investors are now reasonably flexible investment vehicles. Flexibility could be further improved by lifting certain existing prohibitions such as lending money in the context of structured investment transactions. Also, other points considered in paragraph 3.7.2 are worth being clarified by the regulatory authorities, and this will presumably occur over time as the matters arise in the process of approving fund documents. In any event, the above are actually minor aspects if viewed in the wider context of the profound liberalizing efforts made by the legislator and the regulatory authorities over the last few years and the resulting new operational features of these funds. The tax regime of Italian closed-ended funds[23] is another factor that is viewed positively by the market. A short description is as follows:

1. Funds are not subject to ordinary income taxes (IRPEG and IRAP), but only to a 12.5% substitutive tax levied annually on their results. If the results are negative, these can be carried forward and used to offset future positive results. If a fund has less than 100 investors and 50% or less are qualified investors other than individuals, the substitutive tax is levied at a 27% rate on the portion of fund results deriving from investments exceeding 10% or 50% of the share capital with voting power of, respectively, quoted companies or unquoted companies.

2. Funds do not apply any withholdings on proceeds distributed to investors. Tax resident investors:

   ~ if individuals, are not taxed on such proceeds;

   ~ if corporations or other taxable entities, will treat such proceeds as part of their taxable income and benefit from a tax credit of 15% or 36.98% depending on how the distributed income was taxed at the fund level (i.e. at the rate of, respectively, 12.5% or 27%).

3. Non-resident investors are not subject to any Italian taxation on the proceeds distributed by the fund and receive from the SGR a 15% cash reimbursement on the net proceeds distributed to them by the fund. If a fund is subscribed entirely by non-resident investors, the 12.5% substitutive tax is not payable by the fund (and no cash reimbursement is accordingly paid by the SGR).

---

[23] See article 11 of Law 344/93 as amended by article 8, paragraph 3, of legislative decree no. 461 of 22 November 1997 and legislative decree no. 505 of 23 December 1999 as amended by law no. 342 of 21 November 2000.

Financial costs deriving from the prescribed capitalization of SGRs[24] have sometimes been handled by market players in a creative fashion, e.g. under joint venture arrangements between managers/promoters of new investment initiatives and SGRs belonging to banking groups. The delays triggered by the requirement that fund documents must be approved by the Bank of Italy[25] are expected to progressively reduce as the process of standardization of fund documents continues thanks also to A.I.F.I's efforts (see paragraph 3.7.3). When a substantial international fundraising is planned by the promoters of a new investment initiative, Italian closed-ended funds may still be at a certain disadvantage as compared with other legal schemes due to a number of factors. International institutional investors are normally very sophisticated and fund documents tend to be rather complex to be a viable compromise between the conflicting requirements of sponsors, managers and groups of investors often having different negotiating targets or expectations. It is not infrequent for fund promoters and their legal advisors to issue definitive fund documents after long and troubled negotiations with certain investors and their advisors. Although not impossible, this process seems hard to combine with an approval of the fund documents by the Bank of Italy. Indeed, looking at the French experience, the so-called FCPRs with simplified procedures[26] - which share the legal nature of Italian closed-ended funds - do not require any approval of the fund documents by COB[27] if interests in such funds are placed exclusively with qualified investors or persons investing at least Euro 30,000 into the fund or certain employees, executives or individuals acting on behalf of the relevant fund's management company. Only FCPRs' management companies are subject to the supervisory powers of COB, although COB may order the amendment of the fund documentation or discontinuation of their circulation at any time. Tax transparency is another important point. Although the current tax regime is designed to grant non-resident investors a tax position equal to that they would have enjoyed with a tax transparent entity, this is not true under all circumstances. Indeed, the cash reimbursement payable by SGRs is limited to 15% of the net proceeds distributed to these investors even when a 27% substitutive tax applies at fund level. Furthermore, taxation at fund level is effected based on accounts run on an accrual basis, so the taxable base includes any write-up of finan-

---

[24] These costs are substantially reduced in respect of the SGRs governed by the Bank of Italy Regulations of 23 July 2001 (see paragraph 3.7.3).
[25] The approval process can take up to 4 months from filing of the fund documents with the Bank of Italy - see section 39.3 of Legislative Decree 58/98.
[26] Fonds communs de placement à risque bénéficiant d'une procédure allégée.
[27] The Commission des Opéerations de Bourse.

cial assets[28] as well as any realized capital gains. Potential mismatches between fund taxation and cash reimbursement can work to the detriment of the international investors' overall tax position. Moreover, the 15% cash reimbursement applies to reducing the substitutive taxes payable by SGRs, and no cash reimbursement is available when taxes are insufficient to fund this payment. Again, looking at the French experience FCPRs are treated as truly tax transparent entities. Finally, a point which could perhaps be improved – and not only with a view to enhancing the international fund-raising chances of Italian closed-ended funds - is the requirement that investors' interests in a fund all have equal value and carry the same rights[29]. While this is a sensible requirement so far as real investors' interests are concerned, it would be desirable to have more flexibility when the managers wish to invest in the fund. In this event, the managers do not necessarily need to share with the investors the same rights, given the different roles played by managers in an investment scheme. If the managers' interests in a fund could be treated as carrying different rights from those of the investors, it would be possible to structure carried interest payments under fund documents in a more appropriate fashion than under the current system. This would be a further incentive to attracting the best managers to work for these schemes, which would be beneficial for the entire fund industry. Indeed, structuring fund documents so as to enable individual managers to share in the carried interest with the SGR through mechanisms working efficiently from both a financial and a tax standpoint is one of the most challenging tasks facing whoever wishes to establish a new investment scheme in the form of a closed-ended fund. As creative as the mechanisms devised so far may be, none of them seems to work satisfactorily in every possible respect.

3.8 Other Regulatory Requirements

Placing interests in Italian closed-ended funds with qualified investors in most cases involves virtually no regulatory requirements. As a general rule, no offerings of securities may take place unless a prospectus is pub-

---

[28] To be made in accordance with the rules set forth by the Bank of Italy.
[29] See section 36.8 of Legislative Decree 58/98. In recent instances the Bank of Italy has interpreted this requirement in a more liberal fashion, de facto allowing the creation of fund interests carrying different rights for the managers. Yet it is not certain that these fund interests will enjoy the same tax treatment as those in the investors' hands.

lished except in those circumstances where an exemption applies. Exemptions include, inter alia, offerings:

1. exclusively addressed to professional investors;

2. addressed to 200 offerees or less; and

3. involving an investment of Euro 250,000 or more.

The definition of professional investors[30] for the purposes of the exemption under (1) above is somewhat narrower than that of qualified investors applicable under the rules governing closed-ended funds[31]. Therefore, a closed-ended fund reserved to qualified investors does not necessarily fall within the scope of the exemption under (1) above. Nonetheless, the scope of the exemptions under (1), (2) and (3) above is wide enough for SGRs to be able to elaborate appropriate strategies to market their closed-ended funds reserved to qualified investors without being caught by the requirement to publish a prospectus.Under their licenses, SGRs can place interests in the closed-ended funds promoted by them with qualified investors without restrictions concerning the place where offerings are made. Nonetheless, should offerings be made outside the SGRs' principal or branch offices and be addressed to parties not qualifying as professional investors under section 30.2 of Legislative Decree 58/98 (see footnotes 30 and 31), special rules apply. In particular:

1. completion of the agreements under which interests in closed-ended funds are placed with investors in Italy is subject to the condition precedent that the investors do not withdraw their consent within 7 days from their entering into the agreements[32]; and

---

[30] Contained in section 31.2 of CONSOB Regulations No. 11522 of 1 July 1998, as amended implementing section 30.2 of Legislative Decree 58/98.

[31] For example, individuals qualify as professional investors solely if they are able to document that they meet the competence requirements set forth by the Ministry of Treasury Decree no. 468 of 11 November 1998 to hold the position of director, executive or auditor within a licensed investment firm, whereas they are treated as qualified investors whenever they declare in writing to have specific competence and experience in securities transactions.

[32] The contractual documents must include a notice indicating this right of the investors. Failure to include this notice gives investors the right to claim that the agreements are null and void.

2. contacts with the investors may only be made by licensed financial promoters acting for the promoting/offering SGR on an exclusive basis[33].

As discussed above, these rules de facto have a marginal scope of application where closed-ended funds are reserved to qualified investors. Placing interests in investment schemes established under the laws of a jurisdiction other than Italy with Italian investors raises, mutatis mutandis, the issues discussed above insofar as prospectus requirements are concerned. As to licensing requirements, foreign issuers may directly place interests in their investment schemes with investors in Italy, but are precluded from offering such interests outside their principal or branch offices to parties not qualifying as professional investors under section 30.2 of Legislative Decree 58/98. Any such offerings may solely be made by licensed placing agents[34] acting on behalf of the issuers or the promoters of said investment schemes. The special rules discussed above in respect of SGRs apply to such offerings provided that they are made outside the placing agents' principal or branch offices and are addressed to parties not qualifying as professional investors under section 30.2 of Legislative Decree 58/98. More importantly, offering interests in non-domestic investment schemes in Italy are in principle subject to further regulatory requirements which are discussed below. In addition, investments effected in Italy by non-domestic investment vehicles could entail the application of certain regulatory requirements which are also dealt with below.

The first question raised, with regard to the placement in Italy of interests in venture capital investment schemes governed by foreign law, is whether the relevant offer is subject to an authorization procedure. Section 42.5 of Legislative Decree 58/98 provides that any "offer in Italy of units of mutual investment funds not falling within the scope of the directives on undertakings for collective investment shall be authorized by the Bank of Italy [........] provided that their operating schemes are compatible with those prescribed for Italian undertakings". Section 42.6 of Legislative Decree 58/98 provides that the "Bank of Italy [......] shall govern through Regulations the conditions and procedures for granting the authorizations

---

[33] This requirement does not apply when contacts are made by mail or other means of communication not allowing customized communication and immediate interaction with investors - see section 76 of CONSOB Regulations No. 11522 of 1 July 1998, as amended.

[34] These include investment firms and banks licensed to place securities in Italy as well as EU investment firms and banks licensed to place securities in their own jurisdiction and enabled to carry out this business in Italy under the rules implementing the applicable EU Directives on the so-called single passport.

provided by paragraph 5". In addition, section 50.2 of Legislative Decree 58/98 provides that any "offer in Italy of shares in foreign SICAVs shall be subject to section 42". The Bank of Italy adopted Regulations implementing section 42.6 of Legislative Decree 58/98 on 31 December 2001. Pursuant to these Regulations, the Bank of Italy shall authorize the offer if the mutual investment fund or SICAV concerned:

1. has an operating scheme compatible with that prescribed for Italian undertakings;

2. is supervised by an authority in its jurisdiction carrying out controls similar to those applying to Italian undertakings;

3. markets its units or shares in its jurisdiction;

4. meets certain prescribed standards concerning the spread of information to the public and the exercise of the economic rights to which Italian investors are entitled; and

5. employs managers meeting standards of reputation and competence equivalent to those applying to managers of SGRs and SICAVs.

Should the foreign undertakings be established in a non-EU jurisdiction, additional requirements shall apply. As discussed above, venture capital (and private equity) investment schemes are commonly established in the form of fixed-capital investment companies or limited partnerships. The question then is whether these legal structures fall within the scope of the aforesaid rules. Section 1.1, letter j), of Legislative Decree 58/98 defines "mutual investment funds" as "segregate pools of assets, divided into units, pertaining to a plurality of participants and managed on a collective basis"[35]. The term "SICAV" is defined by section 1.1, letter i), of Legislative Decree 58/98 as "an investment company with variable capital whose registered and head offices are located in Italy and whose exclusive purpose is collective investment of the capital raised by offering its shares to the public". Investment schemes established in statutory form as fixed-capital investment companies do not fall within the meaning of "mutual investment fund" or "SICAV" as provided for by Legislative Decree 58/98. Doubts

---

[35] Also provided are the definitions of mutual investment funds which are respectively "open" and "closed" illustrated in letters k) and l) of said section 1.1. The former is "a mutual investment fund whose participants have the right to request, at any time, redemption of the units pursuant to the procedures established by the rules of the fund". The second is "a mutual investment fund in which the right to request redemption of the units may be exercised by participants solely at predetermined maturities".

could instead be cast about limited partnerships in light of the broad definition of "mutual investment fund" referred to above. However, unlike mutual investment funds, limited partnerships are not purely contractual in nature, although they are governed by a contract between the members. UK
limited partnerships (and those in UK-based jurisdictions such as the
Channel Islands and Cayman) rest on partnership law, which establishes a
relationship between persons carrying on business together with a view to
profit. They do not have separate legal personality. The business is carried
on by all of the partners, through the agency of the general partner(s). Assets are held by the general partner(s) on behalf of all the partners on trust.
Trust concepts do not transpose into the civil law legal concepts underlying the definition of mutual investment funds, so the authorization requirement does not seem to apply to these limited partnerships. This is
even more apparent in the case of Delaware limited partnerships (or other,
mainly US, partnerships which follow the Delaware model). These partnerships have separate legal personality. It is therefore the partnership itself which carries on the investment business and is the owner of assets.
The partners simply own interests in the partnership, and do not carry on
the business themselves. Their position is therefore analogous to that of
shareholders in a corporation. One could argue that the differences between the mutual investment fund and the limited partnership schemes are
of no relevance for the purposes discussed herein, the rationale of the rules
subjecting certain offerings to an authorization of the Bank of Italy being
the need of protecting investors with regard to all investment schemes not
harmonized at EU level. Nonetheless, the body of law introduced by Legislative Decree 58/98 does not seem to allow this conclusion. Indeed, Legislative Decree 58/98 could well have rooted the authorization requirement
on legal concepts with a broader scope than those considered above so as
to encompass all possible collective investment schemes. For instance,
"undertakings for collective investment" is a term often used for this purpose consistently with EU Directives[36]. However, Legislative Decree 58/98
does use this term ("Organismi di Investimento Collettivo del Risparmio"
or "OICR") but in a narrower meaning encompassing only "mutual investment funds" and "SICAVs" (see section 1.1, letter m), of Legislative
Decree 58/98.) as defined above. On the other hand, fixed-capital invest-

---

[36] Under Legislative Decree No. 86 of 27 January 1992 (repealed by Legislative
Decree 58/98) an authorization procedure was required in respect of any offer to
the public in Italy of "shares or units of undertakings for collective investment on
transferable securities" not falling within the scope of EU Directives. These undertakings are broadly defined by Directive 85/611 and encompass all types of statutory or contractual vehicles carrying out such business.

ment companies are clearly outside the scope of said definitions, and treating limited partnerships differently can hardly have a rationale if one bears the above remarks in mind. Thus, it is sensible to conclude that the authorization requirement provided for by Legislative Decree 58/98 only applies to certain investment vehicles, i.e. those undertakings which are most frequently used to raise funds from the public and whose shares or units are normally negotiable on a market. This is confirmed by the implementing Regulations issued by the Bank of Italy on 31 December 2001 which have not sought to provide a broader meaning of the legal concepts used by Legislative Decree 58/98.

Another point to examine, in relation to the placement in Italy of interests in venture capital investment schemes governed by foreign law, is whether the Bank of Italy is to be notified of the relevant offer under section 129 of Legislative Decree No. 385 dated 1 September 1993 as amended ("Legislative Decree 385/93"). The aforesaid section provides that the Bank of Italy is to be notified of all "... offers in Italy of foreign transferable securities...":

1. which exceed "...the amount of Lire one hundred billion[37] or the higher amounts established by the Bank of Italy.."[38]; or

2. which relate to securities (x) falling outside certain categories identified by law, or (y) having characteristics different from certain standards "established by the Bank of Italy in compliance with the CICR[39] resolutions".

Section 129 of Legislative Decree 385/93 grants the Bank of Italy the powers to prohibit or defer the relevant transactions in order "to ensure stability and efficiency of the transferable securities market". These powers and the relevant notification requirement do not normally apply to offers in Italy of either limited partnership interests or shares in fixed-capital investment companies governed by foreign law. As regards limited partnership interests, the Istruzioni di Vigilanza[40] define "transferable securities" as "the instruments of collection of funds, negotiated or negotiable on

---

[37] Approximately Euro fifty million.

[38] All transactions relative to the same issuer effected in the preceding twelve-month period are taken into account to establish whether the relevant amounts are exceeded.

[39] CICR (or Comitato Interministeriale per il Credito e il Risparmio) is a governmental committee endowed with supervisory and regulatory powers on banking matters.

[40] Supervisory Guidelines for banks issued by the Bank of Italy - see Title IX, Chapter I, Section I, paragraph 3 (April 1999/January 2002).

a market, intended for a plurality of investors...". Limited partnership interests normally do not meet the negotiability requirement indicated above. These interests are indeed based on a contractual relationship revolving around the Limited Partnership Agreement which provides, as a rule, that their transferability is subject to the approval of the general partner. In granting or denying its approval, the general partner exercises discretionary powers taking into account the creditworthiness of the proposed transferee, the regulatory and tax issues possibly affecting the investment scheme (or certain investors) as a consequence of the envisaged transfer, as well as the commercial strategies pursued by it in handing relationship with the investors. Similar considerations apply to shares in foreign fixed-capital investment companies. In a venture capital investment scheme established in corporate form, investors are required to execute certain documents governing the investment scheme (typically including a Subscription Agreement and a Shareholders' Agreement), and the issue or sale of shares in the investment vehicle to them is governed by said documents. Each investor has contractual rights and obligations under these documents and may only transfer its shares to a third party simultaneously with such rights and obligations. As well as in a limited partnership scheme any such transfer is normally subject to the approval of the issuer and/or the investment manager, and such approval is granted or denied through the exercise of discretionary powers. There are additional grounds supporting this conclusion with regard to fixed-capital investment companies. Section 129.5 of Legislative Decree 385/93 provides that shares are exempt from the notification requirement (and the Bank of Italy's powers) discussed above "provided that these do not represent a participation in open- or closed-ended undertakings for collective investment". No definition of undertakings for collective investment is provided by either Legislative Decree 385/93 or the above-cited Istruzioni di Vigilanza. Nonetheless, in defining such undertakings ("Organismi di Investimento Collettivo del Risparmio"), section 1.1, letter m), of Legislative Decree 58/98 only refers to "mutual investment funds" and "SICAVs"[41]. Even if the definition of "undertakings for collective investment" contained in Legislative Decree 58/98 is not directly applicable in this context, it is sensible to refer to this definition also for the purposes contemplated by section 129 of Legislative Decree 385/93 for reasons of consistency amongst connected provisions. Indeed, the provisions contained in the Istruzioni di Vigilanza dealing with shares representing a participation in an undertaking for collective invest-

---

[41] See this paragraph 3.8 above for a definition of these terms.

ment refer to Legislative Decree 58/98[42]. Accordingly, the duty to notify the Bank of Italy under article 129 of Legislative Decree 385/93 should not apply to the offer in Italy of shares in fixed-capital investment companies governed by foreign law whether or not these instruments qualify as "transferable securities" in light of the contractual provisions governing the investment scheme[43].

The last point to review is the following. An Italian company carrying out a venture capital business must be entered in a register (the "General Register") kept by the Ufficio Italiano Cambi on behalf of the Ministry of Treasury[44]. Solely companies satisfying certain conditions including the exclusive exercise of financial activities and a capital not lower than Euro 500,000 can be filed with the General Register. In consideration of certain thresholds relating to the volume of financial activities and the endowment of capital, companies filed with the General Register (the "Financial Intermediaries") are also required to be entered in a special register (the "Special Register") kept by the Bank of Italy and are subject to supervision of the Bank of Italy[45]. Pursuant to Article 114 of Legislative Decree 385/93, the Ministry of Treasury is entitled to govern the exercise of the activities reserved to Financial Intermediaries on the part of entities having their registered offices abroad. A decree of the Ministry of Treasury dated 28 July 1994 required "companies having registered offices abroad" which exercise the activities reserved to Financial Intermediaries in Italy "with establishment or through another permanent organization including representative offices" to file with the General Register. Filing with the General Register by these companies is subject to similar requirements to those applying to Italian Financial Intermediaries. The foregoing include the exclusive exercise in Italy of financial activities and the establishment of an endowment fund amounting to not less than the minimum capital required

---

[42] Pursuant to the Istruzioni di Vigilanza these shares are exempt from the notification requirement (and outside the scope of the Bank of Italy's powers) discussed above if the relevant undertakings for collective investment have been authorized under section 42.5 of Legislative Decree 58/98.

[43] Under certain conditions, after completion of the relevant transaction a notice is due to the Bank of Italy merely for statistical purposes.

[44] This requirement is set forth by section 106 of Legislative Decree 385/93 and applies to a number of financial activities including the acquiring of equity holdings which - as set forth by a decree of the Ministry of Treasury dated 6 July 1994 - is aimed at their later divestment and is characterized, in the holding period, by "measures aimed at business reorganization or product development or meeting the financial needs of investee companies also by resort to third parties' equity funds".

[45] This requirement is set forth by section 107 of Legislative Decree 385/93.

for Italian Financial Intermediaries carrying out investment business (Euro 500,000). Alternatively, the existence of a sector legal framework equivalent to that provided for Italian Financial Intermediaries is required in the home country of these companies. Furthermore, foreign Financial Intermediaries must be entered in the Special Register when the thresholds applying to Italian Financial Intermediaries are met with reference solely to activities carried out in Italy. The exercising of any activity reserved to Financial Intermediaries (including investment business) without being registered in the General Register is subject to criminal sanctions. It is, at any rate, doubtful that such criminal sanctions could be applicable in relation to foreign entities. From a practical point of view, the questions raised by the above body of law concern (a) the foreign entities which could be subject to the registration requirement with the General Register and (b) the conditions under which the registration is required. With regard to the first point, it seems first of all possible to exclude that the registration obligation applies to limited partnerships, since these are not "companies with registered offices abroad" as set forth by the decree of the Ministry of Treasury dated 28 July 1994. It could perhaps be argued that, whenever limited partnerships are not legal persons within their legal systems, the obligation to register would be imposed on all partners having corporate nature. However, this argument would seem clearly strained, if one were to consider that the requisites imposed by the Ministry of Treasury for registration in the General Register (including the endowment fund) should exist in relation to each individually considered partner, while the investment activity carried out in the frame of limited partnerships would have a unitary nature. As regards investment companies governed by foreign law, which do not usually have offices in Italy (to avoid the risk of a permanent establishment being deemed to exist for direct tax purposes), the only practical question that should be considered relates to those advisors which have registered offices in Italy and carry out ancillary activities consisting essentially in seeking out investment opportunities. The question is whether these advisors can be treated as an "establishment" or as "permanent organizations" of the aforesaid investment companies. The Ministry of Treasury has not given any official interpretation in this connection. In a different ambit, similar problems have arisen in relation to the possibility for EU banks and/or finance companies controlled by same to carry out, in a member State different from their home country, services admitted to mutual recognition under the regime of free provision of services (i.e. not applying the rules on freedom of establishment) when such services are offered continually, in said other member State, through independent intermediaries. The problem was faced in relation to Directive no. 89/646/CEE of 15 December 1989 (the "Second Banking Directive") and to the relevant implementing rules in each member State. In this connection, reference should be made to an interpretative communication from the EU Commission published in 1997[46] and the Supervisory Guidelines issued by the Bank of Italy in relation to the rules implementing the principle of mutual recognition contained in the Second Banking Directive. According to the aforesaid communication from the EU Commission and the Supervisory Guidelines of the Bank of Italy, services rendered through the use of independent intermediaries cannot enjoy the regime of free provision of services (and are, therefore, subject to the rules governing the right of establishment) whenever such intermediaries:

~ operate exclusively for a single EU bank/financial company;

~ have the power to negotiate transactions with third parties;

~ can undertake obligations on behalf of the EU bank/financial company; and

~ work on a continual basis.

Should all of the above conditions be satisfied, the provisions governing the right of establishment - and not those governing the free provision of services - shall apply. As previously mentioned, such criteria are not directly applicable to investment business activities carried out by EU subjects in Italy, but can provide a useful reference for the interpreter wishing to give a more precise meaning to the notion of "establishment" or "permanent organization" contained in the above-cited decree of the Ministry of Treasury. A different sort of position expressed by the UIC through some of its releases should, at any rate, be mentioned. According to the UIC, foreign subjects not provided with establishment or other permanent organization cannot acquire equity holdings in Italy directly, but rather only "through" licensed Financial Intermediaries. In other words, these Financial Intermediaries should operate "in the name and on behalf" of the foreign subject. According to the UIC, this interpretation would satisfy the need of enabling the Italian authorities to monitor financial flows by and for Italian subjects for the purposes of the applicable anti-money laundering provisions. An exception to this principle would only be admitted in relation to cross-border investment activities carried out occasionally and on a limited scale. The UIC's position appears to lack sound legal grounds, since it is de facto aimed at imposing registration in the General Register based (substantially) on the frequency and volume of investment transac-

---

[46] Commission 97/C 209/04 published in the EU Official Gazette no. 209, of 10 July 1997.

tions effected in Italy, rather than on the manners in which foreign subjects operate (as required by the decree of the Ministry of Treasury). On the other hand, the need to operate "through" a licensed Financial Intermediary would subject foreign investment companies to evident tax-related risks connected with the existence of a permanent establishment in Italy. Sometimes co-investment arrangements between a licensed Financial Intermediary and a foreign investment company (not registered in the General Register) has been deemed capable of complying with the position assumed by the UIC without entailing tax-related risks. The foregoing solution, nonetheless, despite having the benefit of offering a level of greater visibility to the activity of acquiring holdings in Italy by foreign investment companies is not, indeed, required by the decree of the Ministry of Treasury nor, if closely examined, perfectly coherent with the position expressed by the UIC (by co-investing with a licensed Financial Intermediary, the foreign subject does not operate "through" the Financial Intermediary but effects its own distinct investment alongside that of the Financial Intermediary).

References

A.I.F.I - Associazione Italiana degli Investitori Istituzionali nel Capitale di Rischio (various years) Il mercato italiano del venture capital e private equity. Rapporti periodici, Milan

A.I.F.I. - Associazione Italiana degli Investitori Istituzionali nel Capitale di Rischio (1996) La valutazione della performance. Guerini & Associati, Milan

A.I.F.I. - Associazione Italiana degli Investitori Istituzionali nel Capitale di Rischio (2000) Guide to Venture Capital. Milan

A.I.F.I. - Associazione Italiana degli Investitori Istituzionali nel Capitale di Rischio (variuos years) Development capital. Guerini & Associati, Milan

A.I.F.I. - Associazione Italiana degli Investitori Istituzionali nel Capitale di Rischio (various years) The Italian venture capital and private equity market. Periodic reports, Milan

A.I.F.I.- Associazione Italiana degli Investitori Istituzionali nel Capitale di Rischio (1999) L'attività di venture capital and private equity. Il Sole 24Ore Libri, Milan

Abbot S, Hay M (1995) Investing for the Future. Ft Pitman Publishing, London

ABI – Prometeia (2000) La gestione dei servizi finanziari alle PMI nell'esperienza delle Banche Europee, I° Workshop Comitato Tecnico, January 26, 2000

Aghion P, Bolton P (1992) An Incomplete Contract Approach to Financial Contracting. The Journal of Finance, n. 1

Anonymous (1983) SBICs after 25 years: Pioneers and Builders of Organized Venture Capital. Venture Capital Journal, October

Anonymous (2001) Financial Snapshot for may 2001: Stocks starts to blossom, Signalsmag, 5 September (www.signalsmag.com)

Arnott R, Stigliz J (1991) Moral hazard and nonmarket institutions: dysfunctional crowding out or peer monitoring?. The American Economic Review

Assogestioni (1999) Mutual investment trusts. Data and statistical guide, Milan

Autorità Garante della Concorrenza - Banca d'Italia (1997) Indagine conoscitiva sui servizi di finanza aziendale. Rome

Bank of England (1999) Practical Issues arising from the Euro. June, London

Bank of England (2000), Finance for small firms: A seventh report. London

Bank of Italy (1994) Il mercato della proprietà e del controllo delle imprese: aspetti teorici e istituzionali

Bank of Italy (1999) Guidelines for banking supervision - Circolare n.229 – 21.4.99

Bank of Italy (1999 and 2000) Annual report

Bank of Italy (2000) Statistical bulletin I, II

Baravelli M (1999) Strategia e organizzazione della banca. Egea, Milan

Barca F (1994) Imprese in cerca di padrone. Proprietà e controllo nel capitalismo italiano. Laterza Rome - Bari

Barney JB, Busenitz LW, Fiet JO, Mosel DD (2001) New Venture Teams' Assessment of Learning Assistance from Venture Capital Firms, Journal of Business Venturing n.18

Barry CB, Muscarella CJ, Peavy III JW, Vetsuypens MR (1990) The Role of Venture Capital in the Creation of Public Companies. Journal of Financial Economics, 27

Basile I, De Sury P (1993) Il ruolo delle banche nell'offerta di servizi d'intermediazione mobiliare. In Banche e Banchieri

Beatty RP, Ritter JR (1986) Investment banking, Reputation, and the underpricing of initial public offerings. Journal of Financial Economics, vol. n. 15

Berger AN, Hannan TH (1997)Using measures of firm efficiency to distinguish among alternative explanations of structure-performance relationship. In Managerial Finance, n. 23

Berger AN, Udell GF (1998) The economics of small business finance: the roles of private equity and debt markets in the financial growth cycle. Journal of Banking and Finance, vol. n. 22

Bester H. (1987)The role of collateral in credit markets with imperfect information. In European Economic Review, N. 83

Bhave MP (1999) A Process Model of Entrepreneurial Venture Creation. In Journal of Business Venturing, vol. 9

Bhidè A (1999) Developing Start-up Strategies. In Sahlman W, Stevenson HH, Robberts M, Bhidè A, The Entrepreneurial Venture. Harvard Business School press, Boston

Billingsley S (1996) Merck & Company: A Comprehensive Equity Valuation Analysis. Working Paper, AIMR, Charlottesville

Black B, Gilson R (1998) Venture capital and the structure of capital markets: banks versus stock markets. The Journal of Financial Economics, vol. 47

Block Z, Mac Millan I (1993) Corporate Venturing. Harvard Business School Press, Boston

Bond & Pecaro Inc. (2000) CyberValuation, InternetBusiness Trends, Analysis and Valuation, New York

Brealey RA, Myers SC (2000) Principles of Corporate Finance. McGraw-Hill, New York

Bruyat C, Julien PA (2000) Defining the Field of Research in Entrepreneurship. Journal of Business Venturing, vol. 14

BVCA (1996) Guide to Venture Capital

Bygrave WD (1999) The venture capital handbook

Bygrave WD, Timmons JA (1992) Venture capital at the crossroad. Harvard Business school Press, Boston

Cairns JC, Davidson JA, Kisicevitz ML (2002) The limits of bank convergence. The McKinsey quarterly journal, n. 2

Canals J (1997) Universal Banking. International Comparison and Theoretical Perspectives. Clarendon Press, Oxford

Capizzi V (2000) Il ruolo degli intermediary finanziari nelle operazioni di finanza straordinaria. In Forestieri G (ed) Corporate & Investment Banking. Egea, Milan

Capizzi V (2000) Le condizioni dell'offerta di servizi di corporate finance. In Zara C (ed) Le banche e l'advisoring nell'ambito delle operazioni di finanza straordinaria, Bancaria Editrice, Rome

Capizzi V (2001) La valutazione delle internet companies. Internal mimeo, SDA Bocconi, february Milan

Capizzi V (forthcoming) The wealth of expertise in corporate and investment banking activities. Enbicredito Research

Caselli S (1999) Family Small and Medium-Sized Firms' Finance and Their Relationship With The Banking System in Italy. International Council for Small Business, 44th World Conference, "Innovation and Economic Development: the Role of Small and Medium Enterprises", 20-23 June

Caselli S (2000) Lo sviluppo del corporate e investment banking in Italia: profili strategici e organizzativi.In Forestieri G (ed) Corporate & Investment Banking. Egea, Milan

Caselli S (2001) Corporate Banking per le piccole e medie imprese. Bancaria Editrice, Rome

Caselli S (2001) Tendenze evolutive del corporate banking. In Anderloni A, Basile I, Schwizer P (ed) Osservatorio sull'innovazione finanziaria. Bancaria Editrice, Rome

Caselli S, Gatti S (forthcoming) The characteristics of the demand and offer of corporate and investment banking services in Italy. Enbicredito Research

Cattaneo C (1995) Mercati dei titoli delle PMI: quali intermediari?. In Banche e banchieri, 1

Cesarini F (1985) Il punto sullo sviluppo del merchant bank. In Banche e banchieri, febbraio

Cesarini F (1994) Rapporti tra banche e imprese in Italia: due punti di debolezza. In Economia e Politica Industriale, n. 83

Cesarini F, Monti M, Scognamiglio C (1982) Il sistema creditizio e finanziario italiano. Relazione della Commissione di studio istituita dal Ministro del tesoro. Istituto Poligrafico dello Stato

Chapman S (1992) The Rise of Merchant Banking. Gregg Revivals, Happshire, London

Christiansen CM (1991) The innovator's dilemma. Harvard Business School Press, Boston

Christofidis C, Debande O (2001) Financing innovative firms through venture capital. European Investment Bank Sector Paper

Ciampi F Squilibri di assetto finanziario delle PMI. Finanziamenti e contributi della Comunità europea. In Studi e Informazioni n.45

Collitti G (1987) Capitali coraggiosi, Ipsoa

Coopers & Lybrand (1999) Corporate finance

Copeland T, Koller T, Murrin J (2000) Valuation – Measuring and managing the value of companies. Wiley, New York

Corbetta G (1997) Le politiche di facilitazione allo sviluppo delle PMI: fabbisogni delle imprese e condizioni di efficacia. In Brunetti G, Mussati G, Corbetta G Piccole e medie imprese e politiche di facilitazione. Egea, Milan

Corbetta G, Bolelli F, Caselli S, Lassini U (2001) Gli spazi di collaborazione fra banca e imprese familiari. Rapporto di ricerca SDA Bocconi – Banca Intesa BCI, Milan

Costi R (1985) Una legge per le merchant bank. In Banche e Banchieri n.3

Cotula F, Filosa F (1989) La contabilità nazionale e i conti finanziari. In Cotula F La politica monetaria in Italia, Il Mulino, Bologna

Credit Suisse First Boston (2001) European Technology – a game of two halves. Internal report, January, London

Csikszentmihalyi M (1991) Creativity: flow and the psychology of discovery and invention. HarperCollins. New York

D'Alessio G (1994) Diffusione della proprietà delle imprese fra le famiglie italiane e trasferimento intergenerazionale: alcune evidenze. In Temi di discussione della Banca d'Italia n.241

Dallocchio M (1996) Finanza d'azienda. EGEA, Milan

Damodaran A (1994) On valuation – Security Analysis for investment and corporate finance. Wiley, New York

Damodaran A (1999) Applied corporate finance – A user's manual. Wiley, New York

Damodaran A (2000) Investment Valuation: Tools and Techniques for Determining the Value of Any Asset. Wiley, New York

Damodaran A (2000) The dark side of valuation: Firms with no earnings, no history and no comparables. Working Paper, Stern School of Business, New York

De Bonis R (1996) La riscoperta del debito e delle banche. In Temi di discussione, Banca d'Italia, Servizio Studi, n. 279

De Cecco M, Ferri G (1994) Origini e natura speciale dell'attività di banca d'affari in Italia. In Banca d'Italia, Temi di discussione, n. 242

De Cecco M, Ferri G (1996) Le banche d'affari in Italia. Il Mulino, Bologna

De Laurentis G (1994) Il rischio di credito. I fidi bancari nel nuovo contesto teorico, normativo e di mercato. EGEA, Milan

De Laurentis G (1995) Mercato potenziale e condizioni di successo del leasing. Verso una teoria della differenziazione finanziaria In Carretta A Gli intermediari finanziari non bancari. Egea, Milan

De Laurentis G (1996) Le basi progettuali del corporate banking. In Il risparmio, n. 4-5

De Laurentis G (1997) La valutazione dei fidi e l'assistenza alle imprese nel corporate banking relazionale. In Baravelli M Le strategie competitive nel corporate banking. Egea, Milan

De Laurentis G (2000) Introduzione dei rating interni nelle banche italiane. In Sironi A, Savona P (eds) I modelli per la misurazione del rischio di credito nelle maggiori banche italiane. Edibank, Milan

Desmet D, Francis T, Hu A, Koller T, Riedel G (2000) Valuing dot coms. McKinsey Quarterly, n.1, Spring

Dessy A, Vender J (2001) Company risk and development capital. Egea, Milan

Dini L (1984) Capitale di rischio per le imprese: il ruolo del merchant banking. Relazione al convegno Risparmiare per investire. ABI-Fideuram, Rome

DRI-WEFA, NVCA (2001) Economic Impact of venture capital. Washington DC, October

Drucker PF (1989) The practice of mangement. Heinemann Business Paperbacks, London

Eiglier P, Langeard E (1991) Servuction. Le marketing des services. McGraw-Hill, Paris

Ernst & Young (2000) Convergence. The Biotechnology Industry Report. Millenium Edition, New York

European Commission (2002) Enterprises in Europe. 8th Report, Bruxelles

European Information Technology Observatory (2000). The Millenium Edition

EVCA - Coopers & Lybrand Corporate Finance(1996) The economic impact of venture capital in Europe. September

EVCA - Coopers & Lybrand (1997) The Economic Impact of Venture Capital in Europe. Zaventem

EVCA - European Venture Capital Association (various issues) Yearbook

EVCA (1999) Private Equity Fund Structures in Europe. Internal publication, Zaventem

EVCA (2001) EVCA Mid Year Survey of Pan-European private equity and venture capital. Press release, Helsinki, 17th October

EVCA (2001) GuideLines. www.evca.com, March

Fabrizi PL (1998) La formazione nelle banche e nelle assicurazioni. Bancaria Editrice, Rome

Fama EF (1991) Efficient capital markets. In Journal of Finance, n. 46

Fenn GW, Liang N, Prowse S (1995) The economics of private equity market. Board of Governors of the Federal Reserve System Staff Studies, 168

Financial Times (1999) The venture capital handbook. London

Financial Times (2001) Telecoms job cuts watch. 27th July

Forestieri G (1980) Struttura del mercato del credito e concorrenza bancaria. Giuffrè, Milan

Forestieri G (1988) Lo sviluppo del merchant banking: i possibili modelli operativi. In Scritti in onore di L Guatri, ottobre

Forestieri G (1993) Mercati finanziari per le piccole e medie imprese. Analisi e progetto per il caso italiano. Egea, Milan

Forestieri G (1998) Gli intermediari finanziari di fronte allo sviluppo del corporate finance. Quaderni di politica industriale, mediocredito Centrale, Rome

Forestieri G (2000) Corporate & Investment Banking. Egea, Milan

Forestieri G (2000) La ristrutturazione del sistema finanziario italiano: dimensioni aziendali, diversificazione produttiva e modelli organizzativi. In Banca Impresa Società, n. 1

Forestieri G, Invernizzi G (1996) Considerazioni conclusive: lo sviluppo di iniziative di merchant banking. In Forestieri G, Corbetta G (a cura di) Le banche italiane dal credito al merchant banking. Mediocredito Lombardo, Milan

Forestieri G Mottura P (1998) Il sistema finanziario. Istituzioni, mercati e modelli di intermediazione. Egea, Milan

412   References

Foster R, Kaplan S (2001) Creative destruction: why companies are built to last underperform the market and how to successfully transform them. Currency/Doubleway, New York

Frasca FM (1995) Il rapporto banca impresa e la nuova normativa sulle partecipazioni. In Bancaria 5

Fried VH, Hisrich RD (1994) Toward a Model of Venture Capital Investment Decision Making. Financial Management, vol. 23

Fruhan WE (1979) Financial strategy – Studies in the creation, trasfer and destruction of shareholder value. Homewood, Irwin

Gardella LA (2000) Selecting and structuring investments: the venture capitalist's perspective. In Association for Investment Management and Research. Reading in Venture Capital, AIMR Charlotteville

Gartner WB, Starr JA, Bhat S (2000) Predicting New Venture Survival: an Anlisys of Anatonomy of a Start-up. Cases from Inc. Magazine, Journal of Business Venturing, vol.14

Gatti S (1996) Il merchant banking: prodotti, servizi e competenze. In Forestieri G, Corbetta G (ed), Le banche italiane dal credito al merchant banking. Mediocredito Lombardo, Milan

Gatti S (1997) Le metodologie professionali di valutazione della marca: analisi dei punti di forza e di debolezza. In Zara C (ed) La valutazione della marca, Etas Libri, Milan

Gatti S (1998) L'offerta dei servizi di corporate finance per le Piccole e Medie imprese italiane: struttura, prodotti, concorrenza ed evidenze empiriche. In Mediocredito Centrale, Quaderni di Politica Industriale

Geisst RC (1995) Investment banking in the financial system. Englewood Cliffs, Prenticehall, London

German Association Of Biotechnology Industries (1998) Valuation of Biotech Companies, Berlin

Gervasoni A (1989) I fondi chiusi: aspetti di gestione e di strategia. EGEA, Milan

Gervasoni A (1996) Il ruolo del sistema bancario. In Finanziare l'attività imprenditoriale. Guerrini e associati, Milan

Gervasoni A (2000) I fondi mobiliari chiusi. Il Sole 24 Ore S.p.A., Milan

Gervasoni A, Perrini F (1994) I fondi d'investimento mobiliare chiusi. Egea Milan

Gervasoni A, Satin FL (2000) Private equity e venture capital. Guerini e associati, Milano

Gervasoni A, Sattin FL (2000) Private equity and venture capital. Investment manual of risk capital. Guerini e Associati, Milan

Gilder G (2000) Telecosm. The Free Press, New York

Goldsmith RW (1969) Financial structure and development. New Haven, Yale University Press

Gompers PA (1995) Optimal Investment, Monitoring and the Staging of Venture Capital. The Journal of Finance, vol. 50

Gompers PA (1999) The venture capital cycle. Cambridge, Mit Press

Gompers PA, Lerner J (1996) The use of covenants: An Empirical Analysis of Venture Partnership Agreements. Journal of Law and Economics, n. 39

Gompers PA, Lerner J (1998) What drives venture fundraising?. Brookings Papers on Economic Activity: Microeconomics, July

Gompers PA, Lerner J (1999) What drives venture capital fundraising?. W.p. 6906, NBER Series, Cambridge, MA

Gompers PA, Lerner J (2000) The venture capital circle. Mit press

Gorman M, Sahlman WA(1989) What do Venture capitalist do?, In Journal of Business Venturing, n.4

Guatri L (1998) Trattato sulla valutazione delle aziende. EGEA, Milan

Gupta, AK, Sapienza HJ (1992) Determinants of venture capital firms' preferences regarding the industry diversity and geographic scope of their investments. Journal of Business venturing, n. 7

Hamel G (2000) Leading the Revolution. Harvard Business School Press, Boston

Hannan TH (1991) Bank commercial loan markets and the role of market structure: evidence from surveys of commercial lending. In Journal of Banking and Finance, n. 15

Hayes S, Hubbard P (1990) Investment banking. Harvard Business school Press , Boston

Heifetz R (1994) Leadership without easy answers. Belknap Press, Cambridge, Massachusetts

Hellmann T, Puri M (1999) The interaction between product marketing and financing strategy: the role of Venture Capital. Research Paper 1561, Research Paper Series, Stanford University, May, Stanford

Hellmann T, Puri M (2000) Venture Capital and the Professionalization of Startup Firms: Empirical Evidence. Research Paper 1661, Research Paper Series, Stanford University, Stanford

Hellwig M (1991) Banking Financial Intermediation and Corporate Finance. In Giovannini A, Mayer C European Financial Integration, Cambridge University Press

Hunt D (1995) What Future for Europe's Investment Banks?. In Mc Kinsey Quarterly Review, n. 1

Jeng L, Wells P (2000) The determinants of venture capital funding: evidence across countries. In Journal of Corporate Finance, n. 6

Jensen MC, Meckling WH (1976) Theory of the Firm: Managerial Behavior, Agency Costs and Ownership Structure. Journal of Financial Economics, n. 3

Kaplan S, Strömberg P (2000) How do Venture Capitalists Choose and manage their investments?. Working Paper, University of Chicago

Kay W (1992) Il merchant banking oggi. Edibank, Milan

Keeton WR (1996) Do Banks Mergers Reduce Lending to Business and Farmers? New Evidence From Tenth District States. Federal Reserve Bank of Kansas City, Economic Review, n. 81

Kellogg D, Charnes JM (2000) Real Options Valuation for a Biotechnology Company. Financial Analysts Journal, n. 3

Koller T (2001) Valuing dot coms after the fall. McKinsey Quarterly, Number 2

Kortum S, Lerner J (1998) Does Venture Capital Spur Innovation?. NBER Working Paper, December

Kuhn RL (1990) Investment banking, the art and science of high-stakes dealmaking. Harper & Row, New York

Leithner & Co. PTY (1999) The internet and value investing. Internal report, Brisbane

Lerner J (1994) The syndication of Venture Capital Investments. In Financial Management, vol 23, n. 3, Autumn

Lerner J (1995) Venture Capitalists and the Oversight of Private Firms In Journal of Finance, n. 50

Lerner J (1999) Venture capital e private equity. A casebook. J.Wiley and sons, New York

Lerner J (2000) Private equity and venture capital – a casebook. Wiley, New York

Lerner J (2001) Venture Capital and private equity

Levin JS (1994) Structuring venture capital, private equity and entreprenueurial transaction. IL.CCH Inc, Chicago

Lewis M (1999) The New New Thing. Norton & Company New York

Liaw KT (1999) The Business of Investment Banking. John Wiley & Sons, New York

Llewellyn DT (1992) Financial Innovation: A basic anlysis In Cavanna H Financial Innovation, Routledge, London

Llewellyn DT (1999) The new Economics of Banking. Société Universitaire Européenne de Recherches Financières, Amsterdam

London Stock Exchange (1999) TECHMARK, The technology market. London

Lorenz T (1985) Venture capital today. Woodhead e Faulkner. Cambridge

Mason CM, Harrison RT (2000) The size of the Informal Venture Capital Market in the United Kingdom. Small Business Economics, n. 15, September

Massari M (1998) Finanza aziendale, McGraw Hill, Milan

Mauboussin MJ, Hiler B (1999) Cash Flow.com – Cash Economics in the New Economy. Credit Suisse First Boston Corporation, Internal Report, March, New York

Mauboussin MJ, Regan MT Schay A, Fisher AM (2000) Wanna Be GE?. Credit Suisse First Boston Corporation, Internal Report, February New York

McBride AS, McBride RG (2001) The vital role of investor relations. Strategic Investor Relations, www.iijournals.com, summer

Mccue J (2000) Telecommunications and the New Economy. Lucent Client Success and Partner Conference, Doral, 30th October - 2nd November, Florida

McNamee M (2001) America's future - Investment Plays. In Business Week, 27th August

Megginson WL, Weiss KA (1991) Venture capitalist certification in initial public offerings. In The Journal of Finance, vol. 46, n. 3

Meyers SC, Miluf NS (1984) Corporate Financing and Investment Decisions when Firms have Information Investors do not have. Journal of Financial Economics, n.13

Millan IC, Zeman L (1987) Criteria distinguishing successful ventures in the venture screening process. In Journal of Business Venturing, 2

Mottura P (1994) Evoluzione della banca verso forme di intermediazione finanziaria innovative e diversificate. In Il Risparmio, n. 3

MSDW (1999) Entrepreneur Workshop - Exit Strategies. www.ms.com, Proceeding of the Conference, 31 May, London

Murray G, Marriott R (1998) Why has the investment performance of technologyu-specialist European venture capital funds been so poor?. Research Policy, 27

Myers SC, Howe CD (1997) A Life Cycle Financial Model of Pharmaceutical R&D. Working Paper POPI 41-97, MIT Sloan School of Management

Norman R (1984) Service managment: strategy and leadership in service business. Wiley & Sons, Chicago

NVCA - National Venture Capital Association (various issues) Yearbook, Annual Economic Impact of Venture Capital Study, USA

Onado M (1992) Economia dei sistemi finanziari. Il Mulino, Bologna

Ooghe H, Manigart S, Fassin Y (1991) Growth patterns of the European venture capital industry. Journal of Business Venturing, n. 6

Peek J, Rosengren ES (1996) Small Business Credit Availability: How Important is the side of lender?. In Saunders A, Walter I, Financial System Design: The Case for Universal Banking. Burr Ridge, Irwin Publishing

Piol E (2001) Il ruolo del venture capital nello sviluppo della Net Economy. Mondo Libero, March

Pivato G, Gilardoni A (1999) Economia e gestione delle imprese. Egea Milan

Porter M (1985) Competitive Advantage. The Free Press, New York

Poterba J (1989) Venture capital and capital gain taxation. In Summers L (ed.) Tax Policy and the Economy, MIT Press

Rabitti A, Bedogni C (1998) Commentario al D.Lgs. 58/98. Ed. Giuffrè, Milan

Rappaport A (1997) La strategia del valore. Le nuove regole di creazione della performance aziendale. Franco Angeli, Milan

Ravid SA (1988) On Interactions of Production and Financial decisions. Financial Management, n. 3

Reid GC (1992) Venture capital investment. Routledge, London

Renoldi A (1992) La valutazione dei beni immateriali. EGEA, Milan

Reston J Jr. (1998) The Last Apocalypse: Europe at the Year 1000 A.D.. Doubleway, New York

Revell J (1985) Rischi bancari e assicurazione dei depositi. In Banca, impresa, società n. 3

Robbie J, Wright S, Chiplin M (1999) Funds providers' role in venture capital firm monitoring. In Management Buy-outs and Venture Capital, Edward Elgar Ltd

Robbie K, Wright M (1997) Venture Capital. Dartmouth Publishing, New York

Rock K (1986) Why new issues are underpriced. In Journal of Financial Economics, vol. n. 15

Rosa C (1987) La sfida italiana del merchant banking.Europia, Novara

Ross SA, (1977) The Determination of Financial Structure: The Incentive Signalling Approach, Bell Journal of Economics, Spring

Rybczynsky TM (1996) Investment Banking: its evolution and place in the system. In Gardner E, Molineux P (eds) Investment Banking. Theory and practice. Euromoney Books, London

Sahlman WA (1990) The structure and governance of venture-capital organizations In The Journal of Financial Economics, vol. 27

Sahlman WA (1999) The entrepreneurial venture. Harvard Business School Press, Boston

Sandri S (1988) Il venture capital nel ciclo di sviluppo delle nuove imprese. Giappichelli, Torino

Santomero AM, Babbel DF (1997) Financial marktes, instruments and institutions. Irwin Publisher, Chicago

Sapienza HJ, Amason AC, Manigart S (1994) The Level and Nature of Venture Capitalist Involvement in Their Portfolio Companies: A Study of Three European Countries. Managerial Finance, vol. 20

Saunders A, Walter I (1994) Universal Banking in The United States. What Could We Gain ? What Could We Lose ?. Oxford University Press

Schefczyk M, Gerpott JT (2000) Qualifications and turnover of managers and venture capital financed firm performance: an empirical study of German venture capital investments. In Journal of business venturing n. 16

Schumacher EF (1973) Small Is Beautiful. Blond & Brigger, London

Schwartz ES, moon M (2000) Rational Pricing of Internet Companies. In Financial Analysts Journal, n.3

Schwizer P (1996) La diversificazione bancaria. Aspetti strutturali e misurazione. Egea, Milan

Silver AD (1994) The venture capital sourcebook. Probus, Chicago

Simon H (1989) Price marketing. North Holland, Amsterdam

Simpson I (2000) Fund raising and investor relations. EVCA Association, December

Smith RC, Walter I (1997) Global Banking. Oxford University Press, Oxford

Soda AP (1994) Un'analisi della disciplina dei fondi d'investimento chiusi. In Banche e banchieri n.2

Stein JC (1989) Efficient Capital Markets, Inefficient Firms: A Model of Myopic Corporate Behavior. Quarterly Journal of Economics

Stewart GB (1991) The quest for value – The EVA management guide. Harper Business, New York

Storey DJ (1991) The Birth of New Firms – Does Ununemployment matter? A Review of the Evidence. Small Business Economics, n. 3

Strahan PE, Weston JP (1998) Small Business Lending and The Changing Structure of The Banking Industry In Journal of Banking and Finance, n. 22

Tobin J (1984) On the efficiency of the financial system. In Lloyds Bank Review, July

Tyebjee T, Bruno A (1984) A model of venture capitalist investment activity. Management Science, 30

Van Osnabrugge M, Robinson R (1999) Financing entrepreneurship: business angels and venture capitalists compared. Harvard Business School working paper

Van Osnabrugge M, Robinson R (2000) Angel Investing. Jossey-Bass, San Francisco

Van Osnabrugge M, Robinson RJ (2000) Angel investing. Harvard Business School

Venture Capital Report (1998) Guide to Private Equity & Venture Capital in the UK & Europe. Pitman Publishing, London

Venture Economics (1988) Exiting Venture Capital Investments

Venture Economics (1996) Pratt's Guide to Venture Capital Sources. SDC Publishing, New York

Venture Economics (2001) VentureEdge. Spring

Ventureone Corporation, PricewaterhouseCoopers (2001) The Pricewaterhouse-Coopers MoneyTree Survey in Partnership with VentureOne. August, San Francisco

Vesper KH (1989) A Taxonomy of New Business Ventures. Journal of Business Venturing, vol. 4

Vitale M (1987) Capitale di rischio per lo sviluppo. Realtà operative e strumenti finanziari: opportunità, condizioni, vincoli. In A.I.F.I., Venture capital, Edizioni Il Sole 24Ore, Milan

Vivarelli M (1994) La nascita delle imprese in Italia. Teorie e verifiche empiriche. Egea, Milan

Wall J, Smith J Better Exits. www.evca.com

Walraven N (1997) Small Business Lending by banks Involved in Mergers. Board of Governors of the Federal Reserve, Finance and Economics Discussion Series, n. 25

Walter I (1998) Global Competition in Financial Services. Ballinger – Harper & Row, Cambridge

Ward J (1987) Keeping the Family Firm Healthy. How to Plan for Continuing Growth, Profitability and the Family Leadership. Jossey-Bass, San Francisco

Wetzel WE Jr (1981) Informal Risk Capital in New England. In Vesper KH (editor) Frontiers in Entrepreneurship Research. Babson College, Wellesley

Wilson JW (1986) The new venturers. Inside the high-stakes world of venture capital. Addison Wesley Publishing Company, Beverly

Winborg H, Landstrom J (2000) Financial Bootstrapping in Small Businesses: Examing Small Business Managers' Resource Acquisition Behaviors. Journal of Business Venturing n.16

List of Contributors

Corresponding Address

Professor Stefano Caselli
Professor Stefano Gatti

Institute of Financial Markets and Financial Intermediaries
"L. Bocconi" University
Via Sarfatti, 25
20136 Milan
Italy

stefano.caselli@uni-bocconi.it
stefano.gatti@uni-bocconi.it

Stefano Caselli

Associate Professor in Banking and Finance at "L. Bocconi" University, Milan, Italy. Professor at Banking and Insurance Department of SDA Bocconi, the Management and Business School of "L. Bocconi University", Milan, Italy. He's developing research activity, managerial education and strategic consulting in: corporate and investment banking, leasing and asset finance, credit risk management, SME's financing.

Stefano Gatti

Associate Professor in Banking and Finance at "L. Bocconi" University, Milan, Italy. Professor at Banking and Insurance Department of SDA Bocconi, the Management and Business School of "L. Bocconi University", Milan, Italy. He's developing research activity, managerial education and strategic consulting in: corporate and investment banking, project financing, corporate finance and company valuation.

Vincenzo Capizzi

Associate Professor in Banking and Finance at "A. Avogadro" University, Novara, Italy. Professor at Banking and Insurance Department of SDA Bocconi, the Management and Business School of "L. Bocconi University", Milan, Italy.

Sonia Deho'
Senior Analyst at Pino Venture.

Jonathan de Lance Holmes
Partner, Linklaters London.

Gino Gandolfi
Associate Professor in Banking and Finance at University of Parma, Italy. Professor at Banking and Insurance Department of SDA Bocconi, the Management and Business School of "L. Bocconi University", Milan, Italy.

Manuela Geranio
Assistant Professor in Banking and Finance at "L. Bocconi" University, Milan, Italy. Assistant Professor at Banking and Insurance Department of SDA Bocconi, the Management and Business School of "L. Bocconi University", Milan, Italy.

Renato Giovannini
Assistant Professor in Banking and Finance at "L. Bocconi" University, Milan, Italy. Assistant Professor at Banking and Insurance Department of SDA Bocconi, the Management and Business School of "L. Bocconi University", Milan, Italy.

Anna Giuiusa
Officer assigned to the Bank's Supervision Department in Bank of Italy. (The author is the only responsible for the opinions expressed in the paper).

Marina Maddaloni
Nuovo Mercato - Market for high growth firms stocks - Italian Exchange

Maria Pierdicchi
Managing Director, Standard & Poor's Italy - Formerly, Director of Nuovo Mercato - Market for high growth firms stocks- Italian Exchange

Elserino Piol
President of Pino Venture.

Claudio Scardovi
CEO of Intervaluenet.

Enzo Schiavello
Partner, Studio Gianni, Origoni, Grippo & Partners, Milan.

Lucia Spotorno
Assistant Professor in Banking and Finance at "L. Bocconi" University, Milan, Italy. Assistant Professor at Banking and Insurance Department of SDA Bocconi, the Management and Business School of "L. Bocconi University", Milan, Italy.

Edmondo Tudini
Ph.D student in Business Administration at "L. Bocconi" University, Milan, Italy. Assistant Professor at Banking and Insurance Department of SDA Bocconi, the Management and Business School of "L. Bocconi University", Milan, Italy.

Daniela Ventrone
Ph.D student in Business Administration at "L. Bocconi" University, Milan, Italy. Assistant Professor at Banking and Insurance Department of SDA Bocconi, the Management and Business School of "L. Bocconi University", Milan, Italy.

Springer Finance

M. Ammann, University of St. Gallen, Switzerland

Credit Risk Valuation

Methods, Models, and Applications

Credit Risk Valuation offers an advanced introduction to the models of credit risk valuation. It concentrates on firm-value and reduced-form approaches and their applications in practice. Additionally, the book includes new models for valuing derivative securities with credit risk, focussing on options and forward contracts subject to counterparty default risk, but also treating options on credit-risky bonds and credit derivatives. The text provides detailed descriptions of the state-of-the-art martingale methods and advanced numerical implementations based on multi-variate trees used to price derivative credit risk. Numerical examples illustrate the effects of credit risk on the prices of financial derivatives.

2nd ed. 2001. Corr. 2nd printing 2002. X, 255, 17 figs., 23 tabs. Hardcover **€ 69.95**; sFr 116.50; £ 49 ISBN 3-540-67805-0

A. Ziegler, University of Lausanne, Switzerland

Incomplete Information and Heterogeneous Beliefs in Continuous-time Finance

This book considers the impact of incomplete information and heterogeneous beliefs on investor's optimal portfolio and consumption behavior and equilibrium asset prices. After a brief review of the existing incomplete information literature, the effect of incomplete information on investors' expected utility, risky asset prices, and interest rates is described. It is demonstrated that increasing the quality of investors' information need not increase their expected utility and the prices of risky assets. The impact of heterogeneous beliefs on investors' portfolio and consumption behavior and equilibrium asset prices is shown to be non-trivial.

XIII, 194 p. 51 illus. 2003. Hardcover **€ 64.95**; sFr 108; £ 45.50 ISBN 3-540-00344-4

Printing: Strauss GmbH, Mörlenbach
Binding: Schäffer, Grünstadt